CYBERSECURITY
Ethics, Legal, Risks, and Policies

CYBERSECURITY

Ethics, Legal, Risks, and Policies

Ishaani Priyadarshini, PhD
Chase Cotton, PhD, CISSP

First edition published 2022

Apple Academic Press Inc.
1265 Goldenrod Circle, NE,
Palm Bay, FL 32905 USA

4164 Lakeshore Road, Burlington,
ON, L7L 1A4 Canada

CRC Press
6000 Broken Sound Parkway NW,
Suite 300, Boca Raton, FL 33487-2742 USA

2 Park Square, Milton Park,
Abingdon, Oxon, OX14 4RN UK

© 2022 Apple Academic Press, Inc.

Apple Academic Press exclusively co-publishes with CRC Press, an imprint of Taylor & Francis Group, LLC

Reasonable efforts have been made to publish reliable data and information, but the authors, editors, and publisher cannot assume responsibility for the validity of all materials or the consequences of their use. The authors, editors, and publishers have attempted to trace the copyright holders of all material reproduced in this publication and apologize to copyright holders if permission to publish in this form has not been obtained. If any copyright material has not been acknowledged, please write and let us know so we may rectify in any future reprint.

Except as permitted under U.S. Copyright Law, no part of this book may be reprinted, reproduced, transmitted, or utilized in any form by any electronic, mechanical, or other means, now known or hereafter invented, including photocopying, microfilming, and recording, or in any information storage or retrieval system, without written permission from the publishers.

For permission to photocopy or use material electronically from this work, access www.copyright.com or contact the Copyright Clearance Center, Inc. (CCC), 222 Rosewood Drive, Danvers, MA 01923, 978-750-8400. For works that are not available on CCC please contact mpkbookspermissions@tandf.co.uk

Trademark notice: Product or corporate names may be trademarks or registered trademarks and are used only for identification and explanation without intent to infringe.

Library and Archives Canada Cataloguing in Publication

Title: Cybersecurity : ethics, legal, risks, and policies / Ishaani Priyadarshini, PhD, Chase Cotton, PhD, CISSP
Names: Priyadarshini, Ishaani, 1992- author. | Cotton, Chase, author.
Description: First edition. | Includes bibliographical references and index.
Identifiers: Canadiana (print) 20210190159 | Canadiana (ebook) 20210190213 | ISBN 9781774630228 (hardcover) |
    ISBN 9781774639207 (softcover) | ISBN 9781003187127 (ebook)
Subjects: LCSH: Computer security—Moral and ethical aspects. | LCSH: Computer security—Law and legislation. | LCSH: Computer security—Law and legislation—United States. | LCSH: Cyberspace—Security measures. | LCSH: Computer security—Government policy—United States.
Classification: LCC QA76.9.A25 P75 2021 | DDC 005.8—dc23

Library of Congress Cataloging-in-Publication Data

Names: Priyadarshini, Ishaani, 1992- author. | Cotton, Chase, author.
Title: Cybersecurity : ethics, legal, risks, and policies / Ishaani Priyadarshini, PhD, Chase Cotton, PhD, CISSP.
Description: First edition. | Palm Bay, FL, USA : Apple Academic Press, 2022. | Includes bibliographical references and index. | Summary: "This book is the first of its kind to introduce the integration of ethics, laws, risks, and policies in cyberspace. The book will advance understanding of the ethical and legal aspects of cybersecurity followed by the risks involved along with current and proposed cyber policies. This book serves as a summary of the state of the art of cyber laws in the United States and considers more than 50 cyber laws. It also, importantly, incorporates various risk management and security strategies from a number of organizations. Using easy-to-understand language and incorporating case studies, the authors begin with the consideration of ethics and law in cybersecurity and then go on to take into account risks and security policies. The section on risk covers risk identification, risk analysis, risk assessment, risk management, and risk remediation. The very important and exquisite topic of cyber insurance is covered as well-its benefits, types, coverage, etc. The section on cybersecurity policy acquaints readers with the role of policies in cybersecurity and how they are being implemented by means of frameworks. The authors provide a policy overview followed by discussions of several popular cybersecurity frameworks, such as NIST, COBIT, PCI/DSS, ISO series, etc. Each chapter is followed by an overall summary and review that highlights the key points as well as questions for readers to evaluate their understanding based on the chapter content. Cybersecurity: Ethics, Legal, Risks, and Policies is a valuable resource for a large audience that includes instructors, students, professionals in specific fields as well anyone and everyone who is an essential constituent of cyberspace. With increasing cybercriminal activities, it is more important than ever to know the laws and how to secure data and devices"-- Provided by publisher.
Identifiers: LCCN 2021016904 (print) | LCCN 2021016905 (ebook) | ISBN 9781774630228 (hardback) | ISBN 9781774639207 (paperback) | ISBN 9781003187127 (ebook)
Subjects: LCSH: Computer security--Moral and ethical aspects. | Computer security--Law and legislation. | Internet governance. | Liability insurance. | Computer crimes.
Classification: LCC QA76.9.A25 P754 2022 (print) | LCC QA76.9.A25 (ebook) | DDC 005.8--dc23
LC record available at https://lccn.loc.gov/2021016904
LC ebook record available at https://lccn.loc.gov/2021016905

ISBN: 978-1-77463-022-8 (hbk)
ISBN: 978-1-77463-920-7 (pbk)
ISBN: 978-1-00318-712-7 (ebk)

About the Authors

Ishaani Priyadarshini, PhD
Department of Electrical and Computer Engineering,
University of Delaware, USA

Ishaani Priyadarshini, PhD, has authored several book chapters for reputed publishers and is also an author of several publications for SCIE-indexed journals. As a certified reviewer, she conducts peer review of research papers for IEEE, Elsevier, and Springer journals and is a member of the editorial board of the *International Journal of Information Security and Privacy* (IJISP). She is a PhD candidate (Department of Electrical and Computer Engineering) at the University of Delaware, USA, from where she also obtained her master's degree in cybersecurity. Prior to that, she completed her bachelor's degree in computer science engineering and a master's degree in information security from Kalinga Institute of Industrial Technology, India. Her areas of research include cybersecurity (authentication systems, cybersecurity ethics, and policies) and artificial intelligence.

Chase Cotton, PhD, CISSP
Professor of Practice, and Director, University of Delaware Center for Intelligent CyberSecurity, USA

Chase Cotton, PhD, is a successful researcher, telecommunications carrier executive, product manager, consultant, and educator in the technologies used in Internet and data services in the carrier environment for over 30 years. Beginning in the mid-80s Dr. Cotton's communications research in Bellcore's Applied Research Area involved creating new algorithms and methods in bridging multicast and many forms of packet-based applications, including voice and video, traffic monitoring, transport protocols, custom VLSI for communications (protocol engines and content addressable memories), and Gigabit networking. In the mid-'90s, as the commercial Internet began to blossom, he transitioned to assist carriers worldwide as they started their Internet businesses, including internet service providers (ISPs), hosting, and web services, and the first large scale commercial deployment of digital subscriber line (DSL) for consumer broadband services. In 2000, Dr. Cotton assumed research, planning, and engineering for Sprint's global Tier 1 Internet provider, SprintLink, expanding and evolving the network significantly during his eight-year tenure. At Sprint, his activities include leading a team that enabled infrastructure for the first large-scale collection and analysis of Tier 1 backbone traffic and twice set the Internet 2 Land Speed World Record on a commercial production network.

Since 2008, Dr. Cotton has been at the University of Delaware in the Department of Electrical and Computer Engineering, initially as a visiting scholar and later as a Senior Scientist, Professor of Practice, and Director of Delaware's Center for Information and Communications Sciences (CICS). His research interests include cybersecurity and high-availability software systems with funding drawn from the NSF, ARL, CERDEC, JPMorgan Chase, and other industrial sponsors. He is currently involved in the educational launch of a multi-faceted cybersecurity initiative at the University of Delaware, where he is developing new security courses and degree programs, including a minor and MS in Cybersecurity. Dr. Cotton currently consults on communications and Internet architectures for many carriers and equipment vendors worldwide.

Contents

Abbreviations .... *ix*
Preface .... *xv*
Introduction .... *xvii*

PART I: Cybersecurity: Ethics and Legal 1
1. Introduction to Cyberethics 3
2. Ethical Issues in Cybersecurity 41
3. Cybersecurity Ethics: Cyberspace and Other Applications 69
4. Introduction to Cyber laws 135
5. Cyber laws in the United States 157

PART II: Cybersecurity: Risks and Policies 239
6. Risks in Cybersecurity 241
7. Cyber Risks and Cyber Insurance 307
8. Introduction to Cybersecurity Policies 329

Bibliography *395*
Index *403*

Abbreviations

ACH	automated clearing house
ACP	access control policies
AI	artificial intelligence
AICPA	american institute of certified public accountants
ALARP	as low as reasonably practicable
AO	authorizing officials
AT&T	american telephone and telegraph
ATO	authorization to operate
AUP	acceptable use policy
AWS	autonomous weapons systems
BA	british airways
BCP	business continuity plan
BGLTQ	bisexual, gay, lesbian, transgender, queer, and questioning
BYOD	bring your own device
CBS	columbia broadcasting system
CCPA	california consumer privacy act
CD	compact disc
CDE	cardholder data environment
CEO	chief executive officer
CERT	community emergency response team
CFAA	computer fraud and abuse act
CFTC	commodity futures trading commission
CI	critical infrastructure
CICA	canadian institute of chartered accountants
CICS	center for information and communications sciences
CIO	chief information officer
CIS	center for internet security
CISA	cybersecurity information sharing act
CISO	chief information security officer
CNIL	commission nationale de l'informatique et des libertés
COBIT	control objectives for information and related technology
COPPA	children's online privacy protection act
COSO	committee of sponsoring organizations

CPU	central processing unit
CRO	contract research organizations
CS	control strength
CSF	common security framework
CSF	cybersecurity framework
CVSS	common vulnerability scoring system
DCS	distributed control systems
DDOS	distributed denial of service
DEX	decentralized exchange
DFAR	defense federal acquisition regulation
DHS	department of homeland security
DMCA	digital millennium copyright act
DNS	domain name system
DoCRA	duty of care risk analysis standard
DoD	department of defense
DPC	data protection commission
DSL	digital subscriber line
DTA	decision tree analysis
DVD	digital versatile disc
EAR	export administration regulations
ECPA	electronic communications privacy act
EDI	electronic data interchange
EEA	european economic area
EHR	electronic health records
EL	environment and legal pest
EMV	expected monetary value
ERM	enterprise risk management
ETRA	enterprise technology risk assessment
EU	european union
FACA	federal advisory committee act
FAIR	factor analysis of information risk
FAR	federal acquisition regulation
FBI	federal bureau of investigation
FCC	federal communications commission
FCRA	fair credit reporting act
FDA	food and drug administration
FERPA	family educational rights and privacy act
FFIEC	federal financial institutions examination council

FISMA	federal information security management act
FMEA	failure mode and effect analysis
FOIA	freedom of information act
FTA	fault tree analysis
FTC	federal trade commission
GAPP	generally accepted privacy principles
GDPR	general data protection regulation
GLBA	gramm-leach-bliley act
GPS	global positioning system
HAZOP	hazard and operability
HCS	harvard computer society
HHS	health and human services
HIPAA	health insurance portability and accountability act
HITRUST	health information trust alliance
HIV/AIDS	human immunodeficiency virus/acquired immunodeficiency syndrome
HR	human resources
ICO	information commissioner's office
ICO	initial coin offering
ICS	industrial control systems
ICT	information and communication technology
ID	identification
IDS	intrusion-detection systems
IEC	international electrotechnical commission
IG	inspector general
IJISP	international journal of information security and privacy
IoMT	internet of medical things
iOS	iPhone operating system
IoT	internet of things
IP	intellectual property
IP	internet protocol
IR	incident response
IRB	institutional review boards
IRMI	international risk management institute
IRPA	individual risk per annum
IRS	internal revenue service
IRTPA	intelligence reform and terrorism prevention act
ISACA	information systems audit and control association

ISMS	information security management system
ISO	international organization for standardization
ISP	internet service provider
IT	information technology
ITAR	international traffic in arms regulations
ITC	information technology and communication
JSTOR	journal storage
LEF	loss event frequency
LSIR	location-specific individual risk
MAE	major accident event
MDM	mobile device management
MIT	massachusetts institute of technology
MOOC	massive open online courses
MTTD	mean time to detect
MTTR	mean time to remediate
NCCIC	national cybersecurity and communications integration center
NCPAA	national cybersecurity protection advancement act
NFC	near-field communication
NGT	nominal group technique
NIST	national institute of standards and technology
NITRD	networking and information technology research and development
NPI	nonpublic personal information
NPR	national public radio
NSF	national science foundation
OCTAVE	operationally critical threat, asset, and vulnerability evaluation
OMB	office of management and budget
OSP	online service providers
PCI DSS	payment card industry data security standard
PCI SSC	PCI security standards council
PCI	payment card industry
PCNA	protecting cyber networks act
PEST (EL)	political, economic, social, technological environment and legal
PFI	personal financial information
PHI	personal health information

Abbreviations

PI	probability impact
PII	personally identifiable information
PIMS	privacy information management system
PLL	potential for loss of life
PLM	probable loss magnitude
PM	project managers
PMBOK	project management body of knowledge
PoW	proof-of-work
PR	public relations
QRA	quantitative risk analysis
R&D	research and development
RAM	risk assessment matrix
RFID	radio-frequency identification
RICO	racketeer influenced and corrupt organizations
RMF	risk management framework
SCA	stored communications act
SCADA	supervisory control and data acquisition
SD	standard deviation
SEC	securities and exchange commission
SEO	search engine optimization
SLA	service level agreement
SME	small and medium-sized enterprises
SMS	short message service
SOA	service-oriented architecture
SOC	service organization and control
SOX	sarbanes-oxley
SQL	structured query language
SSC	security standards council
SSL	secure sockets layers
STD	sexually transmitted disease
SWIFT	structured what-if technique
SWOT	strength weakness opportunity and threat
TCap	threat capability
TCP/IP	transmission control protocol/internet protocol
TEF	threat event frequency
TLS	transport layer security
TSC	trust services criteria
TM	trademark

UCLA	university of california, los angeles
UIGEA	unlawful internet gambling enforcement act
UK	united kingdom
UMC	unfair methods of competition
UN	united nations
US	united states
US-CERT	united states computer emergency readiness team
USD	united states dollar
USSOCOM	U.S. special operations command
VPN	virtual private network
Vuln	vulnerability
WA	wassenaar arrangement
WAF	web application firewalls
Wi-Fi	wireless fidelity
WIPO	world intellectual property organization
WLANs	wireless local area networks

Preface

The world of cybersecurity happens to be very dynamic. As technology progresses, the field witnesses novel innovations in the forms of offensive tools and defensive mechanisms that make it a constant battleground, where the fight between red teams and blue teams ensues habitually and often on a large scale. As a cybersecurity researcher, I have always been intrigued by what is allowed and what is impermissible in the cyberspace battleground. Could there be rules? Is everything fair in cyberspace?

The curious mind raises questions on morality and legality in cyberspace. The world relies on the Internet, and cyberspace incorporates networks. Thus anyone who owns a device that connects to the network is technically a part of cyberspace. Cyberspace neither knows race nor religion, and therefore is for everyone, and as users and part of cyberspace, we have a responsibility towards it. It is necessary to ensure that cyberspace follows a code of conduct where users create a live and let live environment. Often it is seen that cyberspace and the tools and techniques that its harbors are abused, knowingly, as well as unknowingly, and supposedly innocent victims have been convicted of crimes they committed but never realized. Hence, it is necessary to understand the ethical and legal aspects of cyberspace followed by the risks involved along with the proposed policies.

This book is not specific to a particular group of audience but is for everyone who is a part of cyberspace. Whether you own a phone or break into systems for fun, this book is for you. Cyberspace being the habitat of so many users, enlists a plethora of crimes, some of which are yet to be committed in the future. The book incorporates several reports and case studies derived from real incidents that occurred in the past, which justifies the necessity of ethics and laws in cyberspace. I realize the demand for a textbook that sees the relationship between ethics, laws, risks, and policies in cyberspace. This book is the first of its kind to integrate these four domains in cyberspace. While the majority of textbooks target an audience like instructors, students, professionals specific to their topics, this book targets a larger audience and may be of interest to anyone and everyone who is an essential constituent of cyberspace.

This book is also intended to serve as a summary of the state of the art of Cyber laws in the United States. More than 50 Cyber laws have been considered in the book. Also, a number of organizations have come up with their own risk frameworks and security frameworks. Many of these exist as white papers, which may not be easily accessible to everyone. In this book, we have tried to incorporate various risk management and security frameworks for better comprehension. The book provides state of art and defines the scope of cybersecurity in the near future.

Introduction

This book underpins the concepts of ethics, law, risks, and policies in cyberspace, respectively. The book has been divided into two parts, such that the first part takes into consideration the ethics and law in cybersecurity, while the second part takes into account risks and security policies.

Although the first part of the book incorporates five chapters and the second one only three, the magnitude of material and information for both parts is comparable. The book is a compilation of several works that have been discussed before so as to provide readers with clear knowledge on the topic. Each chapter is followed by an overall summary and review that registers the key points in the corresponding chapter. Further, a section dedicated to a few questions from the chapter is also presented for readers to evaluate their understanding based on the chapter content.

We start with a very basic overview of ethics and why it is necessary for cyberspace. Chapter 1 also discussed the need, the principles, and the sources so that the readers can know more in less time. We gradually explore the depths of cybersecurity to which ethics may be applied by highlighting specific cybercrime issues in Chapter 2. Nest, Chapter investigates ethics in cybersecurity in a more extensive way by detailing ethics in each component of cyberspace. Moreover, cybersecurity can be applied to many applications. The chapter encompasses 10 application domains with respect to cybersecurity from an ethical viewpoint.

Chapters 4 and 5 are mainly about the legal aspects of cyberspace. In Chapter 4, we layout an overview of Cyber laws, the need for Cyber laws, and their association with intellectual property (IP). Substantial information on Cyber laws is presented in Chapter 5, in which as many as 50 Cyber laws have been discussed. Specific Cyber laws are also supported with case-studies from real-life incidents that happened in the past to stress on the importance of ethics and law in cybersecurity.

The second part of the book describes cyber risks and security policies. In Chapter 6, we cover substantial information related to cyber risks in the form of risk identification, risk analysis, risk assessment along with frameworks, risk management along with frameworks, and risk remediation. An in-depth explanation of the topics in this chapter, along with the explained industry-standard frameworks, provides the readers with a better understanding of the state of art of cyber risks. Chapter 7 discusses a very important and exquisite topic of cyber insurance which is well connected to cyber risks. Chapter 7 provides an overall idea of cyber insurance followed by its benefits, types, coverage, etc. Finally, the last chapter of the book, as well as the second part, is about cybersecurity policies. The chapter consists of a policy overview followed by several popular cybersecurity frameworks like NIST, COBIT, PCI/DSS, ISO series, etc. The chapter is aimed at acquainting readers with the roles of policies in cybersecurity and how they are being implemented by means of frameworks. Since security loopholes emerge from time to time, these frameworks undergo multiple improvisations and may leave enough to the readers' imagination.

PART I
Cybersecurity: Ethics and Legal

CHAPTER 1

Introduction to Cyberethics

In this chapter, we introduce the concept of cyberethics and its importance in cyberspace. We also present certain terms and their definitions. Several concepts have been explained in the form of illustrations followed by explanations in order to present the readers a better idea about the topics.

1.1 ETHICS IN CYBERSPACE AND CYBERSECURITY

The National Institute of Standards and Technology (NIST) defines cyberspace as the interconnection and association of networks of information technology (IT) infrastructures. The infrastructure comprises computer systems, internet, telecommunications networks, and embedded processors and controllers in critical industries [1].

1.1.1 DEFINITION 1.1: CYBERSPACE

> In 1982, a science fiction writer by the name William Gibson coined the term 'Cyberspace' [2]. Cyberspace may be defined as the national environment in which communication over computer networks occurs. It is a computer network that integrates and incorporates a worldwide network of computer networks. These networks use the TCP/IP (transmission control protocol/ internet protocol) network protocols for facilitating data transmission and exchange.

While cyberspace takes into account the overall functioning of all these components, it also commands safety and reliability due to the huge amount of data traveling across cyberspace. This data may be highly sensitive. Data protection and integrity of computing assets that are a part of an organization's network are critical. Hence, there is a need to defend the assets from any kind of cyber-attacks. Cybersecurity may be defined as the process of preventing any damage, followed by protecting, and restorating the computers, electronic communications services, wire communication, electronic communications systems, and electronic communication [3].

1.1.2 DEFINITION 1.2: CYBERSECURITY

> Cybersecurity refers to the body of technologies, processes, and practices designed for the protection of networks, devices, programs, and data from attack, damage, or unauthorized access. It incorporates securing the networks, applications, information, and operations that take place over cyberspace.

These components may incorporate within themselves some information, and for securing these components and the data within it, it is mandatory to ensure its availability, integrity, authentication, confidentiality, and nonrepudiation which are basically the pillars of cybersecurity [14].

Figure 1.1 describes the pillars of cybersecurity which are essential for securing cyberspace. They are as follows:

1. **Confidentiality:** It refers to protecting the information from disclosure to unauthorized parties. It is associated with the protection of details which should be visible or accessible to people who have appropriate privileges.
2. **Integrity:** It is responsible for ensuring trustworthiness, accuracy, and completeness of the sensitive information. The main objective of integrity is to protect information from being altered by unauthorized or unintended parties and individuals.
3. **Availability:** It is responsible for ensuring that only authorized parties can access the information when at the time of need.
4. **Authentication:** It refers to the process of ensuring and confirming the identity of a user.
5. **Non-Repudiation:** It can be used to ensure that a party involved in a communication cannot deny the authenticity of their signature on a document or the sending of a message that they originated.

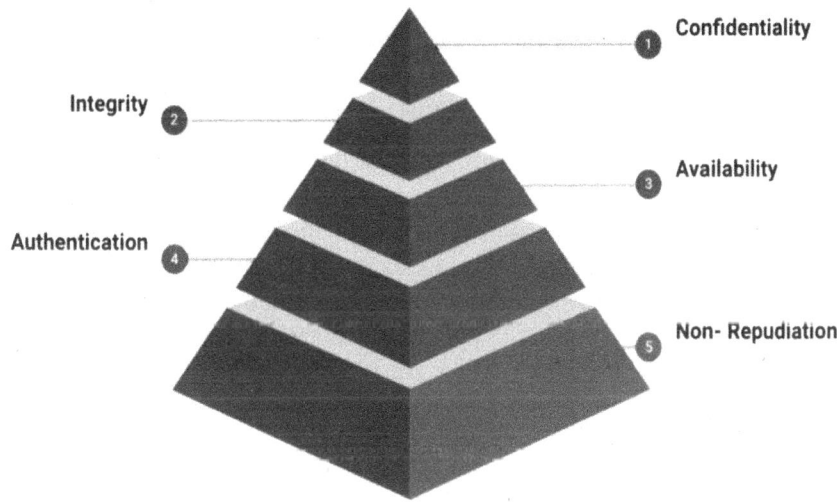

FIGURE 1.1 Pillars of cybersecurity.

Over the last few decades, cyberspace has expanded tremendously and has influenced several other fields like healthcare, finance, technology, business, etc. As cyberspace got associated with these fields, the underlying risks of cybersecurity also made their way into these fields. The interaction of humans with cyberspace has also increased with the advancement of technology. This has also contributed a lot to the cybersecurity risks [15]. As people are connected to each other using such a substantial platform, a lot of activities take place over cyberspace that are morally wrong. Internet crimes have shown an upward trend in graphs over the last decade, indicating the number of crimes that have been increasing day by day [16]. A large chunk of crimes committed in cyberspace may be traced down to employees working within an organization. Sometimes these employees target their previous work places. Spam and phishing emails for ransomware infection are another form of cyber-attacks that are common. Much of these cybersecurity crimes can be attributed to a lack of end user cybersecurity training. The lack of end-user cybersecurity training can be traced down to few observation points:

- First, lack of proper cybersecurity ethics training may make the employee unaware of any breach of ethics in cyberspace. An employee may not realize if he/she has been victimized on the grounds of ethics.

- Second, the employee themselves may at some point of time do something that is ethically wrong.
- Third, due to the lack of cyberethics training, an employee may not know what is to be done in this situation, further deteriorating the situation.

Thus, there is a need to highlight the importance of cyberethics.

Cyberethics presents certain situations in the form of behavior as good or bad, and right or wrong. Cyberspace may be influenced by both people as well as technology. Cyberethics gives an overview of the situations and challenges that arise in the cyberspace due to people and technology. Cybersecurity ethics is a part of cyberethics. While cyberethics deals with computers and networks and programs and the people and organizations that use them, cybersecurity ethics is more concerned about the ethics associated with the practice of cybersecurity, usually what is done by the cyber professionals.

As we know, neither cybersecurity issues know any bound, nor does cyberspace. Thus, there are no physical borders when it comes to cyberspace. Although national and international laws exist, cyberspace is global. Therefore, ethics in cyberspace is also global as well as interconnected.

1.1.3 DEFINITION 1.3: CYBERETHICS

> Cyberethics may be defined as the code of responsible behavior over cyberspace. Laws are the outcomes of Ethics. Ethics are principles that are responsible for guiding a person or society. They are created to decide what is good or bad, and what is right or wrong in a given situation. It is used for regulating a person's conduct and also assists individuals in living better lives by considering basic moral rules and guidelines.

Thus, ethics promotes a sense of fairness and promotes acceptable behaviors. Ethical practices are useful for identifying what unacceptable behavior looks and feels like. Ethical behavior requires courage, i.e., say something if you see something that is morally wrong, and to stand up against what is morally wrong. Ethical behavior also requires humility, i.e., if you are ethically wrong, accept it because wrong decisions can be made by anyone.

The concept of Cyberethics is not confined, rather varies across many definitions. Wikipedia defines 'cyberethics' as the philosophic study of ethics for computers. It also highlights user behavior, what computers are programmed to do, and the effect of this on individuals and the society [4]. Further, Pusey and Sadera defined cyberethics as a set of "moral choices made by individuals using Internet-capable technologies and digital media [5]. Cyberethics is also referred to as a branch of applied ethics that explores the issues related to computer/information and communication technologies morally, legally, and socially. Sometimes it is also mentioned as Internet ethics, computer ethics, and information ethics. The expression "Internet ethics" is quite narrow as it is not enough to investigate the expanse of cyber-related ethical issues that are due to independent internet and networked computers. Therefore, there is a need to explore a new type of ethics which has emerged over the last few decades as a result of the creation of the Code of Computer Ethics. This new kind of ethics may have a binding effect on the professional, especially if the code is incorporated into work ethic and procedure. The Centre for Internet Security has its own definition of cyberethics which is, "the code of responsible behavior on the internet [6]. Thus, in general, cyber-ethics encourage the use of appropriate ethical behavior and acknowledge rights and responsibilities that are associated with online environments and digital media.

Ethics in cyberspace may be described in the following ways:

1. Cyberethics underpins the study of ethics pertaining to computers and networks;
2. Cyberethics also takes into account user behavior as well as what computers are programmed to do;
3. Cyberethics has its effects on individuals and society;
4. Cyberethics focuses on responsible behaviors on the internet;
5. Cyberethics deal with ethics applied to the Online Environment.

1.2 ETHICS, LAW, AND POLICY

As mentioned before, laws are the outcomes of ethics. It is interesting to observe that Cyber law is related to cyberethics. While cyberethics is concerned with providing foundations for ethical behavior in cyberspace, thereby reflecting the ethical standards of human civilization, Cyber law, on the other hand, as a discipline, deals with legislations that are passed in different countries. These legislations are capable of effectively providing sanction, validity, and enforceability to various principles concerning ethical behavior in cyberspace. Thus, Cyber law as a discipline, effectively strengthens the foundations of good ethical behavior. This is a requirement for cyberethics. Moreover, cyberethics can be significantly enhanced by deploying cyber legal frameworks, as the ethical principles on their own do not have any respective standing. Cyberethics only stipulates moral values but until ethical standards concerning ethical behavior in cyberspace are sufficiently backed by appropriate legal provisions and sanctions, they rarely get complete enforceability. Cyber Security Policy is a formal set of rules and must be followed by people who are given access to company technology and information assets. The Cyber Security Policy serves several purposes, like defining what technology and information assets must be protected. It is also responsible for identifying any underlying threats to the assets. The Cyber Security Policy also describes the responsibilities and privileges that users have.

There is an interconnection between Ethics, Law, and Policy in the Cyberspace. While Ethics account for code of conduct and responsible behavior that should be followed in cyberspace, Cyber laws are legislations that focus on the acceptable behavioral use of technology in cyberspace and must be followed. Policies are made to achieve some goals and are therefore usually followed (Figure 1.2).

1.3 PRIVACY AND SECURITY

As the world progresses towards hyperconnectivity, the issue of security versus privacy remains one of the fiercely contested global issues. However, privacy, and security are comparable. Privacy compares to any rights one has to control the personal information and how it is used. Security, on the other hand, considers how personal information is protected. Privacy and Security often overlap in the real world; however, they are not the same, and knowing how they differ may assist further protection. When we deal with personal information, there may be three situations:

- First, both privacy and security are maintained. Your personal information stays with you;

Introduction to Cyberethics 9

- Second, privacy may be compromised, but security is still maintained (Figure 1.3). Your personal information may not be only with you; and
- Third, both privacy and security are compromised.

This usually happens in case of a data breach.

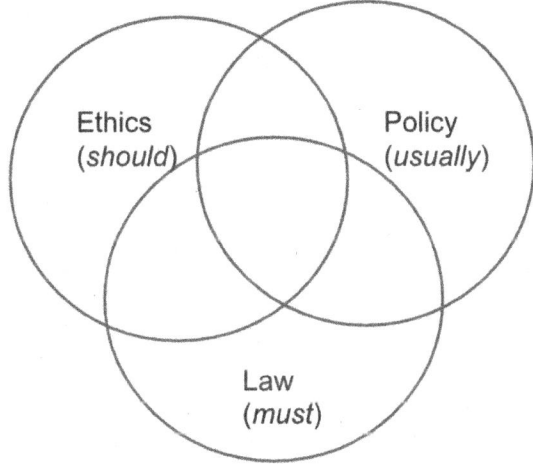

FIGURE 1.2 The relationship between ethics, policy, and law.

Thus, privacy may be more often compromised than security. Privacy is concerned with the collection and use of data about individuals. Privacy is an ethical concern. Privacy breaches disrupt trust and initiate the risk of losing security. It disrespects the code of conduct and violates ethical principles. In Figure 1.3, we see that a group of laborers are working on ensuring security, however, they are seizing privacy [17].

1.3.1 DEFINITION 1.4: SECURITY

> Security is essentially about protection against the unauthorized access of data. It is specifically conducted by deploying security controls to limit who can access the information. Security is primarily concerned with the protection of data while stored, in transit, and during processing, and the related informational assets like servers and mobile devices

FIGURE 1.3 Illustration showing security and privacy.

1.3.2 DEFINITION 1.5: PRIVACY

> Privacy deals with an individual's right to own the data that is originally triggered by his or her life and activities, and for restricting the outward flow of that data. personally identifiable information (PII), personal health information (PHI), Personal Financial Information (PFI), etc., are some private information related to a person.

Figure 1.4 describes privacy and security. Privacy deals with information collection, using, and disclosing personal information in an authorized manner, data quality and access to personal information. Security is primarily concerned with confidentiality, integrity, and availability.

1.4 THE NEED FOR CYBERETHICS

As we know, cyberethics deals with the responsible behavior that should be followed in cyberspace. The constituents of cyberspace are the internet,

Introduction to Cyberethics

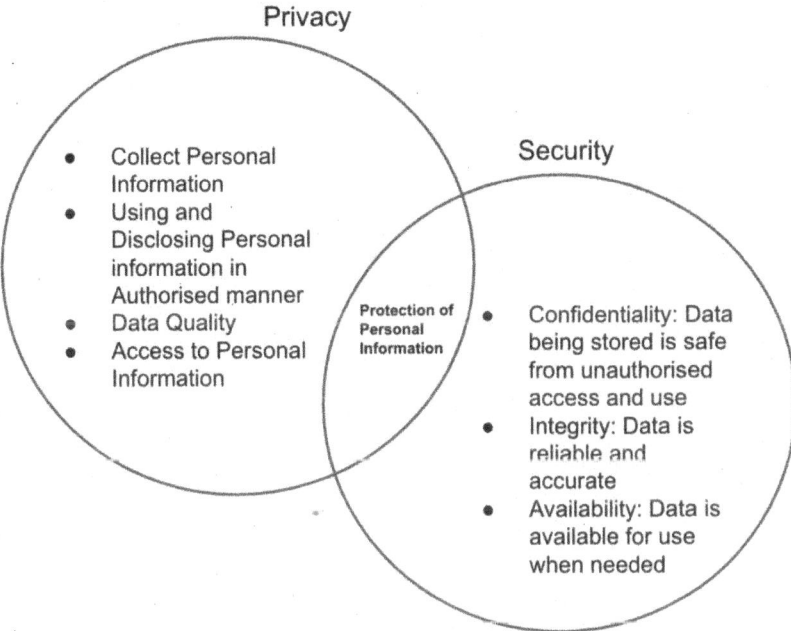

FIGURE 1.4 Differences between privacy and security.

computer systems, telecommunications networks, and embedded processors and controllers in critical industries. All these components are interconnected, but have their own ways of operating. All these are subject to different ethical issues, as they work in different ways. Therefore, the Ethics that must be applied and followed in each of these domains might also differ. We should be concerned because it may have a bad effect not only in cyberspace but also on people who are interacting with these systems as we all do with each other over cyberspace [7]. Some of the situations that require us to follow Ethics in cyberspace are as follows:

1. Social media constitutes a significant part of cyberspace, and the number of users in social media is very high. These are the platforms where people can express their opinions anonymously in the free society. However, these platforms also witness anonymous postings to blogs, websites, and social media that can encourage bad behavior anytime. Bullying, harassment, stalking, gender inequality, human rights and fake death news are very common over social media.

2. Although information in cyberspace can be retrieved globally, it is now standardized. Any information that is available on cyberspace can be accessed unless protected. Consent must be taken before accessing data. It is the moral responsibility of an individual to respect anonymity, confidentiality, and privacy of the information available on cyberspace.
3. Ethical issues may also be attributed to the use and development of technologies. IT not only affects things that are done, but also about how they are thought of. This challenges some of the very basic organizing concepts of moral and political philosophy such as property, privacy, the distribution of power, basic liberties and moral responsibility. There is a need to understand security, privacy issues, and adverse effects of IT in cyberspace.
4. Computer networks are prone to many internal and external hazards internationally, which may not be ethical in numerous situations.

1.5 PRINCIPLES OF CYBERETHICS

Cyberethics underpins the idea of understanding the responsibility that is followed by knowing and using cybersecurity principles.

The Internet Architecture Board [8] has defined the following activities to be unethical and unacceptable:

1. Activities seeking to gain unauthorized access to the resources of the internet;
2. Activities disrupting the intended use of the internet;
3. Activities that lead to wasting of resources (people, capacity, and computer);
4. Activities that destroy the integrity of computer-based information;
5. Activities that compromise the privacy of users.

Moreover, The Computer Ethics [9] published the 10 commandments to manifest the idea of what is ethically acceptable in Cyberspace:

1. A computer shall not be used to harm other people;
2. One must not interfere with other people's computer work;
3. It is unethical to snoop around in other people's computer files;
4. A computer must not be used for stealing;
5. A computer must not be used for bearing false witness;

6. It is unethical to copy or use proprietary software without permission or paying for it;
7. It is unethical to access an individual's computer resources without authorization;
8. It is unethical to appropriate other people's intellectual output;
9. When writing a program or designing a system, one must think about the social consequences;
10. A computer shall always be used in ways that ensure consideration and respect for fellow humans.

Based on the 10 commandments, it is possible to establish a set of Principles for Cyberethics in the form of what should be done morally and what should not be done [125].

Do's:
1. Think about the social consequences of writing a program and designing a system.
2. Always use a computer in ways that guarantees consideration and respect for other individuals

Don'ts:
1. Use a computer for harming other people.
2. Interfere with other people's computer work.
3. Snoop around or access other people's computer files.
4. Use a computer for stealing.
5. Use a computer for bearing false witness.
6. Copy or use proprietary software without paying or obtaining permission.
7. Use other people's computer resources without authorization.
8. Appropriate other people's intellectual output.

1.6 SOURCES OF ETHICS: WHAT MAKES SOMETHING ETHICAL OR UNETHICAL IN CYBERSPACE?

Cyberspace has an impact over different parts of society and cyberethics includes almost all ethics domains [7]. In today's world, for perceiving any problem from the ethical viewpoint, the cyber-dimension of the ethical question must be taken into account. Figure 1.5 shows some sources of ethics.

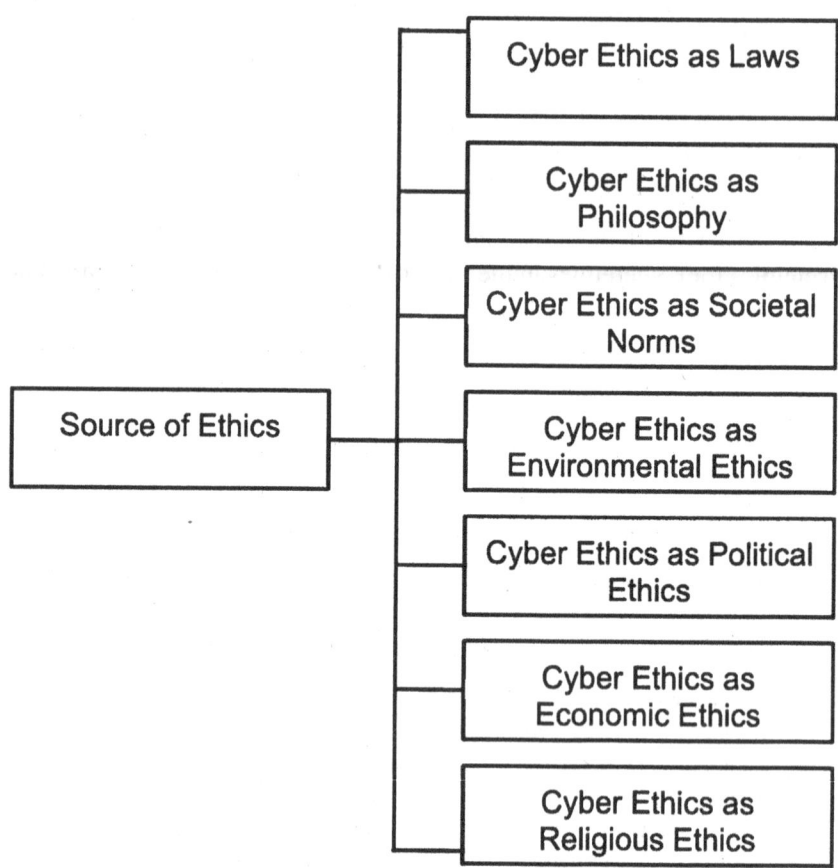

FIGURE 1.5 Sources of ethics.

1. **Cyberethics as Laws:** It refers to the fact that specific laws are derived from ethics. Cyber law and Cyberethics are interconnected and dependent on each other. The increasing significance of cyberspace may further bring complex legal, policy, and regulatory issues. The need for cyberethics to be strengthened by Cyber laws.
2. **Cyberethics as Philosophy:** It explores the philosophical aspect behind ethics that includes life between birth and death. Gaining

unauthorized access to someone's account or compromising the privacy of an account are situations that are unethical and Ethics as Philosophy could avert such situations.

3. **Cyberethics as Societal Norms:** It deals with Community Ethics. These underpin questions of bad and good of social media, how it has an impact on the community life and chances of global communication. It also takes into account abuse in terms of cyber bullying, mobbing, etc. It manifests the core values and virtues that usually have their source in the family.

4. **Cyberethics as Environmental Ethics:** It deals with the impact of cyber technology on human-nature relations. It also highlights the environmental negative impact of energy use as well as the positive impact of environmental advantages of weather forecast, scientific research, etc. It deals with questions like whether it is ethical to jeopardize natural resources in order to carry out research.

5. **Cyberethics as Political Ethics:** It is concerned with changes in political systems. These may be in the form of elections, security, armies with autonomous weapons, need, and limits of regulation of cyberspace on international and national levels, etc. Elections have been known to get manipulated over cyberspace; Ethics from the political perspective may be one of the ways to avoid it.

6. **Cyberethics as Economic Ethics:** It explores the positive and negative impacts of cyberspace. The factors taken into account are economic growth, job creation or job losses, financial investments in cybersecurity research, etc. The financial sector is frequently hit by data breaches and monetary losses. Economic Ethics could be followed while dealing with such situations.

7. **Cyberethics as Religious Ethics:** It looks at the ethical and unethical impact of cyberspace on culture, music, art, dance, language diversity, cultural inclusion or discrimination, religious respect or hate messages through the internet, etc. Social media provides platforms for religious and cultural disputes. Keeping in mind Religious Ethics may be a way of avoiding it.

Thus, ethics may be derived from different domains, which may also be applied to the cyberspace.

1.7 ETHICS IN CYBERSECURITY

In the previous sections, we emphasized on Ethics in Cyberspace. As we know Cybersecurity Ethics is different from Cyberethics. While Cyberethics is more or less concerned with the entire Cyberspace, Cybersecurity Ethics is more or less concerned with the security aspect of it. In this section, we will highlight the Ethics pertaining to Cybersecurity specifically [7].

Cybersecurity practices are concerned with safeguarding data in computer systems and networks (software and hardware). The data that is being kept safe may be sensitive and might hold economic value. Hence cybersecurity practices aim to protect confidentiality, integrity, availability (authentication, non-repudiation), etc. While cyberspace sees users from all respects that communicate with devices and networks, cybersecurity is of utmost importance to cybersecurity professionals. Therefore, protection of information is critical as the data might belong to other people. For example, safeguarding healthcare data is the job of a cybersecurity professional, but the data that is being protected is of a sick patient. Thus, a patient's privacy, health, and failure is somehow related to cybersecurity professionals. Further, the well-being of patients' families and caregivers may be shielded by the practices of a cybersecurity professional.

Critical information like these spread across cyberspace. This information may belong to different domains like finance, agriculture, power, etc.

This brings us to the conclusion that ethics form a significant part of cybersecurity practices. This is because security practices are increasingly required for securing and shielding the ability of human individuals and groups to live well. With the increasing complexity and the underlying challenges in securing online data and systems across a continuously growing landscape of cloud computing services, Wi-Fi-enabled mobile devices, and 'smart' objects, the ethical responsibility to protect data that belongs to others is borne by cybersecurity professionals is a significantly heavy burden.

Let us take a few examples to analyze the importance of best cybersecurity practices:

1. James is a hard-working employee who works in the finance industry. A phishing email opened by him for a bank infects the company network with malware. The malware targets customer profiles and authorizes all transactions.

2. Lee and John had been working late at night on an operating system when they identified a vulnerability, which causes execution of malicious code and switches on the camera and voice recorder without the knowledge of the user. As they agree on informing the manufacturer the next day and posting the bug on their blog, they are approached by a rival of the phone company, who lures them with money if they do not expose the vulnerability.
3. A company is offering a brand-new Porsche to the 500th caller. Derek, a teenage hacker, takes control of the phone network and effectively blocks calls so that he could be the 500th caller.
4. Pam is a history teacher who teaches sophomores straight from the presentation slides. These slides are saved in her email folder. A group of students who do not pay attention in her class and got bad grades for the midterm decided to hack her email. When she opened her folder the next day, she was shocked to see her picture replaced with a dinosaur and the slides having pictures of cartoons.
5. A massive network outage caused by a virus blocked access to weather updates, following which the Wilson family was stuck in the snowstorm for hours.

Which of these hypothetical cases are concerned with ethical questions in cybersecurity? The answer is all of them. We observe that in all cases, some way or the other, the lives of people are impacted by cybersecurity professionals' other actors in the information security space have or have not done. In many cases, cybersecurity best practices may have prevented or limited the harm. Thus, cybersecurity professionals are challenged ethically on multiple levels. A cybersecurity professional has to deal with several ethical questions on a daily basis. They are sometimes technical as:

1. Which security techniques may be more effective, and what resources do they require?
2. How can network intruders be kept away from the system?
3. What levels and types of security risks are tolerable?
4. How can the users and other affected stakeholders be warned of such risks?

In many situations, the questions may not be technical as:

1. How to justify the act of exposing others to a breach or greater risk for their own personal profit?

2. How to balance competing duties to the employer?
3. What levels of care must be taken to perform the role responsibly?

Therefore, a broader and better understanding of cybersecurity ethics is essential for promoting and protecting human flourishing in an increasingly networked society.

1.8 IMPORTANT ETHICAL ISSUES IN CYBERSECURITY

In the previous section, we introduced the concept of ethics specifically for cybersecurity. Cybersecurity deals with the protection of data, hence it is prone to a large number of ethical issues [10]. In this section, we will be taking into account some important ethical issues in cybersecurity. However, before that we need to address two important concerns. First, what makes an ethical significant, and second, what important ethical benefits and harms are associated with cybersecurity efforts.

Ethical actions in cybersecurity is basically about bringing moral discipline to cyberspace from the security perspective. As a part of cyberspace, we are obliged to care about ourselves as well as those with whom we share cyberspace. The reason we mention the significance of ethical issues in cybersecurity is because certain unethical practices may harm others or may have the potential to affect the chances of having a good life for some individuals. In a society, where we need to flourish together, there is a need to abandon acts that may be deemed unethical.

In the previous section, we mentioned how technical choices may become unethical. Similarly, with respect to other technical contexts like engineering, and use of surgical devices, aeronautics, buildings, nuclear power containment structures, manufacture, and bridges, there is a high probability of ethically significant harms that may result due to poor technical choices. Moreover, there are ethically significant benefits of choosing to adopt the best technical practices known to us. Cybersecurity particularly is a good example of such a domain where not following best practices or making poor technical choices may harm people, property, economy, infrastructure, etc. Significantly. Thus, many innocent people may suffer if we overlook public welfare and act carelessly or irresponsibly by choosing bad practices or poor technical choices. On the other hand, if we act ethically by choosing best practices or good technical choices, people will enjoy better lives. Choosing good practices over bad practices preserves and enhances the opportunities that other people have for enjoying a good life. Good technical practice leads to better ethical practice. Consider a situation where a civil engineer willfully or recklessly ignores a bridge design specification, which resulted in the bridge getting collapsed leading to deaths of a dozen people. Such an engineer is not just bad at his or her job but is also guilty of an ethical failure. Even if they are shielded from professional, legal or community punishment, it is still true. As far as cybersecurity practices are concerned, the potential harms and benefits, including matters of life and death, are based on the significance of ethics. However, since cybersecurity efforts are often carried out 'behind the scenes' largely hidden away from customers, clients, and other users, the ethics involved in cybersecurity practice may not be easy to recognize.

Now let us talk about the significant ethical benefits and harms linked to cybersecurity efforts. Understanding benefits and harms is usually linked to our life interests. As humans, we tend to have greater inclination and interests towards health, privacy, economic security, happiness, family, friendship, social reputation, liberty, autonomy, knowledge, etc. As humans, we also look out for respectful and fair treatment by others, entertainment, education, meaningful work, and opportunities for leisure, and creative and political expression, among other things. Each of these fundamental interests can be significantly impacted by Cybersecurity practices. Therefore, it is right to say that cybersecurity has a much stronger need of holding onto ethics than other technical practices like the engineering of bridges. Making unethical design choices for building bridges may lead to destruction of bodily integrity and health. Such damage may have an impact on society and can make it harder for people to flourish. In the realm of cyberspace, unethical choices with respect to cybersecurity may cause different kinds of harm. Cybersecurity failures may not only have an economic impact on our lives, but can also damage our reputation, savings, or liberty. Firms that are affected by security breaches or health records that are sabotaged are some examples which leave our physical bodies intact yet put us at risk of losing our reputations and privacy. Hence, there is a need to introduce effective cybersecurity practices. These practices are capable of generating a vast range of benefits for society, some of which include stable infrastructure, diminished social and economic anxiety, and better investment and innovation.

1.9 SOME ETHICAL ISSUES AFFECTING CYBERSECURITY

In this section, we will take into account specific ethical issues that prevail in the cybersecurity domain. Some of them are discussed in subsections [10].

1.9.1 ETHICAL ISSUES AFFECTING PRIVACY

Cyberspace incorporates oceans of sensitive data, which could be related to the finance industry, healthcare industry, etc. The data in cyberspace is prone to cybersecurity risks, which makes it imperative for the best cybersecurity practices to be followed. Common cyber threats include:

- Identity theft, wherein, PII may be stolen and used to impersonate victims. For example, Using credit card information to make unauthorized purchases.
- Social engineering, in which the adversary uses deception techniques for manipulating individuals into disclosing confidential or personal information to be used for conducting fraud.
- Hacking and other network intrusions are popular techniques for obtaining sensitive information about individuals and their activities. Such information can be used for blackmailing, extortion purposes, and other forms of unethical and/or illegal manipulation.

Such situations affect the privacy of victims and can cause harm to the interests of third-parties. Using blackmail to pressure compromised employees for betraying sensitive client information, trade secrets, or engaging in other forms of corporate espionage or government misconduct are some unethical cybersecurity practices. They are also responsible for adding to the risks of privacy and are further supported by the rampant growth of the global data ecosystem, wherein individuals may not personally control, correct, curate or delete, the storage or release of their private information. Anonymized data, when linked to other datasets, might reveal sensitive information about an individual. Information related to facial, gait, and voice recognition algorithms, along with geocoded mobile data, can be used to identify and gather a lot of information about a particular individual. Since cyberspace is networked, it is difficult to confine data or add boundaries to it. This is a challenge for cybersecurity professionals who are trusted to protect sensitive data against personal and organizational privacy harms. Personal control and containment of sensitive data may not be virtually possible to maintain in networked environments, therefore cybersecurity professionals must be highly skilled and trained with advanced cybersecurity tools. Cybersecurity professionals are therefore responsible for preventing any privacy issues and dealing with certain situations ethically. Use of weak passwords, negligent software updation and patching efforts, outdated encryption tools and lack of incident response (IR) planning contribute to poor cybersecurity practices. These can be unethical, and to a great extent may be responsible for personal and organizational privacy harms.

1.9.2 ETHICAL ISSUES AFFECTING PROPERTY

Abusing data privacy may be followed by property being affected by mechanisms like extortion. However, it is also possible to target property through cyberattacks which may affect intellectual property (IP) and lead to accessing sensitive information like trade secrets, obtain bank account numbers and passwords, and may also cause damage to organization's property. Such harms to property are often targeted by profit-seeking criminal enterprises, by politically-motivated groups, nation-states, detectives, intelligence agents of foreign nations, or by script kiddies and hackers who are seeking to demonstrate their own skills, abilities, and destructive powers.

These harms to property are unethical activities that affect certain individuals who rely upon these properties for their livelihood. While property may not have any ethical value, humans' lives are. Hence, any damage to these properties may be termed as an unethical issue in the cybersecurity domain, even if they may not be punishable by law.

Sometimes, unauthorized destruction of property may be argued to be ethical if it is in the interest of the nation. In 2010, the nation-state used the Stuxnet worm to disrupt Iranian centrifuges being used for enriching uranium. Sometimes, defenders of a network that has been vandalized by a cyberattack, assert an ethical right to 'hack back' in ways that intend to damage the systems of the cyber attacker, so that no further damage may be caused. These situations also generate significant ethical concerns. The release of Stuxnet affected thousands of other computers that were not a part of the Iranian nuclear program. Hacking back may be risky for innocent parties as its effects are unknown. Moreover, spoofing strategies in the past have also led to the misidentification of systems responsible for attacks. Even though all these arguments are valid, cybersecurity professionals must have a default ethical obligation to protect their organization's networks, or those of their clients, from any and all property-targeting intrusions and attacks.

1.9.3 CYBERSECURITY RESOURCE ALLOCATION

There is always a high cost involved in cybersecurity. This is due to the fact that cybersecurity operations consume time, money, and expertise. System resources bear considerable cost and cybersecurity operations incorporate several operations like power efficiency, data storage capacity, system usability/reliability, and speeds related to network and downloading. While cybersecurity operations lead to such high costs, not adopting effective cybersecurity measures may lead to excessive damage and even higher costs. A network that is secure and is practically unusable, or economically unsustainable cannot be justified. Usability and viability concerns are not enough to justify weakening security standards.

Given a situation, an organization wants to make Wi-Fi-enabled music devices, but does not have enough resources necessary for making it effective as well as secure from hackers, then there can be a strong ethical argument that asserts that the organization in the first place should have not been in the business of making Wi-Fi-enabled music devices. Moreover, in this situation, a cybersecurity professional who worked on the devices and the connection or enabled lax security controls with respect to such devices would also be disobeying ethical standards. This is because the professional would not be oblivious to the unacceptable risk of grave harm to others created by his own actions. If there is no clarity on how the issue of resource allocation could be an ethical issue, one must consider the stakes involved in striking the correct balance between competing resources and security. Considering a banking network security administrator exploring a suspicious port scan of the network. The administrator decides to address the possible threat by initiating a new and extremely resource consuming and time-consuming security login procedure. If the administrator does not consider the core function and interests of users of the network, customers' lives could be endangered, especially in departments that require quick network access for making urgent or immediate transactions. Therefore, maintaining a balance between resource utilization and cybersecurity along with various functionalities requires ethics. It demands careful analysis of the benefits, harms, values, and rights involved in the process and the potential impact of the decision on the ability of others to seek and lead good lives.

1.9.4 TRANSPARENCY, DISCLOSURE, AND ACCOUNTABILITY

Ethical issues in cybersecurity practices are linked to transparency, disclosure, and accountability. A significant part of cybersecurity is concerned with risk management. And because of that, since risks also have an impact over people in cyberspace, the default ethical duty to disclose the risks when known is necessary. This ensures that the people who might be affected due to the risks can be better prepared.

In a situation where an organization discovers a critical vulnerability, the customers/clients are informed about the vulnerability in a timely manner so that a patch may be installed or other defensive measures may be taken. Similarly, following a compromise in the user accounts of an organization, the customers are asked to change their passwords or PINs as soon as possible. In many cases, 'timely' notification is a matter of debate, considering the appropriate mode and extent of the disclosure. In a situation where a vulnerability is extremely hard to discover and cannot be resolved quickly by the security team, and also involves a critical network of high utility to customers, there may be some delay in notification before a patch is available and deployed. This is still ethical because premature disclosure may perhaps invite an attack that otherwise would not be there, creating a higher risk of harm to others. Also, there are several general transparency and disclosure guidelines that may be worth exploring. However, given the fact that in cyberspace, different scenarios involve different facts, and place different goods and interests at stake, there may be dynamic rules or instructions that may be applied to guarantee appropriately transparent cybersecurity practice. Thus every case requires careful ethical analysis of a particular scenario and the specific risks, benefits, tradeoffs, and stakeholder interests involved. Only then can a well-reasoned ethical judgment be followed about what is best to do, in light of the particular facts and options.

1.9.5 CYBERSECURITY ROLES, DUTIES, AND INTERESTS

Cybersecurity practices are not confined to one specific role but involve a number of distinct roles and interests. These roles may often conflict with each other, for example, while the red teams are known for exploiting vulnerabilities, blue teams are used to patch them. In many cases, it is difficult to comprehend what ethical duties are, what are the greatest ethical concerns, and whose interests should be protected the most. The variety of roles in cybersecurity practice may also lead to skepticism regarding the ethical standards of the cybersecurity community. This calls for careful ethical evaluations for overcoming such skepticism and for arriving at a justifiable consensus for particular cases. One of the most debatable issues in cybersecurity roles is related to 'white-hat,' 'black-hat,' and 'gray-hat' hackers. For developing and honing their own security skills, cybersecurity professionals occupy multiple and competing roles. This helps them develop their own security skills and knowledge base. It is common for cybersecurity professionals to feel conflicting loyalties to the interests of the public and government agencies. Sometimes they also work for employers or clients, particular subcultures and interest groups within the security community. Many times, cybersecurity professionals have their own personal interests.

In many situations, one cybersecurity researcher may desire publishing a previously unknown technique of undermining a common encryption key management system, in order to improve the community's knowledge base and for spurring research into countermeasures. The same researcher may also have clients in consulting firms who may be placed at greater risk by such a disclosure. Consider a situation where a chief information security officer (CISO) wishes to hire penetration testers for his 'Red Team.' He/she comes across a brilliant, experienced hacker with good skills but with professionalism and ethical values are still underdeveloped. In this case the CISO may want to mentor this person fully into the culture of 'white-hat' cybersecurity practice. This will give the hacker more ammunition to the 'good guys,' however, the risk involved is that one could expose his or her employer to an internal breach. All these issues that deal with ethical choices must be made by cybersecurity professionals. Thus, ethical issues are extremely complex and variable. Hence they may also be subjective. Several responses to cybersecurity issues are clearly ethically wrong; many are clearly ethically dubious. Some also fall short of what is expected from any respected cybersecurity professional. The task of finding cybersecurity solutions that may be called right or justifiable by professional standards is extremely challenging. It requires careful ethical analysis, reflection, and problem-solving. Cybersecurity professionals need to pay attention to the different ways in which their practices can remarkably impact the lives of others. These professionals must also learn how to anticipate their potential harms and benefits so that they may be effectively addressed for the future (Figure 1.6).

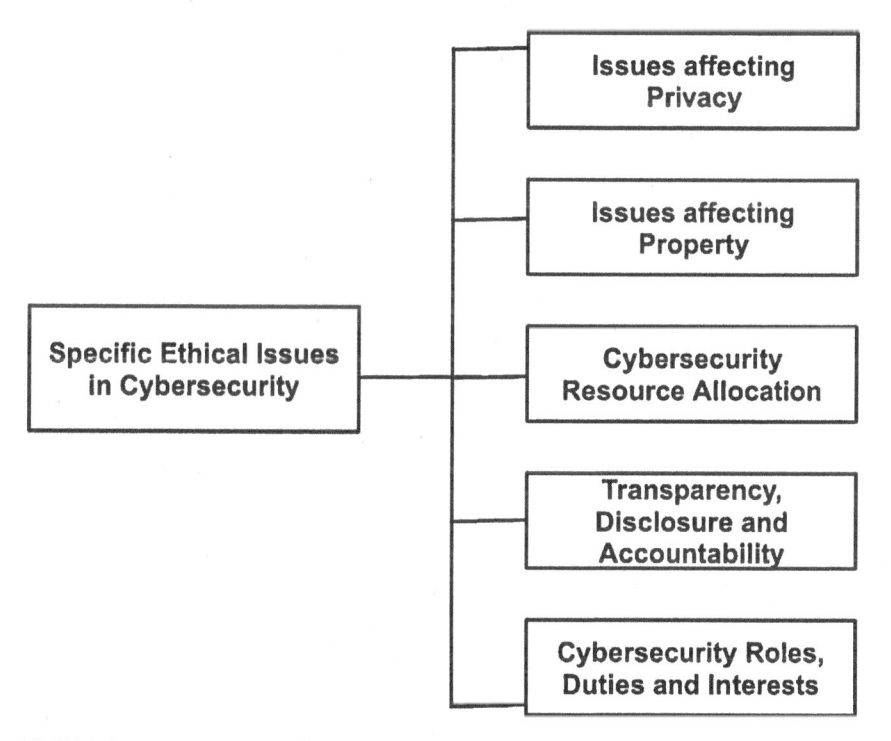

FIGURE 1.6 Specific ethical issues in cybersecurity.

1.10 THE CURIOUS CASE OF BLACK HATS, WHITE HATS AND GRAY HATS

The curious case of Black Hats, White Hats and Gray Hats In the previous section, we listed some specific ethical issues in cybersecurity. We came across the fact that cybersecurity roles, duties, and interests may conflict when it comes to ethics in cybersecurity. One of the most debatable issues related to the roles of cybersecurity professionals is based on cybersecurity practices based on hacking. Security professionals have different aims and interests for getting into systems, and they are usually characterized as people who wear different colored hats, manifesting their roles [11]. We list those here:

1. **White Hats:** Or ethical hackers, are referred to as the good guys in the hacking world. They are concerned with getting into computer networks and systems in order to test or evaluate its security systems. They assist companies in removing viruses or act as penetration hackers. They have adequate knowledge on how to find vulnerabilities and they can also create plans to eradicate them before they are compromised.
2. **Black Hats:** These are commonly known as crackers; these are the bad guys in the hacking world. They hack into computer networks and systems with malicious or criminal intent. They are responsible for most of the cybercrimes over cyberspace. Black hat hackers are primarily responsible for carrying out security operations involved in creating worms, DDoS, vandalizing networks, identity theft, for destroying and wreaking havoc.
3. **Gray Hats:** These are computer security experts who often violate laws or typical ethical standards. While they breach security standards like black hats, they do not have any malicious intent.
4. **Green Hats:** These hackers are usually newcomers who work on improving their skills to get better. They often idolize well-known black hats. This is because they are serious about elevating themselves to the real world of hacking.
5. **Red Hats:** These are also known as the vigilantes of the hacking world. This is because much of their work is about living and playing by their own rules. Although they do want to halt black hats just like white hats, their approach is different. White hats hunt for loopholes and try to build a stronger wall. At the same time, white hats also want to see black hats prosecuted for their crimes, although Red hats want to do more harm to the black hats. Upon finding a malicious hacker, red hats launch a full-scale attack. They upload viruses, access unauthorized information, initiate dosing and also attempt to access the hacker's computer to completely take control over it.

6. **Blue Hats:** These hackers can be malicious but usually, target a person or company. They launch attacks in order to seek retribution for some type of offense. Novice blue hats start out as Script Kiddies. They are driven by events and incidents which converts them from being nonchalant about hacking to being focused. They agree on some kind of hacking catastrophe for their target. While blue hats are not interested in wanting to learn or improve their skills, they want to learn enough only to retaliate.

Thus, we have acquainted ourselves with the different types of cybersecurity professionals who are concerned with getting into a system. We realize that although they perform similar functions, they are placed into different categories. However, if we are given the task of placing a random cybersecurity professional in any of these categories, it may not be that simple. One of the ways of categorizing these individuals is based on their intent which we explain in Table 1.1.

TABLE 1.1 Categorizing Cybersecurity Professionals based on Their Hacking Intent

Category	Intent
White hats	White hackers are influenced to use hacking as a profession. They are responsible for keeping the companies safe for which they work.
Black hats	Black hats are convinced by greed and have little consideration for anything else.
Gray hats	Gray hats are concerned with proving their own importance, seeing themselves as the most elite hackers
Green hats	Green hats aim for proving themselves and hone their skills. They are determined in their pursuits.
Red hats	Red hats are driven to end black hat hackers but do not want to play by society's norms and rules
Blue hats	Blue hats are influenced by retribution and do not think much of the after-effects.

Although different types of cybersecurity professionals exist, over the last few years, black hats, white hats and gray hats have grown quite popular. From an ethical point of view too, these three form the most debatable issue. Moreover, when it comes to ethics, White Hats and Black hats have certain boundaries defined:

- An Ethical hacker or a White hat Hacks with authorization and on behalf of an organization, while a Black hat does not seek authorization.

- While an Ethical hacker shares information and issues for preventing future attacks, Black hats take advantage of the vulnerabilities discovered within an organization's network.
- Ethical hackers' perforate networks and systems. This is done for the purpose of evaluating potential vulnerabilities and exploits to provide actionable solutions, but Black hats steal sensitive data or malicious software because of personal gains or just because they simply can.

So, by now we know the stark contrast between a White hat and a Black hat, but the world of cybersecurity is neither completely white nor completely black. There is another group of hackers known as Gray hats that are a bit of both. So how are the three different apart from their intent? How are their actions different? We answer that in this section:

- When a vulnerability is discovered by a White hat, they exploit it with permission and not disclose its existence until it has been fixed. Black hat, on the other hand, will exploit it without authorization. A Gray hat will exploit it with authorization, but will not tell others how to do so.
- The techniques used for discovering vulnerabilities are also different. The white hat breaks into systems and networks with explicit permission to measure how secure it is against hackers. However, black hats will vandalize a system or network for uncovering sensitive information and for personal gain. Thus gray hats possess the skills and objectives of the white hats but break into systems or networks without permission (Figure 1.7).

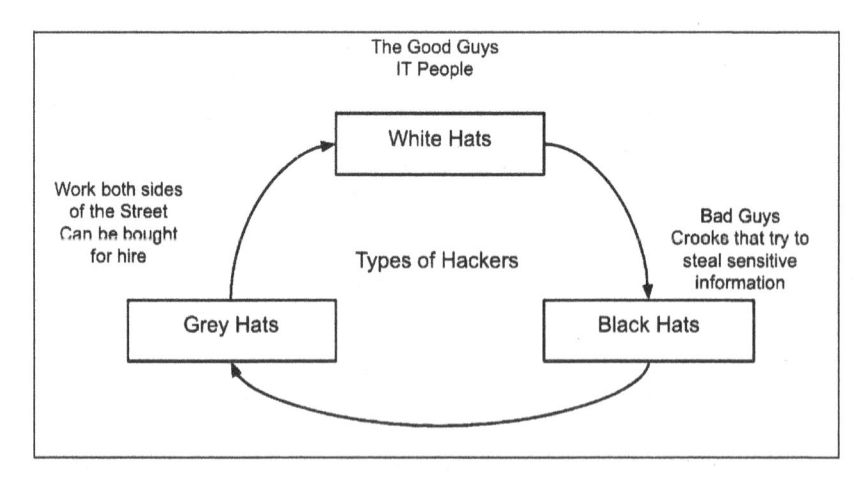

FIGURE 1.7 How black hats, white hats and gray hats function.

Introduction to Cyberethics

> **Gray hats from an ethical point of view:**

> Once a vulnerability is discovered by Gray hats, they do not tell the vendor how exploits work, rather they offer to fix it for a small fee.
>
> When one successfully gains unauthorized access to a system or network, it may be suggested to the system administrator that one of their friends be hired for solving the problem.
>
> Gray hat hackers only arguably violate ethics and law in an effort to explore and enhance security: Legalities are set with respect to the particular ramifications of any hacks they are a part of.
>
> The search engine optimization (SEO) community, labels gray hat hackers as manipulators of websites' search engine rankings. Much of this is attributed to the usage of improper or unethical means however, it is not considered search engine spam.

Some gray Hat Ethics Incidents are as follows [12]:

1. **April 2000:** Hackers referred to as "{}" and "Hard beat" successfully gained unauthorized access to Apache.org. The hackers alerted Apache crew of the issues rather than damaging the Apache.org servers.
2. **June 2010:** Computer experts group known by the name Goatse Security exposed a flaw in AT&T (American Telephone and telegraph) security. The flaw allowed the email addresses of iPad users to be revealed. The security flaw was exposed to media following notification to AT&T. Following that, an investigation was opened by the FBI. The houses of the most prominent members of the group were raided.
3. **April 2011:** Security experts inferred that the Apple iPhone and 3G iPads were capable of "logging where the user visits." The primary company released a statement asserting that the devices were only logging the towers that could be accessed by the phone. The incident was regarded as a minor security issue and may be classified as "gray hat" since the matter was reported even if its objective was malevolent.
4. **August 2013:** Computer security researcher Khalil Shreateh, hacked the Facebook page of Mark Zuckerberg, Facebook's CEO, with the intention of correcting a bug that was discovered. The bug had a tiny flaw that allowed him to post anything to any

user's page on Facebook without their consent. Time and again he had tried informing Facebook about this bug, which Facebook repeatedly neglected, asserting that it was not a bug. The incident prompted Facebook to rectify the vulnerability, which could have been misused as a powerful weapon if discovered by professional spammers. Facebook's White Hat program did not compensate Shreateh since the policies were violated which made this event a gray hat incident.

1.11 ETHICAL CHALLENGES FACED BY CYBERSECURITY PROFESSIONALS

Cybersecurity is one of the domains with increasing popularity. Although several cybersecurity practices are legal, they may not be completely ethical. Unethical or ethically dubious cybersecurity practices may have a negative impact on the reputation of network users, clients, companies, the public, and cybersecurity professionals [10]. Hence, it necessitates the understanding of ethical challenges existing in the cybersecurity space, and how they can be addressed through the ethical 'best practices.

1.11.1 ETHICAL CHALLENGES IN BALANCING SECURITY WITH OTHER VALUES

Cybersecurity incorporates within itself a lot of other values like speed, reliability, network/device usability, and other resource needs of the organization as well as its stakeholders. There is a need to maintain transparent conversations with clients and customers about balancing all these values along with cybersecurity. One should not provide the stakeholders with a false sense of security or make promises about functionalities that cannot be kept. In case of a serious data breach or any other security issues that lead to significant harm to customers or stakeholders, the organization must justify the reason. Cybersecurity professionals must not show a negligent attitude towards such situations, and their actions must reflect consistency, sincerity, and transparency.

Consider a situation where an organization ensures several layers of privacy protection mechanisms for personal data from third-party requests by law enforcement. The question will be if the organization has taken all

Introduction to Cyberethics 31

the security measures to ensure the same? Will the organization keep on monitoring the personal information of clients to ensure whether they are safe? In case there arise inconsistencies, will the organization be able to justify the actions from an ethical point of view?

1.11.2 ETHICAL CHALLENGES IN INCIDENT RESPONSE (IR) ACTIVITIES

Every organization that deals with data has some kind of cybersecurity threat or risk associated with it. Since security risks cannot be 100% mitigated, one cannot ensure a fully secure system. An IR plan assists IT staff in an organization for identifying, responding to, and recovering from cybersecurity incidents. IR plans are primarily concerned with preventing damages pertaining to data loss and theft, and unauthorized access to organizational systems. An ethical perspective would always keep an organization prepared with an appropriate IR plan, for each type of threat or incident that it might face. There must be worst-case scenarios, mitigation strategies to limit or remedy harms if the primary measures fall short. It should incorporate concrete plans, mention about the resources and systems to successfully implement the IR plan.

One of the significant things it needs to mention is how much of all these actions are ethically acceptable if an organization must respond to threats. There might be some ethical gray areas in the IR practice, like justifying the remediation and discouraging the attacks. The IR plan should have enough information about anticipating the risks that may vandalize third parties or tarnish the reputation of the organization. Thus, crossing the ethical boundary requires certain analysis. For example, in the case of a ransomware attack, from an ethical perspective, would it be right to pay ransom to the attackers for restoring the functionality or access? Will that put any burden on the clients and coax them to follow best practices like protecting and backing up their data in order to limit the ransomware vulnerability?

1.11.3 ETHICAL CHALLENGES IN SECURITY BREACHES AND VULNERABILITIES

With networks and devices being interconnected in so many different ways, security breaches are very common now. In the case of a security incident like a breach or a vulnerability, an ethically sound plan must be present for notifying users and stakeholders of the incident. Even if there is a delay of notification, or if the notification is selective and not for all users and stakeholders, it must be justified. Some questions that must be considered from an ethical point of view are whether the reporting is correct and timely, whether an efficient system is investigating and planning an effective response to the incident, is the response that has been drafted seems appropriate, whether the response is susceptible to any uncertainty, disorder or dispute, and if there are good protocols in place for communicating to affected stakeholders so that they can get more information on the incident. They may also be directed to certain measures to protect themselves from further risk. There may be a need to reconsider a delay or selective notification, if it may lead to further damage. It is ethical to disclose the level of the incident, i.e., whether it is at a lower risk, moderate risk or critical. Such measures may be taken considering ethics during a security breach.

1.11.4 ETHICAL CHALLENGES IN NETWORK MONITORING AND USER PRIVACY

Another debatable issue involving ethics is network monitoring and user privacy. Monitoring networks effectively while justifying intrusions upon users and their privacy is a gray area. Monitoring user activity pretty much involves reading their emails, logging their keystrokes, evaluating their physical locations and tracking their websites. Can these activities on personal devices be monitored if they are not owned by the organization and legitimately used for other purposes? Is there a limit on how much the users and network are aware of security monitoring activities? Depending on the profession of users, can the monitoring activity change? Does monitoring information change and increase or decrease for people in different professions?

1.11.5 ETHICAL CHALLENGES WITH COMPETING INTERESTS AND OBLIGATIONS

The harmful results of Security Breaches may be short term or long term, and these may be ethical harms too. The significant interests of all the stakeholders who are affected must be taken into consideration. One group of people must not be favored as compared to another group from an ethical point of view, i.e., never overlook or discount impact in favor of interests of others. In situations where an employee requires performing a privilege escalation for another user, they must be able to justify the reason, and cybersecurity professionals must thoroughly analyze the situation.

For example, granting access to one user might put some other sensitive information at risk. In a situation where an employee wants a delay in patch for protecting the reputation of the organization, it is imperative to analyze if there is any additional risk. There may be situations when cybersecurity professionals are supposed to breach a professional duty of cybersecurity practice. This breach may be in the interest of national security, or any non-professional ethical interest. Such requests need an ethically accepted response. This might also incorporate the cost that must be borne to carry out the highest ethical duties or compensation arrangements that must be made in case there is a conflict of interest between the professional and those who need to be protected.

1.11.6 ETHICAL CHALLENGES IN DATA STORAGE AND ENCRYPTION

Storing data safely and transmitting sensitive data has an ethical side to it. The users, customers, and stakeholders who are somehow related to data must be given accurate information about how data is being stored, encryption techniques, key management mechanisms, and other practices. Third parties who are concerned with encryption tools, conducting security audits, cloud services must be contacted time and again owing to cost, reputation, reliability, infrastructure, and legal protections. Different encryption techniques fall within different ethical boundaries. They have different risks and benefits associated with them which must be mentioned. The security mechanisms must comply with the industry standards, which in turn must be ethically defensible.

Sometimes law enforcement as well as intelligent agencies may request weakening of encryption practices or decryption of certain devices. Such requests must be addressed with some policies that have either been made public or has been made available to specific individuals. Long term data storage might have ethical issues like how long must the data be kept or how often should it be purged for security purposes. Data storage must have a proper security plan which must be regularly updated and improvised to thwart any cyber-attacks or security issues that may potentially harm the sensitive data.

1.11.7 ETHICAL CHALLENGES IN INTERNET OF THINGS (IoT), SMART GRID, AND PRODUCT DESIGN

With the IoT becoming popular, as more and more devices are being added to the network, there is an increased risk for security. Many of these devices deal with sensitive data and now the device being a part of the network has an added risk of intrusion as well as tampering. Smart homes have Bluetooth enabled door locks which may also have security risks. Thus, exploiting IoT vulnerabilities is not very difficult. Since the nodes are mobile in a wireless network, therefore abusing, exploiting, and misusing devices over the IoT network is easy. Thus, it is necessary to upgrade security designs and practices due to such potential scenarios. Further, end user training for safe and secure operation is a must. Communicating a device with the local server directly to the internet arrests the risk of security. Devices may be designed to encourage best cybersecurity practices like strong passwords or regular updation of software.

Smart Grid technologies and other systems that require critical public functions like healthcare, voting, etc., and utilities like power and water, public interests may be taken into consideration for protection. Thus, it is necessary for cybersecurity professionals involved to comprehend the underlying situation from an ethical point of view to deal with such issues.

1.11.8 ETHICAL CHALLENGES WITH ACCOUNTABILITY FOR CYBERSECURITY

Accountability may be used to safeguard data against illegal tampering and also to protect valuable data [13]. In the domain of cybersecurity, ethical cybersecurity practice is a high-level goal, although it is difficult to achieve. There may be a situation that demands ethical cybersecurity practice, but there are no specific individuals tasked with specific actions to achieve it. There will be certain risks and harms, hence the need for accountability. There may be a lot of people involved in an unwanted situation, and it will be extremely hard to hold only one individual accountable for the entire situation. Therefore, an organization must have policies and routines established and enforced to promote ethical cybersecurity practice. Regularly scheduled privilege levels, audits, password management, incident reports, and reviews of access policies, could be a way of ensuring accountability of the organization.

An established process for correcting, repairing, and ensuring an iterative improvement of cybersecurity efforts could contribute a lot to accountability from the ethical point of view. Cybersecurity efforts must be handled by appropriately skilled and responsible individuals who have a sufficient level of training and practice. Inadequate instructions and training, as well as giving stakeholders too much control over data, must be avoided, as it can lead to security issues.

1.11.9 ETHICAL CHALLENGES IN SECURITY RESEARCH AND TESTING

Security researchers are always concerned with unknown security techniques, tools, and vulnerabilities. While there is risk involved in the techniques used or vulnerabilities exploited, it is not completely moral to publish and spread information about these. Developing and releasing automated security tools may have its disadvantages as such technologies are known to become popular very fast. Systems end up being infected by malware and worms and mostly without the consent of the system owner. Thus balancing ethical interests and timescales of security research is hard. Publications may create new risks, hence there is a need to address the issue in an ethical manner. Security research and testing must incorporate individuals that promote a culture of moral growth and maturity. There may be some ethical implications of participating in the private market for zero day exploits as an exploit seller, buyer, and developer. Many companies sell exploits to governments, which are then used against political activists.

1.11.10 UNDERSTANDING BROADER IMPACTS OF CYBERSECURITY PRACTICE

The domain of cybersecurity has grown extraordinarily over the last few years. It would be appropriate to assert that cybersecurity practices today may have an impact on others now, as well as in the future. Cybersecurity professionals must be capable of understanding and anticipating these impacts, which may vary depending on individuals, identities, cultures as well as interest groups. Further, neglecting the same or minimizing our efforts towards cybersecurity ethics may have a negative impact in the future. Many times, privacy law experts, political groups, privacy advocates, demographic minorities, and civil rights advocates, pitch in their ideas about security practices.

Cybersecurity practices must not breach the legal or moral rights of any individual, nor should they confine their fundamental human capabilities, or vandalize their fundamental life interests. They should not impede upon the moral and intellectual habits, values, or character development of any individual or affected parties. Cybersecurity practices have the potential of damaging the professional reputations of people and organizations, cybersecurity professionals must keep that in mind (Figure 1.8).

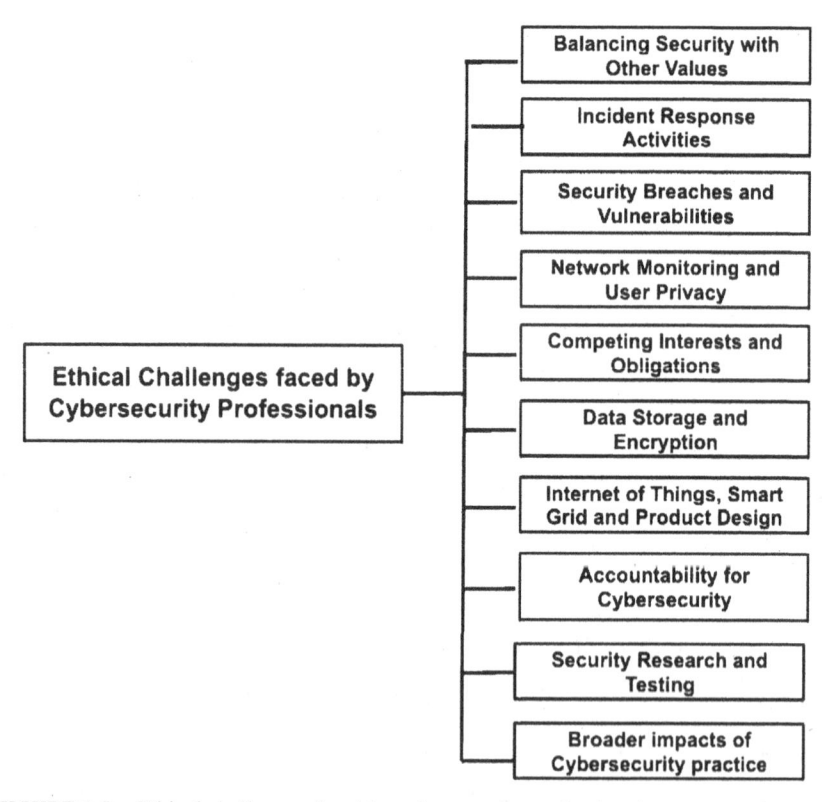

FIGURE 1.8 Ethical challenges faced by cybersecurity professionals.

Introduction to Cyberethics 37

In this chapter, we discovered the significance of ethics in cyberspace and cybersecurity. We explored how ethics, laws, and policy are related to one another and also focused on the need for ethics and the principles behind it. We discussed some important ethical issues in cybersecurity, both generic as well as specific. Finally, cyberspace is a dynamic domain which witnesses a lot of hands-on work; therefore, we familiarize ourselves with the different ethical challenges that most cybersecurity professionals face.

1.12 SUMMARY AND REVIEW

- Cyberspace as the interconnection and association of networks of information technology (IT) infrastructures.
- Cybersecurity is concerned with different kinds of technologies, processes, and practices aimed to protect programs, networks, data, and devices, from attack, damage, or unauthorized access.
- Cyberethics may be defined as the code of responsible behavior over cyberspace. Laws are the outcomes of ethics.
- Ethics, law, and policy in cyberspace are related. Ethics account for code of conduct and responsible behavior that should be followed in cyberspace. Cyber laws are legislations that focus on the acceptable behavioral use of technology in cyberspace and must be followed. Policies are aimed at achieving some objectives and goals and are hence usually followed.
- Security is primarily concerned with protection against the unauthorized access of data.
- Privacy deals with an individual's right to own the data generated by his or her life and activities, and to restrict the outward flow of that data.
- We should be concerned about cyberethics because it may have a bad effect not only in cyberspace but also on people who are interacting with these systems as we all do with each other over cyberspace.
- Principles for cyberethics lay down some commandments to manifest what should be done morally and what should not be done.
- What makes something ethical or unethical in cyberspace is greatly influenced by the source of ethics, namely, laws, philosophy, societal norms, environmental ethics, political ethics, economic ethics, and religious ethics.

- While cyberethics takes into account the entire cyberspace, cybersecurity ethics is more or less concerned with the security aspect of it.
- Ethical issues are the heart of cybersecurity practices.
- Ethical issues exist in the cybersecurity domain. Good technical practice is also ethical practice.
- Specific ethical issues in cybersecurity underpin privacy, property, resource allocation, transparency, and cybersecurity roles.
- Cybersecurity professionals may have similar role, but maybe known by different names: black hats, white hats, gray hats, green hats, red hats and blue hats.
- An ethical hacker or a White hat Hacks with authorization and on behalf of an organization, while a Black hat hacks without authorization.
- Cybersecurity professionals face different kinds of ethical challenges.

QUESTIONS TO PONDER

1. **Case Study 1:** David and his group had been working on their Senior Year project for a year. The group project is due tomorrow, failing the submission will cost each of them a letter grade. Yesterday, the group gave the final copy to SAM so that She could print it out and turn it in. Unfortunately, Sam had to leave the town last minute due to some unavoidable business. The group is now at a risk of getting a lower grade if the project is turned in late. Few days ago, David had seen Sam's list of passwords that she usually saves in her classroom desktop which is not password protected. David knows that he could get into the system, get a hold of her password to print the project and turn it in. What could he do in such a situation, and how would his actions vary ethically?

2. **Case Study 2:** Eric's university organizes an annual Capture the Flag event, which pretty much involves cybersecurity operations. Scoring the maximum number of points will give him $5000 USD and an internship with Google. The competition features a scoreboard showing the scores of the top 10 teams. Eric somehow gets access to the systems the teams are allotted to and attempts to identify all the vulnerabilities found by the top teams. Thus, he gains maximum points using them. Has he acted ethically? Why/Why not?

3. **Case Study 3:** Ryan and Jane were in a relationship for a long time before Jane discovered that Ryan was cheating on her and broke up with him. An angered Ryan posted a picture of Jane on Facebook, which makes it look as if she is consuming alcohol. Jasper is their common friend and was at the party when Jane was posing with alcohol so he knows that she was not consuming any. Jane is very upset and Ryan refuses to take the picture down. Jane asks Jasper to help her in getting into Ryan's Facebook account to remove the picture. What would be ethically right or wrong in this case?
4. **Case Study 4:** Professor Emma has set up her own server for students to do an assignment for her Algorithms class. The better graphs and results you show, the more points you get. The scores stay anonymous until the end. Matt is the class topper and has been doing well in his midterms and final exams. Only during the last project, he found out that the server has a glitch and one can see the scores of other students. He wants to tell his professor about the glitch but doubts that she might think he had been cheating for the whole semester. What should he do?

KEYWORDS

- **cyberethics**
- **cyberspace**
- **electronic communications systems**
- **information technology**
- **National Institute of Standards and Technology**
- **telecommunications networks**

CHAPTER 2

Ethical Issues in Cybersecurity

In the previous chapter, we introduced the concept of ethics in cybersecurity as well as cyberspace. We likewise inferred that there are a few standards related to ethics, and these are determined from several sources. We specifically focus on cybersecurity, and therefore we highlighted ethical challenges in cybersecurity and the ethical challenges faced by cybersecurity professionals. In this chapter, we will be focusing on the ethical issues that are present in cybersecurity as well as cyberethics frameworks.

2.1 ETHICAL ISSUES IN CYBERSECURITY: OVERVIEW

As we know, ethics take into account the moral principles that govern a person's behavior. Without clear ethical standards and rules, it may not be easy to tell apart cybersecurity professionals from the black-hat criminals against whom they seek to protect systems and data. Thus, ethics are a critical part of cybersecurity. The world of cybersecurity is full of ethical challenges that were discussed in the previous chapter, some of which are as follows:

1. **Vulnerability Disclosure:** When and how should researchers or professionals inform the public about a newly found vulnerability.
2. **Encryption Issues:** What should be done when law enforcement requests for encrypted data.
3. **Automated Security Tools:** If it is ethical to release tools that may assist in automating attacks on a large number of systems in the cyberspace and beyond.
4. **Sale Restrictions:** What could be the responsibility of cybersecurity professionals for trying to prevent the sale of products they have developed to autocratic governments that may possibly be used to mistreat their citizens?

5. **Incident Responses (IR):** What is an appropriate level of incident detail to share with customers and other stakeholders?
6. **Roles and Responsibilities:** What are the roles in your IT department and what are the responsibilities associated with each of those roles?

While these are some ethical challenges that require a certain level of analysis for carrying out appropriate actions, there are some issues that already exist in the cybersecurity domain that challenge basic ethics and are of utmost concern to anyone who is a part of the cyberspace (and cybersecurity domain).

2.2 TYPES OF ETHICAL ISSUES IN CYBERSECURITY

The domain of cybersecurity has spanned across several industries related to medicine, education, commerce, government, entertainment, and society at large [14]. While it is largely responsible for securing networks, infrastructures, systems, etc., it also invites potential risks. Moreover, it presents and makes a few issues identified with morals, and contains by and large three fundamental kinds of moral issues: personal privacy, access rights, and harmful actions [18]. In this section, we will delve into these issues more closely and explore the ways in which they affect society and technological changes.

2.2.1 PERSONAL PRIVACY

In the previous chapter, we defined privacy as an individual's right to own the data, which is usually generated by his or her life and activities. Privacy also restricts the outward flow of this data. Cybersecurity, in terms of privacy, allows the exchange of information over cyberspace to different individuals and organizations, to any location of the world, at any given time. Thus, there is an added potential of information disclosure and privacy violation of individuals and organizations due to data diffusion worldwide. Thus the real challenge is to ensure data privacy and integrity. Since this falls within the boundary of ethics, certain precautions may be taken to ensure the accuracy of data, as well as protecting it from unauthorized access or accidental disclosure to inappropriate individuals.

2.2.2 ACCESS RIGHTS

Access rights may be defined as the authorizations an individual client or a computer application holds that permits them to peruse, compose, alter, erase or in any case get to a system record, change arrangements or settings, or include or remove applications. An organization's network or the administrator is capable of defining permissions for files, servers, folders, or specific applications on the computer. As cyberspace also encourages commerce, access rights are an important issue in cyberspace. Security breaches are normal and large organizations and government associations are at the danger of security assaults. Therefore, the implementation of proper security policies and strategies is mandatory. Setting up permissions and access rights confines security attacks to a certain extent.

2.2.3 HARMFUL ACTIONS

Harmful actions can prompt injuries or negative consequences. Some negative consequences pertaining to the same are undesirable loss of information, loss of property, property damage, or unwanted environmental impacts. Cybersecurity attacks may lead to such harmful actions. This principle is responsible for prohibiting the use of computing technology that may harm users, the general public, employees, and employers. Harmful actions may also lead to the destruction or modification of files and programs. The repercussions could be a critical loss of assets or superfluous use of resources, like the time and effort required to purge systems of malware.

2.3 SPECIFIC ETHICAL ISSUES IN CYBERSECURITY

In the previous section, we highlighted the various types of ethical issues based on their impact in the Cybersecurity domain. In this section, we will talk about some specific ethical issues that have cropped up in cybersecurity over the last few years. These issues are prevalent in cybersecurity, we also talk about some cases that underpin these ethical issues. The following are specific ethical issues in cybersecurity.

2.3.1 COPYRIGHT INFRINGEMENT

Copyright encroachment might be characterized as the utilization or creation of copyright-ensured material without the consent of the copyright holder. It occurs when a third party breaches the rights afforded to the copyright holder. The right could be in the form of an exclusive use of a work for a set period of time. The most common copyright infringement in the entertainment industry is in the form of music and movies as they suffer from significant amounts of copyright infringement. When new works are developed by individuals and companies, copyright protection is often considered. This ensures that individuals and companies can exclusively profit from their efforts. So as to utilize those works, different parties might be conceded consent through licenses. They may also purchase the works directly from the copyright holder.

> **Definition 2.1: Copyright Infringement**

> The United States Copyright Office defines *Copyright Infringement* as follows [22]: Copyright infringement happens when a copyrighted work is imitated, dispersed, performed, freely showed, or made into a subsidiary work without the authorization of the copyright proprietor.

Copyright protection varies from country to country. With technology advancing at such a rapid rate, copying a product or information has become relatively easy. Many companies procure a significant portion of their revenue by replicating the works of other companies and individuals. It might be hard to demonstrate copyright in a global setting along with enforcement of copyright claims from international companies since it may be seen as a threat to national productivity by domestic courts. This is why some international organizations like the European Union (EU), try to coordinate regulations and enforcement guidelines among its member countries.

Although technology has made copyright infringement very easy, at the same time, the growth and popularity of the Internet has conveniently created new hurdles for copyright holders [2]. Nowadays, it is very easy for companies around the world to access copyrighted materials. Additionally, the creation of new technologies has outperformed the regulatory environment's ability to ensure that copyrights apply to new formats. The music industry often faces copyright infringement issues due to newly emerging online music-sharing websites like Napster. To claim copyright infringement, companies may provide specific files and also seek damages from internet service providers (ISPs) and individual users. Accessing content from a particular website, to modify it, or just reproduce it on another website is very easy and common, and has posed challenges pertaining to individual rights and protection. Any person with a computer who has access to the network can become a publisher. Thus, copyright infringement is one of the most popular ethical issues in the domain of cybersecurity [22]. It challenges the code of responsible behavior of an individual or a company over a copyrighted material that solely belongs to someone else.

2.3.2 PIRACY

Piracy happens because of the unapproved duplication of copyrighted substance that is then sold at significantly lower costs. In the course of the most recent couple of years, with the ease of technology, piracy has gotten more common. For instance, CD writers are accessible at exceptionally low costs, and music piracy is common in the entertainment industry. Music piracy involves unapproved replication of music tapes that flood the market as soon as a new release is launched. Music organizations' incomes are hit hard by the overwhelming number of pirated discs and cassettes. This is because cassettes and discs are accessible at generously lower costs compared with that at stores. Owing to the unethical practices of music piracy, many organizations work towards identifying sources of music piracy. Once the sources are identified, they are often confronted with the help of the police. However, since convictions are few and the penalties not harsh enough, piracy continues to prevail.

In cyberspace, software piracy is a global issue. Software development relies on significant financial investment; therefore, software companies rely on profits to continue improving and building software. Software piracy occurs when a software program is illegally copied, downloaded, and/or installed. Illicitly utilizing, disseminating or replicating programming, and wrongfully transferring or downloading on the web music are on the whole demonstrations of piracy. Software piracy may have several forms, such as:

1. **Counterfeiting:** It is the illicit duplication, dispersion, and additionally offer of copyrighted material with the plan of impersonating the copyrighted item. In the case of packaged software, one can easily, one can without much of a stretch, discover fake duplicates of the reduced discs consolidating the product programs. It might likewise be conceivable to discover related labels, manuals, packaging, permit understandings, registration cards, and security highlights.
2. **Internet Piracy:** This sort of piracy happens when the product is downloaded from the Internet. Regularly sites make programming accessible for free download or in return for other Internet auction sites that offer fake or out-of-channel software. Distributed systems are additionally fit for enabling authorized transfer of copyrighted programs.
3. **End-User Piracy:** This kind of piracy occurs when an individual mass-produces copies of software without permission. Using a licensed copy of software to install a program on multiple computers may be thought of as a copyright infringement. Other forms of end-user piracy are acquiring restricted software without license, copying discs for installation or distribution, acquiring academic software without permission, exploiting upgrade offers without having a lawful copy of the version to be upgraded, and trading discs in or outside the working environment.
4. **Client-Server Overuse:** This situation takes place when there are many users on a network trying to use or access a central or specific copy of a program at a given time. If a local-area network is used to install programs on the server for many users at the same time, they all should have the license. Overusing of resources occurs in a situation where there are more users than allowed by the license.
5. **Harddisk Loading:** Businesses selling new computers with illegal copies of software loaded onto the hard disks for making the purchase of the machines seem attractive lead to hard disk loading.

It is worth mentioning that software piracy and cybersecurity threats go hand-in-hand. This is due to the fact that software may be used conveniently to distribute malicious computer code. This computer code may incorporate malware which may compromise individual computers and entire networks, putting companies, governments, and consumers at risk. Privacy and Copyright Infringement seem similar; however, they are not.

2.3.2.1 THE DIFFERENCE BETWEEN PIRACY AND COPYRIGHT INFRINGEMENT

Piracy may be referred to as copyright infringement, but copyright infringement may not completely be synonymous to piracy. There are many instances that may be considered for highlighting the difference between piracy and copyright infringement. There may be instances where actions seem like copyright infringement but cannot be termed as piracy. For example, in a situation where a commodity may not be identically copied, but ideas from it could be heavily borrowed. One could argue that someone did not copy their entire book, but stole a significant portion of their ideas. If it were not for the original product, someone else's new product would not have been made. Yet the products may not be identical.

Because of software piracy, there could be job losses as well as financial losses. The increment in piracy might be credited to the obscurity and finesse with which illicit duplicates of programming can be made and circulated. In this way, every individual who makes illicit duplicates might be in a roundabout way adding to the financial misfortunes brought about by piracy. Much the same as individuals who compose books have the sole option to sell them, individuals who produce the product additionally have rights to benefit from it. Duplicating programming denies the legitimate proprietors of software programs of money-related benefits. The unscrupulous action is sometimes supported by the claim that pirates reserve the privilege to make unlawful duplicates of software in light of the fact that the software may be buggy, excessively costly, or not frequently used by the pirate. However, the claim is flawed. For example, a car may be too expensive and not worth the money, but that does not mean that anyone has the right to steal it. Someone who does not watch a television show may not be entitled to steal it. Pirating software may prove to be very costly. If several copies of the software are sold, the software manufacturers must raise prices, thus making legitimate users pay more due to piracy. Thus, piracy is not as victimless a crime as it may seem since it affects software developers, distributors, and end-users.

2.3.3 INTELLECTUAL PROPERTY (IP)

Intellectual property (IP) may be thought of as ownership of an idea or design by the person who came up with it. It may also refer to creations of the mind like inventions. Many literary and artistic works as well as designs and symbols could become IP. Sometimes names and images used in commerce are also considered IP. In order to protect an idea that may not be stolen by someone, one or more of the four different types of IP may be considered. There are four types of IP:

1. **Trade Secrets:** These are valuable information that are not publicly known and of which the owner has taken sensible and rational steps to maintain secrecy. These may include information related to business plans, customer lists, ideas related to research and development (R&D) cycle, etc. Usually, trade secrets are not registered with a governmental body since all that is needed is the establishment of information to treat it as a trade secret. Access to trade secret information must be given to only those with a need to know. In a situation where an impersonator steals trade secret information from its owner, the owner has the right to sue the imposter for expropriation of trade secrets. However, in a situation where the owner willingly gives information pertaining to trade secrets to another entity without limitation, there is no misappropriation. Hence, the owner cannot sue the individual. It is also possible for information to lose its status as a trade secret. This can occur in a situation where there is a lack of appropriate efforts to keep the information secret or if the information is no longer a secret.

2. **Trademarks:** These protect brands. Trademark is the name of the product linked to a product or a service. A trademark may be used by customers to identify a product or the source of a product, like a name associated with the product. Typically, it refers to the words that may be used to refer to the product or service. Brands and trademarks incorporating words, are called wordmarks. Sounds, colors, smells, and anything else that brings the product and/or its owner to the minds of a consumer can be used as a trademark. Some common trademarks are wordmarks, logos, and slogans. To ensure proper protection of a trademark, one must conduct a search to find out if similar marks are being used. On the off chance that not, at that point a trademark application might be documented to get the trademark enlisted.

3. **Copyrights:** These are means to protect original works of authorship that may have been authored and written down on a

piece of paper or stored electronically. It may have been preserved in some tangible format. Some copyrightable works are books, movies, photos, videos, articles, diaries, and software. It is worth mentioning that copyright does not protect ideas or useful items. This is usually done by patents. Software are functional items, yet they can be protected by copyrights because of the creativity involved in selecting, ordering, and arranging the various pieces of code in the software.
4. **Patents:** It is one more type of licensed innovation. Patents grant their owners the legal right to disallow others from making, utilizing, selling, and bringing in a creation for a constrained time of years. This is done in return for distributing an empowering public disclosure of the creation. Patents may be utility-based as well as design-based.

The Internet is a great platform to showcase artistic and literary creation. It is an ideal platform for creative and academic inventors for disseminating their work since it ensures ease of transmission from creator to viewer, and re-transmission from viewer to viewer. It is also an ideal medium for commercial creators to reach wider audiences. Technology has made it possible for people to innovate, edit, alter, distort, and redistribute the words, sounds, and images on the Internet. Thus, it is common for authors to find their creations altered and posted in various places. This brings the question of respect and integrity, and also raises questions on the ethics of an individual in cyberspace. Since the creators hope that others will respect their original work and will not make unauthorized copies, they are also interested in protecting the integrity of their work by keeping others from tampering or modifying it (Figure 2.1).

2.3.4 COMPUTER FRAUD AND MISUSE

Computer fraud may be defined as is the act of using a computer to access or alter electronic data. It usually occurs when a person uses the Internet through a computing device to obtain something valuable to the individual. This information may come from a personal computer, government facility or a company. There are four elements of computer fraud:

- For computer fraud to occur, the person may be able to access and enter a protected computer which is not owned by them;

- With computer fraud, the individual may either access the device without authorization or exceed the timeframe of authorization given by the company;
- The individual must be capable of accessing with knowledge of engaging in or with every objective to deceive someone or an entity; and
- Driven by the objective and by further accessing the computer, if the individual acquires, copies or steals something valuable.

Different types of Intellectual Property

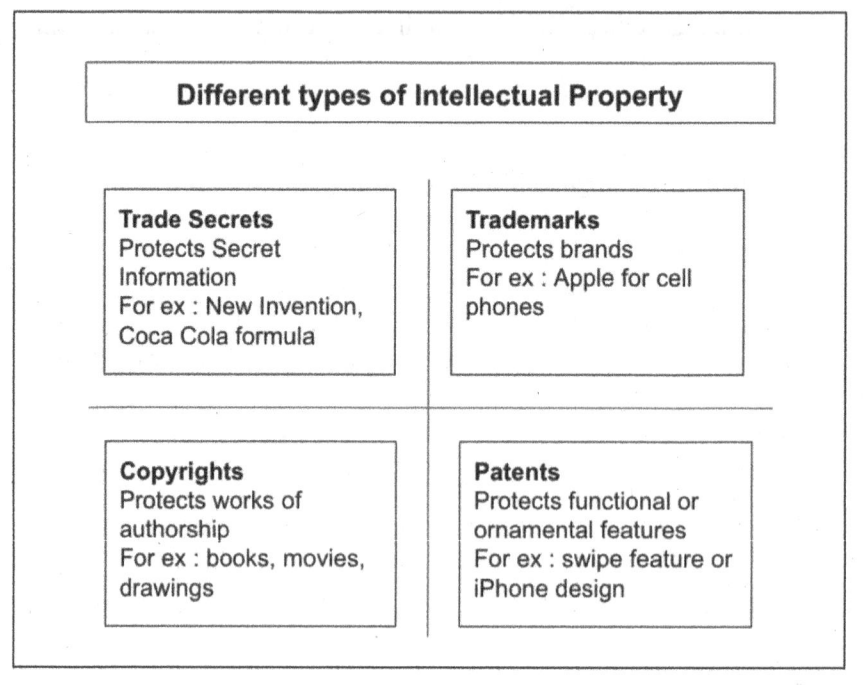

Trade Secrets
Protects Secret Information
For ex : New Invention, Coca Cola formula

Trademarks
Protects brands
For ex : Apple for cell phones

Copyrights
Protects works of authorship
For ex : books, movies, drawings

Patents
Protects functional or ornamental features
For ex : swipe feature or iPhone design

FIGURE 2.1 Types of intellectual property.

The data in question is usually either personal details of a specific individual. This is significant, because the perpetrator could penetrate the system security in the future or for personal instances of fraud that may involve identity theft and using the victim's credit to purchase property. In some instances, the perpetrator may steal computer or system information and hold it hostage until the company owner or management meets certain demands that usually involve a large monetary settlement. This brings ethics or moral character of the individual into question. Social engineering methods like distributing hoax emails, phishing, illegally accessing personal information, using spyware and malware for data mining, are some examples of computer fraud. Sensitive information like credit card details or Social Security numbers may be compromised due to computer fraud [14, 16, 23].

2.3.5 PLAGIARISM

Plagiarism, an act of fraud, simply involves taking someone else's work and/or ideas and using them as your own without giving any credits to the legal possessor or receiving permission before taking the material. It involves directly cloning or replicating and then presenting an existing production without proper citation or referencing. In plagiarism, copied work is often passed off as one's own work, without permission from the original holder. Paraphrasing without citing, copy, and paste plagiarism, patchwork plagiarism and self-plagiarism are some examples of plagiarism. Some forms of plagiarism are as follows:

1. **Direct Plagiarism:** This involves word-to-word transcription of someone else's work without attribution. Direct plagiarism does not involve any quotation marks. Exclusion of quotation marks while performing direct plagiarism is unethical.
2. **Self-Plagiarism:** It occurs when an individual hands over his or her own preceding work, or combines parts of preceding works, without acquiring permission from all authorities involved.
3. **Mosaic Plagiarism:** It is said to occur when an individual borrows phrases from a source unaccompanied by quotation marks. It may also be accompanied by finding synonyms for the author's language while still keeping to the general structure and meaning of the original work constant.
4. **Accidental Plagiarism:** It is said to occur when individuals disregard citing their sources, or misquote their sources, or accidentally reword a source by using similar words, groups of words, and/or sentence structure without attribution.

Since plagiarism is concerned with the works of others, it may be deemed as an issue of trust. There is a need to respect honor codes, so that we gain the solace of realizing that what we read is spoken in the voice of the writer and what we compose will not be distorted as another person's unique work [20].

2.3.6 CYBERBULLYING

Cyberbullying is referred to as the use of electronic communication for bullying a person. It basically includes sending messages of a threatening, intimidating, or compromising nature to the person in question. Cyberbullying has been known to occur through SMSs (short message service), text, and apps, or online in social media, forums, or gaming. These are the platforms where people can view, participate in, or share content. Hence, it is easy to send, post, or share negative, harmful, false, or mean content about someone else. Cyberbullying may also involve sharing of personal or private information about someone else causing discomfort or dishonor. With cyberspace being a home to millions of users, cyberbullying is a common issue that is concerned with cyberethics. Cyberbullying is extremely unethical due to several reasons. It not only affects the victim but also the victims' family and friends if something goes very wrong. Since cyberbullying also deals with the deliberate use of communications and new media, it is exceptionally simple to irritate, undermine, hurt, embarrass, and defraud another with the expectation to cause mischief, distress, and intimidation. We know that harmful actions towards an individual call for ethical issues. Cyberbullying is capable of leaving psychological scars on victims, filling them with low confidence and a higher danger of sorrow, uneasiness, and different stress-related issues. In rare cases, victims go to self-destruction.

2.3.7 IDENTITY THEFT

Identity theft may be defined as the fraudulent act of obtaining the personal or financial information of another person. It is performed by assuming that person's name or identity. It is usually done to make transactions or purchases. There are different types of identity thefts like criminal, medical, financial, and child identity theft. Criminal identity theft witnesses a criminal misrepresent himself as another person during arrest for the purpose of steering away from summons. A criminal identity theft may also involve prevention of discovery of a warrant issuing

Ethical Issues in Cybersecurity 53

real name to avoid an arrest. In clinical identity theft, somebody may mimic as a person to get free clinical consideration. In money-related wholesale theft, somebody may imitate to get credit, products, services, or advantages. Financial identity theft is very common. In child identity theft, someone may use a child's identity for various forms of personal gain. Since children do not have information associated with them that could cause hindrance for the adversary, this information is very easy to gain. Using information like Social Security, one may obtain residence, find employment, obtain loans or avoid arrest on outstanding warrants. One of the biggest issues of identity theft victims is due to the lingering effects of having identity stolen. If a thief opened new accounts or committed a crime in the victim's name, it may take a long time to correct public records and credit. Damaged credit may hurt a victim's finances in the long run. Since many applications for apartment rentals, car rentals and house mortgages involve a credit check, if the victim's credit does not look good due to identity theft, they may be refused housing. They may also receive higher interest rates on mortgages and loans.

2.3.8 RANSOMWARE ATTACKS

Ransomware is a type of malware that is equipped for encrypting files and documents. Once the victim's files are encrypted, the attacker demands a ransom from the victim to restore access to the data after payment. Explicit directions with respect to payment to get to the decryption key are followed and the expenses may range from a couple 100 dollars to 1000. The cost is usually paid in cryptocurrencies. There are several ways and system loopholes that ransomware can rely on to access a computer. Phishing spam is one of the common techniques wherein the victim receives attachment in an email. This attachment often masquerades as a file they could trust, which once downloaded and opened, can take control over the victim's computer. Moreover, built-in social engineering tools are capable of tricking users into allowing administrative access. Other forms of ransomware, like NotPetya, are capable of exploiting security holes for infecting computers without needing to trick users. Victims are prompted on a lock screen to purchase a cryptocurrency, like Bitcoin, for paying the ransom fee. After the ransom is paid, customers receive the decryption key and attempt to decrypt files. There may be varying degrees of success with the decryption key as decryption is not guaranteed. It may also be possible for victims to never receive the keys. A few attacks stress on introducing malware on the computer systems significantly after the ransom is paid by the victim. Ransomware mainly targets business users, since businesses are willing to pay more for unlocking critical systems and resuming daily operations than individuals (Figure 2.2).

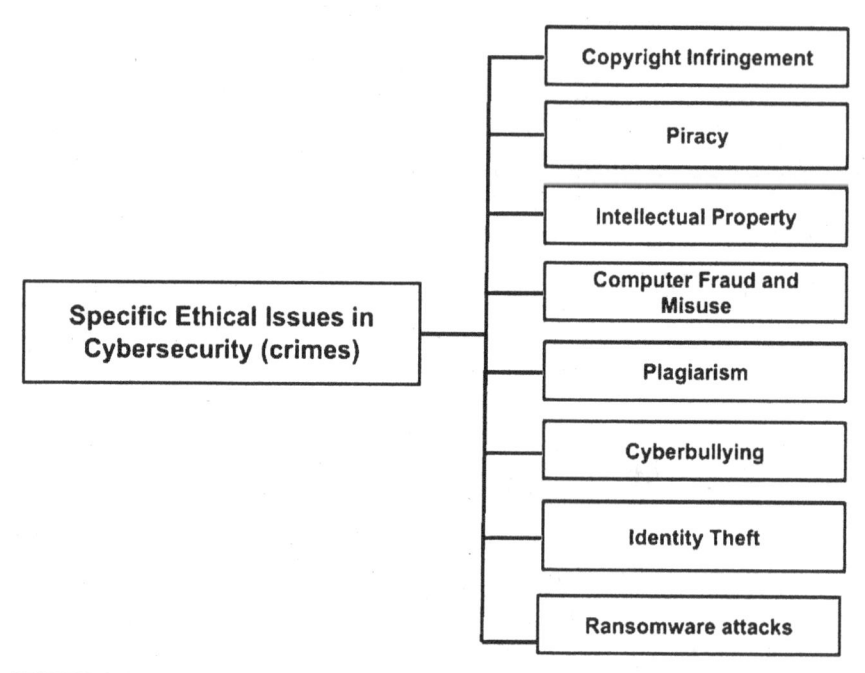

FIGURE 2.2 Specific ethical issues in cybersecurity (crimes).

2.4 FAIR USE DOCTRINE

Fair use is a principle (United States) that authorizes restricted use of copyrighted material without the need to acquire permission from the copyright holder. It is a legal doctrine that assists in promoting freedom of expression by authorizing the unauthorized use of copyright-protected works in various situations. The important categories that are mentioned under fair use are: parody commentary and critique news reporting teaching scholarship and research. Few examples of proper fair use are:

- A film reviewer dispersing clips of a movie throughout a film review;

- A professor copying a single chapter from a book and giving it to his students;
- A researcher quoting passages from another article to support a claim;
- A musician taking the melody of a popular song and recording new lyrics.

There are four factors for determining whether a work is fair use or not. They are described as follows:

1. **The Purpose and Character of the Use:** This incorporates whether such use is of business nature or is for philanthropic instructive purposes. This fair use factor is concerned basically with the function for which the duplicated material is being utilized. Copyright law favors empowering grants, education, training, and critique. In this manner, an adjudicator is bound to make an assurance of fair use if the litigant's utilization is noncommercial, instructive, logical, or verifiable. In any case, an instructive or logical use for business purposes may not be pardoned by the fair use doctrine. Further, groundbreaking uses may almost certainly be viewed as reasonable, since they include something new. Including something new with a further reason or distinctive character, does not substitute the previous work.

2. **The Nature of the Copyrighted Work:** This is the second factor in the fair use assurance. It considers whether the replicated work is educational or engaging in nature. If a substance is replicated from an authentic work, for example, a life story, it might be delegated reasonable use. However, if the substance is replicated from an anecdotal work, similar to a romance book or horror film, it might not qualify as reasonable use. Likewise, it is essentially imperative to consider whether the work that is replicated is distributed or unpublished. Utilizing a more inventive or innovative work is less inclined to help a case of a reasonable use as for a verifiable work Utilizing an unpublished work is more averse to be viewed as reasonable.

3. **The Amount and Substantiality of the Portion Used in Relation to the Copyrighted Work as a Whole:** In this case, both the quantity and quality of the copyrighted material that was used are taken into consideration. If the use incorporates a significant portion of the copyrighted work, it is less likely to be considered fair use.

However, if it uses only a small amount of copyrighted material, it is more likely to be called fair use. Sometimes, the use of an entire work may be considered fair under certain circumstances. In many contexts, using even a small amount of a copyrighted work may not be labeled fair because the selection is an important consideration. The questions taken into consideration may address how much of the original work was taken by the infringer, i.e., a sentence of a book or an entire chapter, a minute detail of a painting, or an entire painting.

4. **The Effect of the Use Upon the Market:** The value of the copyrighted work on the market is substantially important in determining fair use. It is important to understand to what extent, the unlicensed use harms the existing or future market for the copyright owner's original work. To assess this factor, one must consider whether the use is hurting the current market for the original work and if the use may cause substantial harm if it were to become widespread.

Source material may be acknowledged by means of citations. Citing the author or publication may be a considered fair use; however, it may not protect against a claim of infringement. For example, if a feature film is recorded in a theater and the DVDs (digital versatile discs) of that recording are sold, it does not infer that the copyright owner is being acknowledged and may still be considered as an impermissible infringement. Nevertheless, it is important to credit the original source. Crediting the recognition will lessen the distress of witnessing the work being reproduced without authorization for the original authors. Crediting the source also makes it clear to the public one is not claiming that the original work is theirs. The best option is to ask for permission from the copyright holder before using their work, even if one feels that it is fair.

2.5 THE ORIGIN OF CYBERSECURITY ETHICS

Knowledge, information, and communication are the primary drivers of development in globalized, multicultural, knowledge-based societies, which are some key aspects of cyberspace. Ethics in information and knowledge has seven core values, namely equity, freedom, care, and compassion, participation, sharing, sustainability, and responsibility.

Ethical Issues in Cybersecurity

These values are exemplified on core topics of the information society, such as principles, participation, people, profession, privacy, piracy, protection, power, and policy [21]. The origin of cybersecurity ethics may be attributed to the following subsections (Figure 2.3).

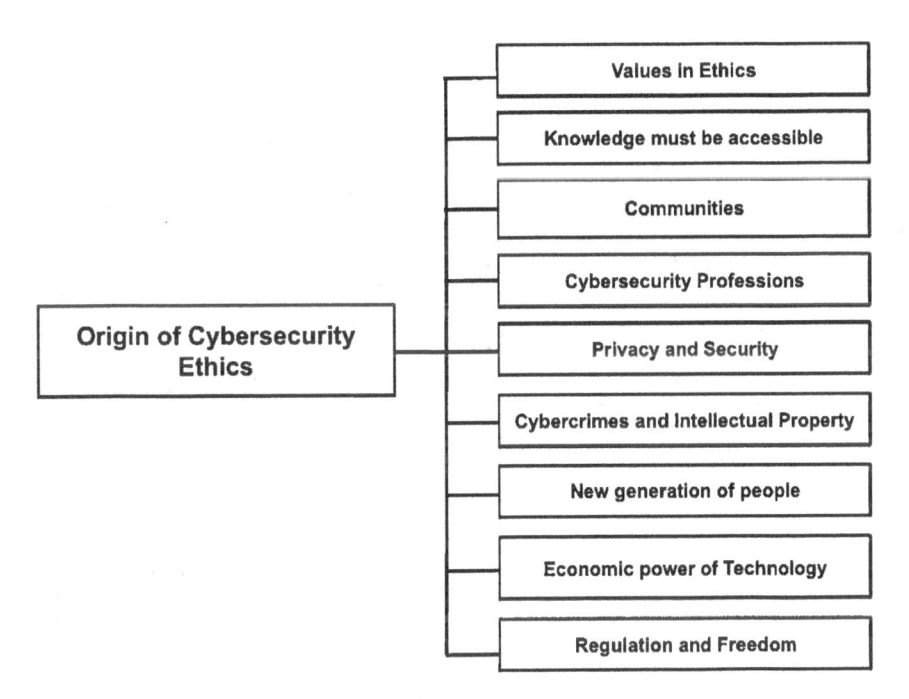

Origin of Cybersecurity Ethics
- Values in Ethics
- Knowledge must be accessible
- Communities
- Cybersecurity Professions
- Privacy and Security
- Cybercrimes and Intellectual Property
- New generation of people
- Economic power of Technology
- Regulation and Freedom

FIGURE 2.3 Origin of cybersecurity ethics.

2.5.1 VALUES IN ETHICS

Knowledge societies over cyberspace that are based on ethical values must be innovative, sustainable, coherent, and integrative. It may also take into consideration other rational opportunities and political or financial interests. Cyberspace is a global forum, and hence due to the globalized multicultural world, the values must be global and respect the diversity of contextual values. Cybersecurity ethics is capable of promoting public awareness of such fundamental values and principles. The universal consensus on human rights is supported by such ethics. They are also concerned with human rights, which legally binds expression of this ethical vision. Cybersecurity Ethics is responsible for fostering trust among human beings and also for strengthening care and action for global environmental protection.

Fundamental values for the knowledge societies in the cybersecurity domain are as follows [21, 124]:

- Equity/value is totally founded on human dignity just as equality. Justice develops when individuals develop a profound regard towards one another. Reasonable and equivalent chances of access to data are important for common understanding.
- Freedom of access to data, of articulation, of accept and of choice is important for human nobility and development. Freedom, value, and obligation balance one another.
- Care and sympathy is the capacity for compassion, regard, and backing of the other. It prompts solidarity.
- Cooperation is the privilege and capacity to take an interest in cultural life and in choices of concern.
- Sharing prompts, empowers, and continues connections between people and fortifies networks. The ITCs (information technology and communication) empower in an uncommon manner the sharing of data and information.
- Sustainability as a drawn-out point of view for green technologies.
- Responsibility is accountability for one's own actions. The degree of responsibility needs to relate to the degree of intensity, limit, and capacity. Those with more assets bear more responsibility. Every one of these qualities are interconnected and balance one another.

2.5.2 KNOWLEDGE MUST BE ACCESSIBLE TO EVERYONE

Cyberspace promotes access to information, communication, and education. This is because knowledge must be thought of as a basic right and public good. Open access for free or for affordable costs enables participation of all in the development of societies. The cybersecurity domain has witnessed a large number of opportunities granted by information and communication technologies. Another methodology of increasing access to knowledge may be in the form of massive open online courses (MOOCs).

Participation should underpin the following ideas:

- Governments and international organizations must reinforce free and fair access to knowledge for everyone who is a part of cyberspace.
- Governments must include support for open access repositories, including training and support, as well as infrastructure for everyone who is a part of cyberspace.
- Regulators must help the improvement of territorial center points that list open access stores, recognizing full-text repositories and those offering just metadata.
- Public and private entertainers must create open access and open distributing activities as a team with organizations that incorporate worldwide perceivability, availability, new ranking mechanisms, building impact factor measurements, and local value attribution.

2.5.3 COMMUNITIES

People: Community, identity, gender, generation, education people, and systems are the key actors of information, communication, and knowledge over cyberspace. For interacting in the cybersecurity domain, there must be sufficient understanding about how to refine, contemplate, and absorb information and knowledge. There is also a need to know how to use the information for advancement and not chaos and confusion, for identity building and not identity-loss, and if knowledge is primarily used to win over others in very competitive markets or to oppress others. The knowledge society in the cybersecurity domain incorporates the following aspects:

1. **Value-Based:** Societies are visualized where people, gatherings, and establishments share information fairly, equally, freely, and for the benefit of caring sustainable communities. These values are derived from communities, environment, culture, economic system, political system, etc.
2. **People-Centered:** Technology is a key driver in the rampant growth in cybersecurity, however technology is not a goal. It must be used to serve people. Information society over cyberspace needs to be people-centered.

3. **Identities Oriented:** Information and communication technology (ICT) trends increase individualism and individual media consumptions. Therefore, it is important to balance the needs and rights of individuals and communities. Information must be constantly deconstructed and reconstructed. Also, there is a need to change and balance stability for building strong identities.
4. **Education Focused:** This deals with responsible use of information and communication. Ethics with respect to information ethics is forever needed and on all levels, ranging from producers to consumers for dealing with information in a responsible way. Hence there is a need to increase awareness since dealing with information, communication, and knowledge can be very challenging. Therefore, it is imperative to highlight the significance of education for the ethical use, as well as personal transformation of such information to knowledge.
5. **Gender Oriented:** Gender fairness in access to data, correspondence, information, and decision making is a significant element of a comprehensive and human focus society.
6. **Generation Sensitive:** Computer literacy of more established people is significant for their investment in the public arena and for intergenerational trade.

People should take responsibility for the following [21]:

- Educational institutions to increase cyberethics as well as cybersecurity ethics in the curricula.
- Educational institutions must care for ethical aspects of fast-growing e-learning, distant learning and mass online courses (all part of the cybersecurity domain).
- There is a need for media providers and educational institutions to maximize efforts in transforming information into adapted and digested knowledge.
- Public and private media institutions must care for the cultural and lingual diversity of programs.
- There is a need to build empowering capabilities of women and girls to use cybersecurity for education.
- Substantiate and incorporate native people's values and knowledge.
- Policymakers must assure freedom of expression while steering away from moral harm and violating the integrity of individuals.

2.5.4 ETHICS OF CYBERSECURITY PROFESSIONS

Professions that are specific to information, communication, processing, creation, dissemination, control, renewal, preservation, archiving, policy-making, and cybersecurity have a special ethical responsibility in implementing core values. Bloggers, cybersecurity engineers, analysts, scientists, social media personnel, security researchers, IT hardware and software developers, curricula developers, politicians, philosophers, and several other professionals in cybersecurity have an impact on private and public opinions. However, in the modern information society, professionals must strengthen values, virtues, and rights themselves through their professional work. This includes: transparent, accountable, fair, corruption-free, and honest, qualitative cybersecurity operations with respect to integrity; taking into account integrity of people and establishments; considering ethical standards with respect to economic profit maximization, audience rating, and entertainment goals. Professional codes of ethics are significant for enhancing the ethical responsibility of content providers in the information society.

Cybersecurity professionals must take into account the following [21]:

- Associations and network of professionals (e.g., security researchers, analysts, policymakers) for ensuring the promotion and enhancing the ethical codes with respect to production, distribution, and archiving of information, communication, and knowledge.
- Affiliations and system of experts to create and advance separate codes for consumers.
- Governments to guarantee a lawful system that offers space for defilement free and legit cybersecurity practices.
- Preparing establishments of media experts to incorporate morals courses as compulsory in the educational plans.
- Preparing establishments to give training in advanced security to cybersecurity experts, both disconnected and on the web.

2.5.5 PRIVACY AND SECURITY

Privacy is a human right. Views on privacy rights differ between countries and political structures. Threats to privacy are continually emerging-particularly from the business and security segments and social networks. Sensible parity should be struck among privacy and security needs. Governments must work towards protection of public security. However, cyber-warfare can threaten public security. Hence, companies in their pursuit of profit, must also consider the privacy of individuals. Privacy is threatened as much by the private sector as by government action or negligence; however, both must be accountable to individuals and organizations for their actions.

> When considering privacy, the following must be taken into account:
> - Organizations should enforce reasonable privacy safeguards for their employees.
> - Organizations to create software and gathering information to guarantee more attention regarding the ethical element of business, including a genuine regard for the piracy of people.
> - Web intermediaries to be more transparent about solicitations they get from governments for information access.

2.5.6 CYBERCRIMES AND INTELLECTUAL PROPERTY (IP)

Piracy has the potential to become an existential threat to prevailing business models for novel content creation and use. Piracy usually occurs when potential users see information or material as too expensive and rights protection as excluding the poor. As we know, most pirates do not steal ideas and information electronically. While stealing ideas and information is quite common, with digital media proliferating and the Internet growing continuously, it has become remarkably easier. New technology has intensified the IP piracy problem. However, it has also provided solutions for the same. Plagiarism these days is handily distinguished utilizing suitable devices and programs. Piracy, if not contained, would represent an existential danger to the current plan of action of distributors and producers. Thus, there is a need to recognize and plan a middle course for executing it in an ethical and powerful way with the goal that all stakeholders feel they have gained.

Ethical Issues in Cybersecurity 63

> When it comes to piracy, the following may be considered:
> - Guaranteeing that copyright requirement activities depend on comprehensive, multi-stakeholder processes that reflect straightforward and responsible procedures.
> - Empowering exploration and discussion on a fair lawful framework to secure IP and to support access to data for all.
> - Supporting relaxation of patents where affordable duplicates of items (for example, drugs) are fundamental for saving lives.
> - Underlining the requirement for it to be legitimately conceivable to do digitally what can be lawfully done in hard copy, including a person's entitlement to possess digital materials and to give or move their proprietorship to other people.

2.5.7 THE NEW GENERATION OF PEOPLE

The Internet is a popular platform for connecting computers, smartphones, and tablets, and young people are connecting with each other and the wider society in different ways [10]. While the digital world promises to bring people together, there are certain issues that come with it. There may be a situation where children, young people and young adults may face specific risks and hazards related to the Internet. Some of these issues may be in the form of cyberbullying, a lack of anonymity and potential addiction to online networks. The digital world may also witness dangers of addiction to online games and social networks. Risks must be managed through requirements of general estimates, for example, transparency, the requirement for explicit consent in sharing of data, and the option to pull back such consent. There must be measures to protect children and young people. Further, clients must have rights to data, freedom of expression and association, privacy, and non-discrimination.

> When it comes to protection, the following may be considered:
> - Internet and social networking platforms must ensure comprehensible and accessible privacy mechanisms.
> - Organizations must support research with respect to cybersecurity (social media) by children, young people and young adults.
> - Ensure investigation and enforcement mechanisms for dealing with cyber-based criminality, which may incorporate exploitation and abuse of minors.

2.5.8 ECONOMIC POWER OF TECHNOLOGY

Key actors in the cybersecurity domain may be in the form of investors and managers of companies and institutions. These actors have specific responsibilities in using their entrusted power. Greater the power, more will be responsibilities and accountability. Power can either be used responsibly or may also be abused. The profits expected from these investments must be optimized and not maximized. Profit must not be treated like a goal, rather it should be considered a means to provide these services for the sustainable development goals. People in cyberspace have their own responsibility in interacting with other users, systems, and information.

> When it comes to power, the following may be taken into consideration [21]:
> - Organizations must pay heed to the qualities and principles of socially responsible investments in all ventures identifying with cybersecurity advances.
> - Media proprietors and other pertinent private-sector enterprises to guarantee that their inclusion in and strategy of cybersecurity organizations depends on ethical qualities and responsibilities regarding the particular effect of the sector for society.
> - People in cyberspace must be entitled to freedom, equal access, peoples' participation, respect of diversity and sustainable development.
> - Producers and consumers in the cybersecurity sector to utilize their individual capacity to advance ethics in cybersecurity society.

2.5.9 ETHICS OF REGULATION AND FREEDOM

Educated users in cyberspace must guarantee that administrative estimates bolster freedom of expression, freedom of association in cybersecurity-related advancements, and the option to look for, get, and impart data and thoughts. Quick innovative turn of events, ethical norms and regulatory frameworks must be more synchronized. The need to maintain crucial qualities, regard human rights and fundamental freedom of others, dealing with the abusive uses of cybersecurity-related technologies, through unlawful and illegal activities raise the issues of regulation and regulatory frameworks. There is a balance to be maintained between Internet freedom, which may risk exacerbating inequality and unequal access, and encouraging an equity of access that requires regulation.

> When it comes to policy, the following may be considered:
> - Promote the regulation and freedom of the cyberspace.
> - Guarantee a multi-stakeholder approach dependent on transparency, responsibility, and representativeness, to incorporate the exercises of transnational enterprises, to address net neutrality and disparities in Internet access.

In this chapter, we acquainted ourselves with specific ethical issues related to cybercrimes like copyright infringement, piracy, plagiarism, cyberbullying, ransomware attacks, etc. Other than that, we explored the idea of fair use doctrine and interpreted the nine Ps of cybersecurity ethics. In the next chapter, we will delve into a little more detailed version of cyberethics.

2.6 SUMMARY AND REVIEW

- There are three kinds of ethical issues in cybersecurity: personal privacy, access rights and harmful actions.
- Copyright encroachment might be characterized as the utilization or creation of copyright-ensured material without the consent of the copyright holder.
- piracy might be characterized as the unapproved duplication of copyrighted substance that is then sold at considerably lower costs.
- All piracy is copyright infringement but not all copyright infringement is piracy.
- Intellectual property (IP) alludes to the responsibility for thought or plan by the individual who concocted it.
- Computer fraud is the act of using a computer to take or alter electronic data, or to gain unlawful use of a computer or system.

- plagiarism is characterized as the act of legitimately duplicating and afterward introducing a current creation without precise reference or referencing, and additionally making the item look like one's own, without authorization from the original producer.
- Cyberbullying might be characterized as the utilization of electronic correspondence to menace an individual, ordinarily by sending messages of an intimidating or undermining nature.
- Identity theft is the crime of getting the individual or financial data of someone else for the sole purpose of assuming that person's name or identity to make transactions or purchases.
- ransomware is a type of malware that encrypts the victim's files. The attacker at that point requests a ransom from the victim to re-establish access to the information upon payment.
- Fair use is a doctrine in the law of the United States that licenses constrained use of copyrighted material without having to initially procure the consent from the copyright holder.

QUESTIONS TO PONDER

1. Take the example of a can of Coca-Cola®. Identify the trademark, trade secret, copyright, and patent.
2. In the event that Open Access in the future requires installment by creators or establishments, what sort of solutions and preferential treatments are to be offered for foundations in developing countries that cannot manage the cost of membership charges and assets for publication?
3. How is Open Access identified with copyright issues?
4. I bought a single license piece of software, and now my whole class needs to obtain the product to load into their systems. Would it be able to be load onto a few machines? In the event that I am not loading the software into their systems yet simply handling it out, am I blameworthy?
5. How might I tell if my software is genuine/authorized?
6. I am a creator and copyright holder myself. Is fair use a threat to my copyrights?

KEYWORDS

- **cyberethics**
- **cybersecurity**
- **environmental impacts**
- **infrastructures**
- **securing networks**
- **vulnerability disclosure**

CHAPTER 3

Cybersecurity Ethics: Cyberspace and Other Applications

In the previous chapters, we acquainted ourselves with the basic concept of ethics in cyberspace. Now that we are aware of the significance of ethics in cyberspace, we will specifically target the sphere of cybersecurity ethics. Further, as cybersecurity is an advancing field and makes its presence felt in various other domains and applications, we will also highlight ethics in those domains and applications.

3.1 CYBERSECURITY ETHICS: A DETAILED STUDY

Let us recall the definition of cyberspace from Chapter 1. According to the National Institute of Standards and Technology (NIST), cyberspace incorporates the Internet, telecommunication systems, computer systems, embedded processors, and control systems. Each of these domains have specific functionalities and operates in various ways [14]. Thus each of these have their own ethics and standards. Each of these components are responsible for establishing cyberspace and their applications traverse the globe. Therefore, a detailed study on ethics for these components is of great significance.

3.1.1 CYBERSECURITY: INTERNET ETHICS

One of the most remarkable and active components of cyberspace is the Internet. In fact, the majority of the communication among devices and users takes place over the Internet. Since the Internet harbors millions of users and devices, exchange of information over the platform must be performed in an ethical way.

> **Definition 3.1: Internet**

> The Internet might be characterized as a method for interfacing a computer to any other computer anywhere in the world by means of dedicated routers and servers. At the point when two computers are associated over the Internet, they can send and receive a wide range of information like text, designs, voice, video, and computer programs.

Many times, ethics over the Internet is broadly referred to as Internet Etiquette or Netiquette. As Etiquette corresponds to the code of polite behavior in society, netiquette is concerned with code of good behavior on the Internet. Netiquette takes into account various aspects of the Internet, like email, social media, and web forums. It also underpins multiplayer gaming, online chats, website comments, and other ways of online communication. There are no specific netiquette rules or guidelines, however the overall idea is to respect others online:

1. **Remember the Human:** When we interact in Cyberspace, we must not forget that we might be interacting with a human who is on the other side of the interactive screen. One must refrain from displaying derogatory behavior and making the other person uncomfortable. Often, what we write or send may be stored somewhere that might not be accessible to us. There is a good chance of it being used against us.
2. **Following the Same Standards of Behavior Online as in Real Life:** Although, the chances of getting caught sometimes seem slim, one must be aware of the fact that breaking a law is bad netiquette. Netiquette mandates that people do their best to act within the laws of society and Cyberspace.
3. **Knowing That You are in Cyberspace:** Netiquette varies across domains. What may seem perfectly acceptable in one domain may be dreadfully rude in another. When entering a cyberspace domain that is new, it is important to investigate and spend some time tuning in to the chat or perusing the files and documents. One must attempt to get a feeling of how the individuals who are there act before participating.
4. **Appreciating the Other People's Time and Bandwidth:** Bandwidth may be defined as the information-carrying capacity of the wires and channels through which every user and system is connected in Cyberspace. There is a specific amount of data that may be carried in

a given time. Bandwidth also refers to the storage capacity of a host system. When the same is accidentally posted to the same newsgroup multiple times, both time (of the people who check all the copies of the posting) and bandwidth (repetitive information sent over the wires and getting stored) are wasted. Therefore, one must not spam others by sending large amounts of unsolicited emails.

5. **Making Yourself Look Good Online:** The Internet lets us reach out to people we would otherwise never meet. One might be judged by the quality of their writing. It is not ethical to use offensive language; there is also no need to be confrontational in situations of conflict. If cursing is indispensable, it is preferable to use amusing euphemisms or classic asterisk fillers.

6. **Sharing Expert Knowledge:** The objective of the Internet is to grow and share information across destinations. It is polite to share the results of questions with others over various platforms on the Internet. Many people explicitly broadcast different kinds of resource lists and bibliographies. These may also include legal resources and popular books.

7. **Helping Keep Flame Wars Under Control:** Flaming is the act of people expressing a strongly held opinion without holding back any emotion. It is the kind of message that instigates people and makes people respond. Netiquette prohibits any kind of flame wars. It may be in the form of a series of angry letters, many of them directed towards specific individuals or groups. They may control or oppress the tone and demolish the companionship of a discussion group. Additionally, it is unfair to the other members of the group. One should avoid posting inflammatory or offensive comments online.

8. **Respecting Other People's Privacy:** A lot of people in Cyberspace do not respect the privacy of others. Failing to respect other people's privacy is not only bad netiquette but also comes with other costs and may also be followed by loss of jobs. One must respect others' privacy by not sharing personal or sensitive information, which may include pictures, and videos that another person may not want to be published online.

9. **Not Abusing Power:** A few people in Cyberspace have more control than others. They might be professionals or experts, and executives in an organization. Knowing more than others, or having more control than they do, does not give one the right to exploit others. For instance, sysadmins should never read private messages.

10. **Be Forgiving of Other People's Mistakes:** People make mistakes, even in Cyberspace. While minor errors do not cause any significant issues, if an issue bothers an internet user, he/she must think before reacting. Having good manners does not give anyone the right to correct everyone else. If one must inform someone of a mistake, it must be pointed out politely, and preferably by private email rather than in public. One must give people the benefit of the doubt by assuming they do not know any better. Netiquette discourages arrogant behavior. Like spelling flames always contain spelling errors, notes pointing out Netiquette violations are often examples of poor netiquette.

Since we do not see or hear the people who we are communicating with, the Internet provides a sense of anonymity when communication is taking place online. However, that is not a reason to have poor manners or post inappropriate comments online. Many users want to hide behind their keyboards, systems, or smartphones while posting online. However, they are still the ones publishing the content. It is important to remember that posting offensive remarks online may lead to serious consequences if the veil of anonymity is lifted, as there might be several questions raised about the indecent comments made. In summary, good netiquette is useful for everyone in Cyberspace, and it ensures a healthy environment.

3.1.2 CYBERSECURITY: TELECOMMUNICATION SYSTEMS ETHICS

Telecommunication systems is yet another important aspect of Cyberspace. Telecommunication systems have advanced tremendously over the last few decades.

➢ **Definition 3.2: Telecommunications**

> Telecommunications is referred to as the transmission of information. This information may be in the form of signs, signals, messages, words, writings, images, and sounds. The information is usually transmitted through means of a wire, radio, optical or other electromagnetic systems. Telecommunication occurs when there is an interchange of information between communication parties which includes the use of technology.

A telecommunications system operates as a cluster of nodes and links to facilitate telecommunication. Telecommunication incorporates three important aspects, namely information, communication, and technology, and they are related to one another. Information is produced due to an interpretation process. Usually, a receiver interprets data during a specific time-space based on the total amount of knowledge and experience of the receiver. Communication may be defined as the interaction between elements and the repercussion of the effect that the performance of one element has on the performance of another element. Technology is a field of knowledge and skills that includes design, construction, and use of created artifacts, and the theoretical science of technology itself.

When technology-based environments are considered, and the complexity that emerges in the cross between humans, between technology and humans, and between information, communication, and technology [25], a cultural focus emerges where human attitudes, values, and behavior sets the conditions for communication and interaction. Ethics and ethical contemplations are significant elements for where to wind up in the range among conflict and agreement. Hence, ethical considerations characterize human attitudes and behavior towards the environment. Inside the ethical circle of information, the issue of suitable or right conduct related to information, creation of information, and transfer of information is considered primarily.

In a changing technological infrastructure, ethical considerations have an effect on the outcome of human communication. Common norms and rules must be respected to enable humans to live and act together in Cyberspace. Rules or norms can be both explicit and implicit. Society depends on technology. This dependence brings to the fore ethical considerations of its design, use, and implementation. One way of dealing with ethics when designing ICT environments is through a framework for ethical guidance in development projects and the design of information systems [26]. The framework includes a method with four steps:

- Identify the different needs that can be associated with the situation;
- Identity where conflicts may occur;
- Solve the problems that can cause conflicts; and
- Let the common needs to be included in a requirement catalog and perform a requirement analysis for the situation to be designed.

3.1.3 CYBERSECURITY: COMPUTER SYSTEMS ETHICS

All the operations in Cyberspace take place due to the interaction of computer systems over the network.

➤ **Definition 3.3: Computer Systems**

> Computer systems are shaped by an arrangement of interconnected systems that share a central storage system. It also supports various peripheral devices like printers, scanners, and routers. Every system associated with the system can work autonomously, however, it can communicate with other external devices and computers.

Cyberspace continues to grow as more and more computer systems are added. More computer systems mean increased usage as well as more communication among users. Therefore, it is mandatory to take care of ethical considerations in Cyberspace with respect to computer systems. Some ethical considerations are as follows [126]:

1. **Contributing to Society and Human Well Being:** While designing or implementing systems, it is mandatory that computer experts guarantee that the results of their endeavors will be utilized in socially mindful manners, will address social issues, and will stay away from destructive impacts to wellbeing and welfare.
2. **Avoiding Harm to Others:** The use of computing technology should not harm users, the general public, employees, employers. Harmful actions incorporate intentional destruction or modification of files and programs. This may have severe consequences in the form of serious loss of resources and unnecessary expenditure of human resources (HR). Additionally, in order to purge systems of computer viruses, a significant amount of time and effort may be required. Sometimes well-intended actions may lead to harm unexpectedly. In such a circumstance, the responsible individuals are committed to fix or alleviate the pessimistic outcomes however much as could reasonably be expected. A mainstream approach to dodge unexpected damage is to deliberately consider likely effects on every one of those influenced by choices made during design and implementation.

3. **Being Honest and Trustworthy:** Honesty and trust go hand in hand. For an organization to function effectively, trust is very important. Honest computing professionals do not make intentionally bogus or tricky cases about a system or system design. Rather, they give complete honesty of all appropriate system constraints and issues.
4. **Being Fair and Taking Action Not to Discriminate:** Discrimination between different groups of people may be the result of use or misuse of information and technology. All individuals are entitled to equivalent opportunity to participate in, or take advantage of, the utilization of computer resources irrespective of their race, sex, religion, age, disability, national origin, or other such comparative elements. These standards, however, do not legitimize unapproved access and utilization of computer resources. Further, they do not give an adequate motivation to infringement of some other ethical objectives of this code.
5. **Honoring Property Rights Including Copyrights and Patents:** Violating copyrights, patents, trade secrets, and the terms of license agreements is prohibited by law in most circumstances. The violations are contrary to professional behavior even if the software is not so protected. Copies of software must be made and distributed only with proper authorization. Unauthorized duplication of materials must be discouraged.
6. **Giving Proper Credit for Intellectual Property (IP):** Computing experts must ensure the integrity of protected innovation. Hence it is not ethical to assume acknowledgment for other's ideas or work, even in situations where the work has not been unequivocally ensured by copyright, patent, and so on.
7. **Respecting the Privacy of Others:** This principle is significant for electronic communications, including electronic mail. It denies strategies that capture or monitor electronic client information, including messages, without the consent of clients or approval identified with system activity and support. Client information must be treated with strictest secrecy, aside from in situations where it is proof for the infringement of law or authoritative guidelines. In such cases, the nature or contents of the sensitive information must be disclosed only to proper authorities under the right circumstances.

8. **Honoring Confidentiality:** Honesty and confidentiality are interconnected. It is necessary to honor confidentiality or, implicitly, when private information is not directly related to the performance of one's duties. Computer professionals must regard all commitments of secrecy to employers, customers, and clients except if released from such commitments by necessities of the law or different standards of this code.

3.1.4 CYBERSECURITY: EMBEDDED PROCESSORS ETHICS

An embedded system may be thought of as a combination of computer hardware and software, which is either fixed in capability or programmable, designed for a specific function or functions within a larger system. Embedded system hardware can be microprocessor- or microcontroller-based.

> **Definition 3.4: Embedded Processor**

> An embedded processor is primarily a microprocessor designed into a system that is responsible for controlling electrical and mechanical functions. An embedded processor can be programmed specifically for the specific operations. It may support many different CPU (central processing unit) architectures.

Since embedded processors are deep-seated in critical hardware, such systems can be remarkably tricky to safeguard. Hence cybersecurity and cyber resiliency must be considered toward the start of the plan and design process. Embedded processors are known to be powerless against cyber-attacks because of a few reasons, which may go from the electronic system's capacity to withstand cyber threats to the side effects of processor pipeline designs to empower cyber-attacks. In addition, fixes may bring about inadvertent impacts, including expanding power utilization and influencing processing or timing, all of which can constrain hardware performance. Therefore, ethical considerations must be kept in mind while designing such embedded processors. This is important since each person involved in the working of an embedded system acts assuming that the other team members have done their job properly, e.g., a captain sails a ship accepting the originator/designer/maker have planned, created, and made appropriately; that the maintenance team have carried out their responsibility and that the ship control works perfectly, etc. It is the ethical obligation of all team members to release their obligations with the most noteworthy honesty and meticulousness. Each individual included must adhere to the most elevated moral norms to forestall the odd practices inside his/her obligation [27]:

1. **Communicating Information Responsibly:** When there are many team members, there is a collective obligation that requires the architects/engineers/makers, clients, and software writers to interface capably, particularly without ambiguity, to guarantee the safe operations of the embedded processors. Protocols with respect to communication must be created for this reason, and correspondence must be kept up in any event until the activity of the processors is steady. Furthermore, there must be appropriate infrastructure set up to convey back diagnostic information to the engineers to improve security. This features another significant domain for which ethical policies should be formulated and the amount of data that must be disclosed. For instance, manufacturers may not be interested in disclosing certain details of the embedded system with the objective of not losing their competitive edge. There may likewise be irregularities in the correspondence interfaces, between artificial intelligence (AI)-based programming and people, because of which data may get lost, stay unattended or is misjudged. Consider the instance of traditional safety-critical systems, Leveson cites poor information flow data as one reason for software-related mishaps. Hence the requirement for communication among various developers expects us to research the moral parts of correspondence with emphasis on what data should be 'disclosed' for moral reasons, while securing different secrets. The biggest challenge involved in the communication infrastructure is that there should be an adequate transfer of pertinent information for the system to operate in a secure manner.

2. **Attributing Responsibility to Software:** Embedded systems must be accompanied by suitable (self) regulatory mechanisms that will guarantee their moral conduct. It is as yet a generally discussed theme and a few ethicists decline any chance of machine morals/ moral machines while some others safeguard a similar line of argument that we present. Notwithstanding, disregarding the

responsibility of software can leave the process of tending to the policy vacuum incomplete. Profoundly keen software agents must be programmed to understand and apply morals so as to ensure that they learn and evolve like humans. A liberal methodology towards the moral obligation of software agents empowers social maintainability. Despite what might be expected, if the software agents related to embedded systems are given immunity from responsibility, mishaps that happen because of wrong decisions taken by software agents will stay a dubious issue. Likewise, different engineers who are not the slightest bit answerable for such mishaps chance being punished. The risk of prosecution may likewise make the designers hesitant to receive the propelled abilities given by the field of AI in building more secure and more productive systems [121]. This hesitance may end up being unfavorable since it might additionally sabotage the objective of social sustainability. Airplane mishaps suspected to be brought about by pilots bring up issues regarding whether software agents would be more secure. In such a manner, it is significant that attributing the responsibility of failures to completely autonomous systems gains acceptance among communities related to government, law, users, etc.

3.1.5 CYBERSECURITY: CONTROL SYSTEMS ETHICS

Control systems are an integral part of the automation industry. Different control loops that regulate these processes may be in the form of industrial control systems (ICS). Supervisory control and data acquisition (SCADA) and distributed control systems (DCS) are some examples of control systems.

> **Definition 3.5: Control Systems**

> A control system is made of a set of mechanical or electronic devices that controls other devices or systems using control loops. It consists of many elements that may be combined together to form a system for producing desired outputs.

The advancement of Control systems has significantly increased the cyber-attack surfaces. The challenges may be in forms like increasing cyber threats, lack of visibility and control, difficulty in securing control layer protocols and unsecured networks. Therefore, addressing the issues from an ethical point of view is mandatory:

1. **Machine Design Ethics:** Machine ethics are to be considered from the earliest starting point of the research work and the meaning of the necessities, past classical safety studies. For each stage of the machine design process, it is advised to realize a risk assessment and to implement redundant technological barriers to contain possible machine ethical risks. It is important to consider how the designed control system could be misused by error or rendered harmful (hijacked) using cyber-attack technologies, for example. Time should be taken as well into consideration for imagining extreme conditions under which the partners might be sued and announced lawfully liable for a dishonest conduct of the control system and then to adapt the implementation process to limit these risks.
2. **Human-Centered Approach:** Researchers must design the control system using a human-centered approach rather than a techno-centered one. This will guarantee utilizing the possible capacity of people when confronting the unexpected while at the same time confining automatic and deliberate abuses of the control system.
3. **Design Process:** It is encouraged to define a ventured design process, containing distinctive approval stages. Although it is a classical design way, adding exaggerated tests at each of the steps could be yet another approach. This approach would be beneficial for testing the robustness of the designed elements facing exaggerated risks, before the final targeted implementation of the control systems.
4. **Safety, Responsibility, and Liability:** Safety aspects should be one of the primary concerns such that control systems and humans can work together without accidents. Since control systems increasingly operate in close proximity to humans, safety is very important. Control systems are not only large and complex, but also intelligent and self-learning. But it is equivalently important to understand that these systems may fail, and neither the systems nor humans can be blamed for it. Besides, it is much harder to locate the underlying reason and attribute liability. It is equally hard to find who is accountable for the failures and malfunctions. Therefore, these questions need to be addressed from an ethical point of view.

5. **Privacy Concerns:** Control Systems require immense measures of information to work successfully, and this poses several privacy questions. For instance, in order to build smart home security systems, smart home systems need to keep track of the times when residents are away. This information may be private, but valuable for burglars. Robots spying on the working habits of their human co-workers may be yet another privacy concern since they may manipulate humans to work harder. The ethics behind reviewing medical professional secrecy concerning the health data stored on connected parts of medico-technical systems also raises many privacy concerns and questions. Thus, collecting data on an individual's lifestyle and physical parameters may contribute towards improving their health, but it is not ethical to take advantage of the data shared on medico-technical systems (Figure 3.1).

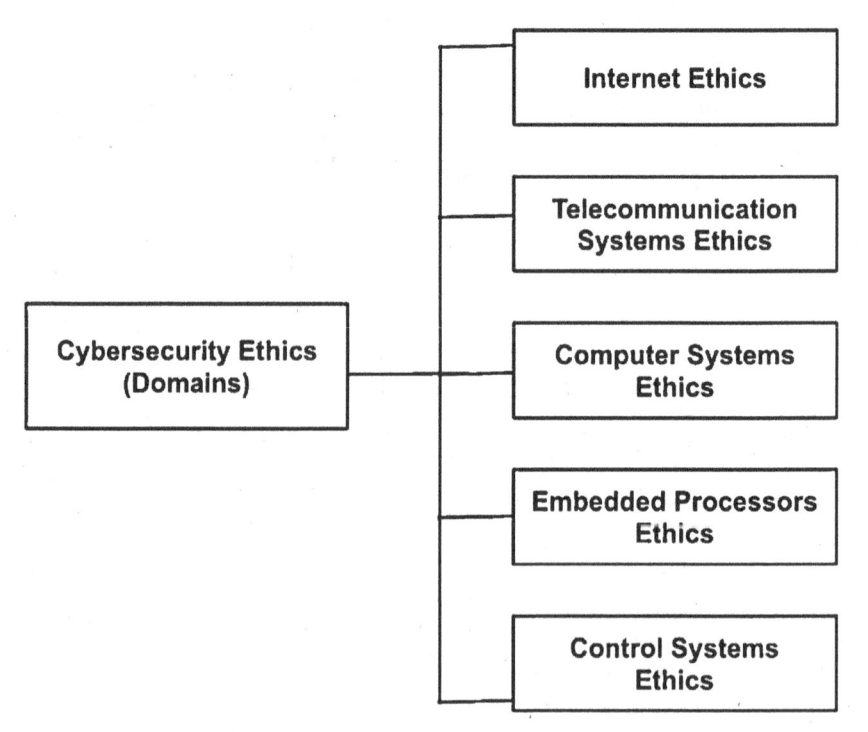

FIGURE 3.1 Cybersecurity ethics (domains).

3.2 CYBERSECURITY ETHICS IN APPLICATIONS

In the previous section, we highlighted the ethical concerns in Cyberspace. Cyberspace incorporates several applications. In this section, we will list the cyber ethical issues some of these applications face (Figure 3.2).

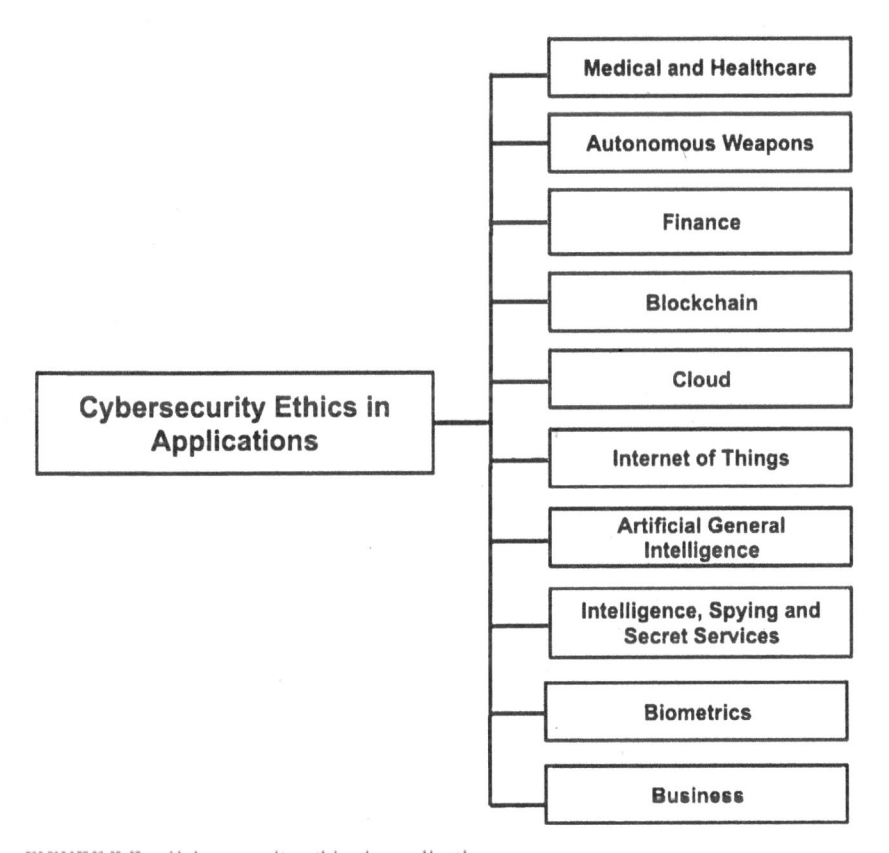

FIGURE 3.2 Cybersecurity ethics in applications.

3.2.1 CYBERSECURITY ETHICS IN MEDICAL AND HEALTHCARE INDUSTRY

Healthcare cybersecurity is a challenging issue in the healthcare industry. Electronic health records (EHRs), include a plethora of sensitive information about patients' health and medical histories. Due to this, the hospital network security is a potential target for hackers. Through EHRs, it is very easy to share essential information among physicians and other healthcare professionals, and insurance companies. A cybercriminal's dream is to easily obtain files that contain an individual's social security number, address, email address, credit card data, and even medical history. Healthcare organizations are vulnerable to cyberattacks, mainly because their security is often much more relaxed than other organizations. The healthcare industry is vulnerable to cyber-attacks due to the following reasons:

1. **High Prize:** Healthcare records fetch for a high price since they incorporate sensitive details like credit card data, email addresses, social security numbers, medical history records, and employment information. Cybercriminals use the data to launch spear-phishing attacks, commit fraud and steal medical identities. Healthcare data may also be used for stealing research and development (R&D), disrupting the supply chain and manipulating stocks.
2. **Ransomware:** Cybercriminals rely on ransomware to make even more money off of the healthcare organization itself. Hackers mostly demand the money to be paid in bitcoin currency, and some have even demanded a ransom as they vandalize the systems and gain control of entire hospital computer systems.
3. **Major Vulnerabilities-Employees and Encryption:** One of the greatest cyber threats to a healthcare organization is its own employees. Over half of the data breaches came from people within the organization. Another major threat is the lack of encryption as the healthcare sector has one of the lowest overall rates of data encryption.
4. **Counteracting Attack Attempts:** Due to the rapid changeover to EHRs, and the fact that healthcare organizations increasingly face sophisticated digital changes, the need for cybersecurity protocol grows more each day. Patients with compromised records usually have no choices in these circumstances. Numerous medicinal services associations have wound up paying the government fines, and settling in civil suits. Smaller organizations lacking experts and staff to prevent these attacks are extremely vulnerable. Therefore,

it is important to train all staff members on proper information technology (IT) protocols. If a user can identify a phishing attack, attacks can be avoided.

Since the data in the medical industry is very sensitive, it is mandatory that information is protected using a combination of ethical as well as a best practices approach.

3.2.1.1 CASE 1: IDENTIFYING COMMON HEALTHCARE SECURITY THREATS

When looking at potential threats, one must consider:

1. **Staff or Insider Threats:** Healthcare employees have easy access to patient files and medical records. Several incidents in the past highlight employees stealing sensitive information. This data may not exclusively be utilized in identity theft, but may also be used to blackmail people. There are several ways of stealing records. In many cases, employees have been accused of accessing confidential financial documents and using patients' credit card numbers for fraudulent purchases. At times, workers have also been reported to steal face sheets, which includes demographic and social security information. This information is also sensitive and may be used to commit a variety of crimes.
2. **Malware and Phishing Attempts:** These are a common means to steal login credentials for compromising an entire system. This may be done using sophisticated malware and phishing schemes which embed malicious scripts on a computer. In the case of malware, it just takes one apparently valid connection to bring a terrible cyber presence into the system, which may not be very easy to thwart. Thus, it is important to prepare staff with the goal that regular phishing attacks might be perceived. A common scam involves emails from authentic-looking sites requesting login information. It is interesting to note that reputable companies never ask for login information through emails. Once a user provides the information, the hacker on the other end can easily get access to the user's system. There are various sorts of viruses that are fit for

mining records and automatically sending them back to the initial host. At times, they may even leave a backdoor open for getting to the machine later.
3. **Vendors:** One of the common mistakes that healthcare providers do is work with vendors without estimating the accompanying risk. Let us take an example of a hospital that hires a cleaning company. The employees may gain access to computers. Although patient information must be locked in ways that the average employee cannot view, it may be a challenging task to safeguard all points of access as cleaning and maintenance are integral to maintaining a healthy work environment.
4. **Unsecured Mobile Devices:** Many healthcare facilities that allow mobile logins do not require the devices to meet security standards. This makes their systems helpless against malware and hackers because all of the organization's planning and security do not have an impact on staff communication devices. This situation aggravates once staff dispose of the equipment in an upgrade. The system data or passwords may in any case be available, making it easy for criminals to access the devices. Thus, the association must set policies or ban user devices altogether.
5. **Lost and Stolen Mobile Devices:** Lost or stolen devices pose an enormous threat. If a mobile device is used to access a facility's network, it becomes a liability as soon as it is lost or stolen. Given that, if it falls into the wrong hands, the user can easily access the system using previously stored data. For malicious users having access to the network, it may be very difficult to detect their presence or reseal the breach.
6. **Online Medical Devices:** These often lack security mechanisms. This makes them obvious objectives for hackers. Medical devices such as infusion pumps only provide information to the doctor and patient involved. However, with the Internet of medical things (IoMT) continuing to grow, such gadgets are known to trade the data to outside sources and are also capable of interacting with the world outside the health centers. This sensitive data may be easily intercepted or manipulated, leading to privacy issues. Further, hackers may access and gain control over most things associated with the system.

7. **Unrestricted Access to Computers:** Computers not confined to restricted areas may be easily accessed by unauthorized personnel. When systems are associated with sensitive patient data, unauthorized staff or others in the area could rapidly discover sensitive information. Further, successful phishing attempts on general-access computers may allow hackers to access more sensitive areas of the network. Therefore, it is important to ensure that any computer that holds patient information must be placed in a secure location.
8. **Inadequate Disposal of Old Hardware:** Deleting information from a system does not mean that the information can never be accessed. Disposing hard drives properly could cost a lot. Old terminals and other equipment that is utilized to get to a system may rely upon credentials which a criminal may attain if hard drives are not properly disposed of. Even after drives have been deleted or reformatted, it is still possible to rescue this information. This means that anything the user saved is still vulnerable.

3.2.1.2 CASE 2: ADDRESSING DATA SECURITY ISSUES IN HEALTHCARE

A few critical ways an organization and its employees can prevent cyber threats include:

1. **Educating Employees:** It is necessary to make employees understand their role in cybersecurity and its influence on patients' lives. This creates an atmosphere that values and respects security. Regular briefing sessions and frequent communication on the state of the organization's security may ensure cyber safety of an organization. Further, staff training sessions and regular cybersecurity meetings could promote cybersecurity awareness among employees.
2. **Establishing Procedures:** Creating a plan summarizes definite guidelines for handling information and networks both physically and virtually. It also ensures that the protocols are being followed. Once the expectations are laid out, the process becomes standardized, which allows more comprehensive oversight for network security

monitoring. It is imperative to create proper penalties for inability to follow the strategies. This discourages careless conduct that may compromise the capacity to remain in consistency with HIPAA. It may also underscore the value one places on keeping patient information secure.

3. **Require Software Updates:** Loopholes in outdated software or other unsecured access points often allow cybercriminals to sneak into systems. To combat this, it is necessary that software updates on machines are regularly enforced. Following certain best practices like utilizing two-factor authorization and automatically regulating monthly password updates that require strong passwords may ensure system security. Employees may be assisted by automatically setting company machines to periodically require changes so that employees only have to come up with a new password or click to allow updates. While it may be incredibly difficult to enforce on staff personal devices, educating employees on the importance of updates is crucial.

4. **Set Strict Personal Device Regulations:** Healthcare providers must establish strict protocols regarding the use of mobile devices. This likewise incorporates the removal of equipment that incorporates sensitive information from an earlier time. Mobile device management (MDM) software permits IT administrators to secure, control, and implement arrangements on tablets, cell phones, and different gadgets. This guarantees representatives do not break noteworthy policies, and information remains safe.

3.2.1.3 CASE 3: DEALING WITH SECURITY BREACH IN A HEALTHCARE ORGANIZATION

In case of a patient information being violated the following steps may be considered:

1. **Reporting the Breach:** If an unsecured or compromised network activity is experienced, it must be reported to the U.S. Department of health and human services (HHS). Reporting times may differ depending on the number of people that have been affected by the breach.

2. **Sharing the Information:** The patients must be helped to understand and recognize signs of fraud, like unrecognized medical bills and unsubstantiated claims from insurance providers. Such detailed information is provided by The Federal Trade Commission (FTC). This information may be passed along to any interested patients. It may also include their rights under the fair credit reporting act (FCRA) and how to proceed with any claims.
3. **Reexamining the Network:** If an attacker gained access to your organization's network, the incident must be investigated. This would ensure that any weaknesses that allowed threats are secured. Network professionals may be enlisted for the same. They are capable of finding the current gap as well as assess for future problems for creating safeguards against future attacks.

3.2.1.4 CASE 4: MINIMIZING SECURITY THREATS IN A HEALTHCARE ORGANIZATION

To limit security dangers in a healthcare association it is pivotal for information to be encoded with the goal that outsiders cannot get to data during transmission or when information is stored:

1. **Understanding the Network Map:** Using innovation that gives a review of the gadgets and capacity on the system is critical. This enables users to see exactly what information is vulnerable in which ways. It also gives information about when new or unauthorized devices join the system. This layout helps establish the access and restrictions for each device on the network and cuts down on inappropriate staff conduct.
2. **Updating the Software:** It is equally important to ensure that all software and operating systems information are up to date. These updates may consolidate basic patches which are equipped for discouraging potential cybercriminals. These attackers take advantage of previously found weaknesses in software. If software is not updated regularly, criminals may take advantage of the loopholes left behind by earlier versions.
3. **Virtual Private Network (VPN) Encryption:** Network privacy may be enforced by encrypting the network connection. It may

also block potential hackers. A VPN encodes data in such a manner that other viewers cannot see what goes out or comes in on the computer system. So even if a connection is being monitored, the observers may not receive anything unless they already had access to the system.
4. **Conducting Regular Audits:** Regular audits must be conducted by system administrators to enhance cybersecurity. Two-step authentication must be enforced for users looking to adjust information or entering new data for verifying their identity. All users must be encouraged to create strong passwords and change them frequently after a certain number of weeks. Periodical review of access credentials ensures that previous or transferred employees do not have access to sensitive patient data.
5. **Setting Strict Access:** Setting severe accesses and making the base measure of data accessible for access is one more technique for securing sensitive information. This would also cut the possibility of staff misuse.
6. **Thinking Like a Hacker:** Since cybercriminals manipulate a network, it is important to obstruct their efforts. While it may seem difficult to ensure healthcare security without having enough background knowledge about healthcare data security measures, it is still crucial to identify any potential gaps.
7. **Using Professional Services:** There are many ways through which health organizations can limit potential threats. For managing data security measures in healthcare, a network security expert or a specialized outside agency, may corroborate professional network safety and support. This would allow staff to focus more directly on medical-related tasks and not specifically on cybersecurity.

3.2.1.5 CASE 5: IMPROVING DATA SECURITY IN HEALTHCARE ORGANIZATIONS

For considering data security, the following may be considered:

1. **Considering All Phases:** Although cybercrimes are a significant threat, records theft may happen at any stage in the recordkeeping process. Therefore, it is crucial to develop technical, administrative, and physical safeguards, so as to ensure protection against

record thefts. Various technical aspects of security may ensure the prevention of inappropriate access to computers and potential backdoor setups. Certain administrative duties like handling of training materials and ensuring that terminated employees no longer have access to the network may be used to establish safeguards against employee misuse. Similarly, adjusting how records are physically handled as well as device locations may significantly reduce opportunities for abuse as well.

2. **Using the Crosswalk:** It is intriguing to take note of that the health insurance portability and accountability act (HIPAA) cybersecurity crosswalk, connects each HIPAA security rule to a related NIST cybersecurity classification. Utilizing this system, it might be conceivable to address HIPAA security issues through the data given by NIST. This ensures alignment and compliance with all regulations (Figure 3.3).

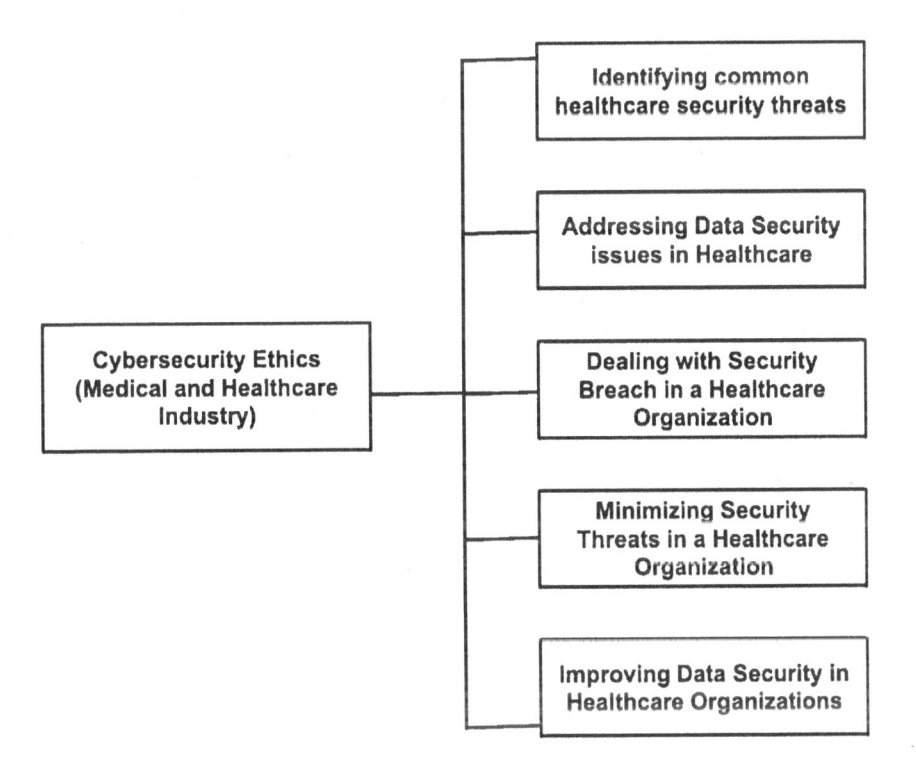

FIGURE 3.3 Cybersecurity ethics in medical and healthcare industry.

➢ Digital certificates may help safeguard healthcare data:

Identity is a significant part of security. If something has an identity, it can be more secure. Along these lines' individuals, devices, administrations, applications, and everything that interfaces with the web or are found in the Internet must have an identity. This identity may assist in encrypting communications and transactions, authenticating to a service, authorizing proper access and proving their integrity. Digital Certificates can ensure identity and trust. Certificates empower numerous security use cases that must be tended to in the HIPAA technical safeguards:

- **Web and Server Security:** Proving that public and private sites and servers are legitimate. At the same time, there is a need to protect and encrypt data submissions and carry out transactions with SSL/TLS Certificates.
- **User and Device Authentication and Access Control:** Executing solid Authentication without burdening end users with hardware tokens or applications and guaranteeing just affirmed clients, machines, and gadgets (including mobile) can access approved systems and administrations.
- **Document Signing:** Digital signatures that use trusted Digital Certificates replace signatures and create a tamper-proof seal that protects patient records and other documents which must be kept secure and private.
- **Email Security:** It is important to digitally sign and encrypt all internal emails so that it mitigates phishing and data loss risks. This might be done by checking the origin of messages such that recipients can recognize real as opposed to phishing messages.

➢ Other technical safeguards:

Technical safeguards may be defined as the technology, policy, procedures, and guidelines that protect electronic protected health information (e-PHI) and control access to it. According to the HIPAA Security Rule, technical safeguards include:

- **Access Control:** All covered entities must implement technical policies, procedures, and guidelines that allow only authorized persons to access e-PHI.
- **Audit Controls:** Every covered entity must actualize hardware, software, and procedural techniques for recording and inspecting access and other movement in information systems that contain or use e-PHI.

> - **Integrity Controls:** All covered entities must implement policies and procedures to ensure that e-PHI is not tampered or destroyed. Several electronic measures can be considered to confirm that e-PHI has not lost its integrity.
> - **Transmission Security:** All covered entities must implement technical security measures to safeguard against any unauthorized access to e-PHI that is being transmitted over an electronic network.

3.2.2 CYBERSECURITY ETHICS IN AUTONOMOUS WEAPONS

The rise of cybersecurity could see cyber warfare [23]. Cyber Warfare alludes to the utilization of computer technology and innovation for disrupting the operations of a state or association. This is done by the intentionally assaulting information systems for key or military purposes. Cyberwarfare incorporates techniques, tactics, and procedures which may be involved in a cyberwar. Autonomous weapons could play a major role in cyber warfare.

> ➢ **Definition 3.6: Autonomous Weapon Systems**

> Autonomous weapons systems (AWS) are characterized by the U.S. Department of Defense (DoD) as a weapon system that, when actuated, is equipped for choosing and engaging targets without any more intercession by a human administrator.

The AWS proposes the chance of dispensing with the human administrator from the war zone which raises several ethical questions.

The questions refer to moral decision-making by human beings involving an intuitive, non-algorithmic capacity that may not be captured by even the most sophisticated of computers. Some of the questions are as follows [127]:

- Is the intuitive moral perceptiveness on the part of human beings ethically desirable?
- Is the automaticity of a series of actions capable of making individual actions in the series easier to justify, as arguably is the case with the execution of threats in a mutually assured destruction scenario?

- Should the legitimate exercise of deadly force should always require meaningful human control?
- What could be the nature and degree of a human oversight over an AWS?
- Should the definition center around the system's capacities for self-governing target selection and engagement, or on the human administrator's utilization of such abilities?
- Should the human administrator's pre-commitment expectation have a conclusive bearing on the system's definition as an AWS?
- What are the suggestions for legal liability?
- Who should bear the legitimate risk for choices AWS makes?

In the past, some ethical considerations that have been taken into account are as follows concerning autonomous weapons.

Many governments are keen on adopting the official policy of humans being responsible for making the final decisions regarding the use of lethal force, even by otherwise autonomous weapon systems [28]. However, some federations believe that completely autonomous systems that can recognize and connect with targets are totally free of human control [29].

In both the cases, national-level arrangement choices legitimately impact the advancement of every nations' individual autonomous weapon systems. Several robotic weapon systems are still being developed. Hence there is a need to explore which approach will prove most advantageous on the future battlefield.

Future front lines will observe battling robots. Major world powers are pursuing variations with respect to autonomous fighting vehicles. Applicable policies and resultant developmental strategies are constantly being worked out and suggest that reality will be some hybrid of their respective visions. Many federations' insistence on human control of lethal effects ensures continued efforts by human-machine teaming that requires robust communication infrastructure, which may be vulnerable to manipulation. Others' evident comfort with autonomous lethal decision-making may lead to more agile systems that nevertheless create significant moral and legal hazards. The vision of future battle is quite unpredictable, the inevitability of fighting robots on future battlefield demands an ethical approach. Therefore, five principles for ethical use of autonomous weapons are as follows:

1. **Responsibility:** There is a need for human beings to exercise appropriate levels of judgment and be responsible for developing, deploying, and using AI. Although the principle sounds straightforward, it is also the most disputable. The use of the term human judgment has greater depth with respect to human control.
2. **Equitability:** The military must take action to avoid unintended bias that may inadvertently cause harm. This is the larger concern around bias in AI, since it is not clear if it can be resolved. There is no neutral data, hence equitability is very important. Consider a situation of a military conflict. Knowledge about the front line might be originating from one party in the dispute, who may be influenced by some other party in the conflict. A case of expected bias is elimination of the foe. An unintended bias may be collateral or civilian damage against targets in the war zone that were discriminated against in the information or knowledge assortment process.
3. **Traceability:** It is an important principle that deals with responsibility for an attack made by an autonomous system. The concept of algorithmic transparency and explainability, refers to a system explaining how it made certain decisions based on a machine-learning model or algorithm. This is a worry with AI all in all, however, it is essentially significant with regards to military matters. Maintaining accountability within the chain of command is crucial, but attribution is a crucial issue when it comes to who is responsible for an attack. A situation where it is unclear who is responsible for an attack should not lead to war.
4. **Reliability:** Reliable behavior of AI is critical. The military does not need an AI application that behaves abnormally. Rather, they require reliable techniques and tools that perform as they are expected to. Therefore, these systems must be tested thoroughly to understand its effects and side effects before using it to carry out any operations.
5. **Governability:** This standard tends to the biggest fear of automation of war and weaponry. A few scientists accept that the systems may become sophisticated and become unmanageable. There is a need for human or automated disengagement or deactivation of deployed systems that may be used to demonstrate unintended escalatory or other behavior (Figure 3.4).

```
                    ┌─── Responsibility
                    │
                    ├─── Equitability
Five principles for │
ethical use of ─────┼─── Traceability
autonomous weapons  │
                    ├─── Reliability
                    │
                    └─── Governability
```

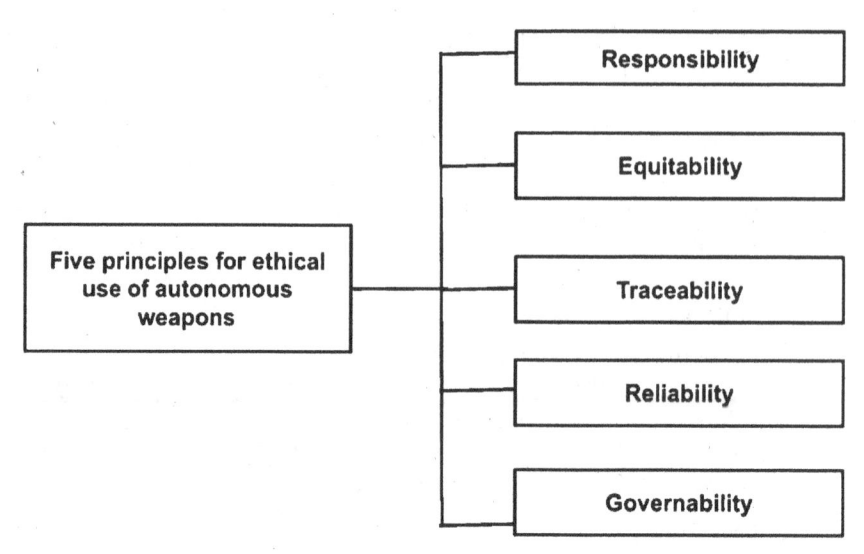

FIGURE 3.4 Five principles for ethical use of autonomous weapons.

While several issues are raised related to automatic weapons, countries will likely continue to develop cyberweapons. The question is if such weapons can be designed to be more ethical [30].

Controllability is a serious issue with cyberweapons. Therefore, ethical weapons must use a variety of methods that ensure focused targeting. System propagation and infection through viruses and worms must be minimized. Attacks must focus on limited sets of important targets. These targets must be clearly identified and confirmed during the attack. The identification must not be restricted to only Internet addresses. Civilian systems like commercial infrastructure should clearly identify themselves as civilians so attackers can avoid them.

It should also be possible for ethical attacks to be stopped easily. In a situation where an adversary surrenders, it is unethical and against international law to continue attacking and vandalizing the target.

Subsequently, all attacks must be under control through some techniques like an emergency channel so that they can be stopped if needed. Although it is difficult since the effects of the attack may delay communication, still it is important. Identification of the attacker may be considered a key feature of ethical attacks. Therefore, a piece of code associated or some data pertaining to the cyberattack should identify who is responsible for the attack. Adding digital signatures to code or data is one way to do it.

Many technologies that do the same are available, like encryption with a private key of a public-private key pair or a hash of the contents being signed. Hashing implies that the information cannot be altered after it is agreed upon since the public key can confirm it. Unattributed attacks may not be effective since attacks are used to force a country to do something, and the victim may not know what to do unless they know the attackers. Thus in such situations, responsible attackers let themselves be identified. Signatures permit the attacker to recognize vandalized systems by recognizing their own signature and avoid attacking them again and again. A constraint of signatures is that they may make it simpler for defenders to determine they are being attacked. This may matter when the attacks are subtle. However, it will not be easy to recognize signatures. Signatures are a function of the contents and therefore will differ when attached to different files or data. It is possible to sign normal code and data to prevent unauthorized modification, Therefore the existence of a signature merely does not imply an attack. It is important to conceal signatures which might be done using steganography. Damage perseverance with cyberweapons is a significant issue; therefore, considering weapons results are easily repairable at the end of hostilities could be ethically feasible. There is a greater likelihood of this being possible in Cyberspace than in conventional warfare. Consider a situation where an attack could encrypt important parts of a target system so that it cannot be used until the attacker supplies a key to undo it. Encryption and decryption being the reverse of each other do not lose any information. Hence, the attack would be totally reversible. This may well fill in as the ethical option in circumstances where it is inconceivable for a victim to reestablish a system from backup. Essentially, an attack could retain significant messages from a victim. If the messages are saved by an attacker, they could be provided to the victim at the termination of hostilities, thereby reversing the attack.

3.2.3 CYBERSECURITY ETHICS IN FINANCE

Financial institutions incorporate wealth management firms, investment brokers, and credit unions. These are very attractive to cyber-attackers as they host a variety of sensitive data like bank account numbers, social security numbers, etc. Attacking financial institutions may lead to loss of sensitive data, which may in turn have a devastating effect on the institution's reputation. Therefore, significant amounts of time and money to repair may be required. While banks have strict regulations and guidelines regarding security, not all these institutions fall under the same rules and regulations. Most security breaches are due to human errors. If one of the employees clicks on a malicious link in an email or allows an attacker to change the client's address without proper verification, it may lead to system compromise. Therefore, all organizations must promote cybersecurity awareness and make security best practices mandatory for their employees.

Banks and financial institutions can ensure cybersecurity by combining best practices as well as cyberethics. Several key security questions arise for a financial firm, which may not be handled completely in a technical manner. There are some moral contemplations that must be considered for such questions. Financial institutions must consider some key security questions for securing their organization:

- Is there an integrated information security strategy that underpins IT risk management?
- Is the association elevating careful risk evaluations to secure client data?
- Are representatives being kept updated on the strategy changes, new data security risks, and best practices?
- Is the association teaching its customers about data security risks and best practices?
- Is the organization effectively managing and monitoring system configurations?
- How is the organization staying abreast of the latest threat and vulnerability information?

The following questions can be answered using various techniques described as follows:

1. **Establishing a Formal Security and Ethical Framework:** Currently, several core security frameworks assist financial institutions in managing cyber risk more effectively. These include:
    i. The NIST cybersecurity framework (CSF) incorporates best practices in five significant domains of information security, i.e., identify, protect, detect, respond, and recover.
    ii. The Federal Financial Institutions Examination Council (FFIEC) IT examination handbook presents a comprehensive list of security guidelines. These guidelines provide information

on a variety of security approaches extending from application protection and end-of-life management to vendor management and the rule of least privilege.
2. **Arming the Employees with Knowledge:** A significant portion of malware propagates through online social engineering schemes that manipulate unsuspecting users into either giving their personal information or making it easy for hackers to gain control of their systems. Fileless, or zero-footprint, malware is one such malware. Malware are effective in bypassing firewalls because they exploit existing applications as opposed to endeavoring to sneak a payload through a web filter. A client may get an email from an obscure sender containing an apparently legitimate document, downloading which may lead to certain legitimate scripts running certain tasks. A macro at the same time may issue a command to a remote server to download malware. Subsequently representatives are the primary line of resistance against such threats. It is significant for employees to figure out how to spot phishing plans. Attachments without context or vague titles, even when sent from an existing contact, may include malicious content. Teaching these identification techniques and some other security best practices like using password managers and logging out of your devices before leaving them unattended may remarkably curb the risk of user-driven compromise.
3. **Performing Continuous Threat Monitoring:** In Finance, threat monitoring is critical. This is because the real harm is usually done when the association is negligent. Most of the data breaches are furtive in nature. Once the attackers hammer the organization's network, they attempt to cover their tracks in order to be persistent. Stealing login credentials through a phishing campaign is usually the first step of vandalizing a financial institution followed by, masking their activities using a series of advanced tactics. Once the attackers get into the system, the risk multiplies exponentially as they can now move laterally to other subsystems and gain more sensitive information. This is often followed by the creation of backdoors through which they can extract data that may be used in future attack campaigns. This data may also be sold on the dark web. This may lead to catastrophic consequences for financial firms.

4. **Assessing and Managing Vulnerabilities:** It is impossible for any organization to address all vulnerabilities, even if they have the best IT teams and technology. Hence there is a need to conduct vulnerability assessment. Vulnerability assessment provides visibility across your environment, which lets an organization know if the software and systems have weaknesses. Organizations usually prioritize the most critical vulnerabilities so as to mitigate them first. Vulnerability management is also one of the most effective techniques to reduce the attack surface. The only limitation of vulnerability management is that it needs to be done consistently. Performing vulnerability scans intermittently (as often as possible), will make it hard for opportunistic attackers to discover their ways into the system.
5. **Managing Third-Party Risks:** Organizations may take several steps to minimize third-party risks. This may incorporate establishment and verification of security posture for vendors and partners. Business associates may be required to maintain security best practices through service agreements. Often organizations segment their network and limit third-party access to critical assets. Monitoring the network for anomalies using a threat detection and response solution is another way of managing third-party risks.
6. **Creating a Strong Cybersecurity Culture, Starting at the Top:** A solid cybersecurity culture goes past a worker mindfulness program. This may be done by positioning cybersecurity as everyone's business and not just an IT problem. Therefore, all stakeholders must view themselves as a critical part of a strong security posture.
7. **Devising Comprehensive Incident Response (IR) Plans:** IR should never be treated like a last-minute process. Since security breaches can occur anytime, organizations should always be prepared. The IT organization should have a comprehensible technique that can be quickly implemented to quarantine, block, or eliminate malicious network traffic if a security breach is reported (Figure 3.5).

3.2.4 CYBERSECURITY ETHICS IN BLOCKCHAIN

Blockchains are digital records of transactions. It gets its name from the structure, in which individual records, also called blocks, are linked

```
                    ┌─ Establishing a formal security and ethical framework
                    │
                    ├─ Arming the Employees with Knowledge
                    │
                    ├─ Performing Continuous Threat Monitoring
Cybersecurity Ethics in ─┤
     Finance        ├─ Assessing and managing vulnerabilities
                    │
                    ├─ Managing third party risks
                    │
                    ├─ Creating a Strong Cybersecurity Culture, Starting at the Top
                    │
                    └─ Devising Comprehensive Incident Response Plans
```

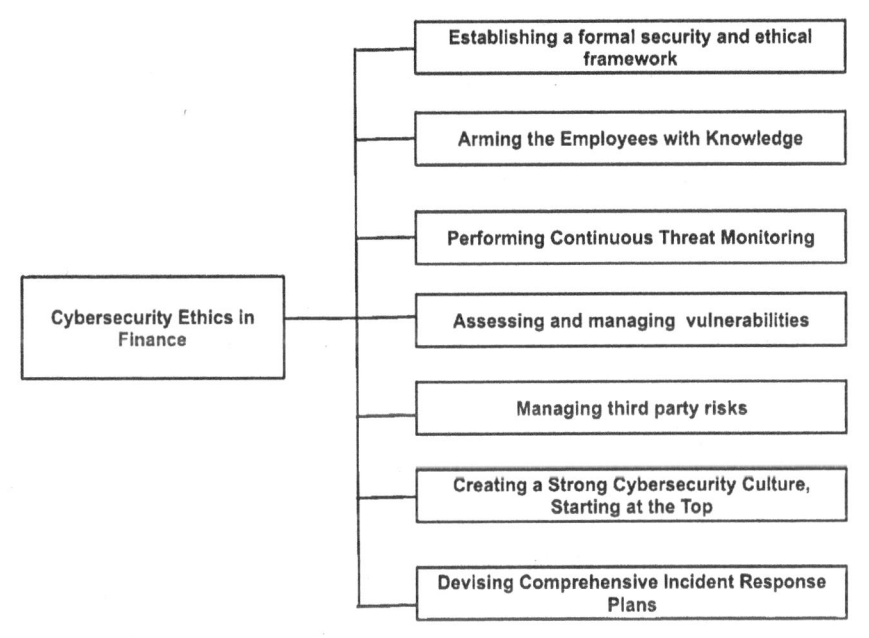

FIGURE 3.5 Cybersecurity ethics in finance.

together to form a single list, which is called a chain [117]. Blockchains can be used for recording cryptocurrency-based transactions, like Bitcoin. Transactions that are added to the Blockchain must be validated through multiple computers on the Internet. These systems are configured to monitor specific types of blockchain transactions and form a peer-to-peer network. Before adding a transaction to a block, it must be ensured that the systems work together to validate the transaction. This decentralized nature of the computers and the network ensure that a single system cannot add invalid blocks to the chain. Given a situation where a new block is added to a blockchain, it must be first linked to the previous blocks using a cryptographic hash. This hash is generated from the contents of the previous block. Due to this, the chain is never broken, and each block is permanently recorded. Since all the subsequent blocks need to be altered first, it is significantly difficult to alter past transactions in a blockchain. This makes blockchain technology inherently secure. Since data is distributed, blockchain technology removes any single point of failure. Blockchain technology is also based on cryptographic proofs and game theory consensus mechanisms, which is why it is impossible to hack.

Although blockchain technology boasts of several safety features, it does not mean that security issues are non-existent. Although Blockchains are robust, they are not immune to security issues. Some of the security issues related to blockchain technology are as follows:

1. **Fifty-One Percent of Attacks:** This is a well-known security issue in Blockchain, in which one, or several, malicious entities acquire majority control of a blockchain's hash rate. With the majority hashrate, it is possible to reverse transactions and perform double-spends. This also prevents other miners from confirming blocks.
2. **Exchange Hacks:** These are one of the costliest blockchain security issues. Cryptocurrency exchanges are lucrative honeypots for hackers because of their ability to incorporate massive crypto holdings and sometimes poor security practices. Since many exchange platforms are inherently centralized, they make the decentralized benefits of blockchains obsolete.
3. **Social Engineering:** This is yet another blockchain security issue that an organization and its employees must be aware of. Social engineering can be performed in many different forms, but the primary aim is to obtain private keys, login information, or cryptocurrency. Phishing is a typical type of social engineering where a malicious attacker conveys an email, message, or even sets up a site or Internet-based social media mirroring a trusted organization. Frequently, they request credentials under the guise of a giveaway or critical issue to compel a sense of urgency. If the information is given to the hackers, they can access the systems and may likewise steal money.
4. **Software Flaws:** Big-name blockchains like Bitcoin and Ethereum have proven their resilience to many types of attacks. However, the applications built on top of them are still susceptible to bugs and invite security issues. Therefore, any software using blockchain technology must undergo rigorous testing and review.
5. **Malware:** This blockchain security issues may range from malicious crypto mining software to code that could shut down a company's servers. Cryptojacking is a malware associated with Blockchain and cryptocurrency. Cryptojacking may be defined as the unauthorized and unnoticeable takeover of a computer's resources to mine cryptocurrency. While cryptohackers do not steal

money from their victims legitimately, the malware they infuse is equipped for prompting execution issues increasing electricity usage, which may further open the door for other hostile codes. Since Cryptojacking is secretive, it is difficult to anticipate how much money is lost, however, if estimates are to be believed, it is somewhere in the multi-millions.

Thus, blockchain technology is susceptible to several cybersecurity risks. However, it is a reliable technology due to several reasons. One of the important features about the Blockchain is that it can never be hijacked since the records are indelible and cannot be falsified and the motivation for true decentralization lies in four areas:

- Anonymity and transaction obfuscation, which makes it difficult to track down and attack the executing parties legitimately.
- Decentralization, which makes it impossible to co-opt or attack the system as a whole through a central entity.
- Strong encryption, which makes it impossible for a powerful outsider to see what goes on.
- Distributed organization, which creates organizational structures that are resistant to interference.

Despite being reliable, Blockchain must address some ethical concerns that we list as follows:

1. **How Ethical is Blockchain Anonymity and Privacy:** Since the individuals behind transactions in Blockchain are anonymous by nature, some companies have taken initiatives to help companies and individuals identify bad actors which may otherwise lead to cyber-crimes and fraud track transactions. These companies are capable of revealing the structure of cyber-criminal organizations. When funds are transferred into fiat currencies, there is the possibility of revealing the true name or an IP address. These transactions have enabled law enforcement and Blockchain investigation firms for identifying and arresting groups and individuals behind ransomware attacks. But due to the ability to identify individuals using the Bitcoin Blockchain and the volatility of the crypto-currency, cyber-criminals have moved away from reliance on Bitcoin to truly anonymous Blockchain cryptocurrencies such as Monero so that their identities can remain hidden. The biggest

question in such a scenario is whether it is ethical to remain anonymous on Blockchain since Cyber-criminals can hide behind the mask of privacy and continue their nefarious activities. Another question that comes into the play is whether it be more ethical to maintain decentralization and not allow anonymous transactions. The anonymity of the Blockchain may also lead to issues with abuse of the technology by bad actors who may knowingly take advantage of the privacy afforded to trade illegal weapons, drugs, even weapons of mass destruction. The want for privacy is unquestionably appealing, but one of the most important questions related to blockchain technology is whether or not it is truly needed from an ethical standpoint.

2. **There is No Possibility to be Forgotten or Erased:** Blockchain innovation is in head-on encounter with the option to be overlooked. Several questions are raised about ethics with Blockchain moving from cryptocurrency to social media platforms and business efficiencies. Some questions raised from the ethical point of view are: Is it ethical to collect information about children on the Blockchain when they have no say in the placement of their information on the Blockchain and no ability to have it removed? Researchers found unknown persons storing links to images and lists of websites of child abuse within Bitcoin's Blockchain. What could be done from an ethical point of view if it was possible to store those references forever, ever embedded in the technology when the information is destructive and highly illegal? As per the law enforcement agencies, for Blockchain, there are advantages of actualizing Blockchain for chain of proof and data sharing among organizations, but what if it is applied to criminal records? Is it ethical to have records that can never be erased? Should the possible mistakes of an individual's youth follow them throughout their lives with the ability to never be forgotten? These are some problems that society may possibly face if Blockchain is actualized in the recordkeeping of law implementation.

3. **Blockchain Ethics in Voting:** Many governments encourage the usage of Blockchain technology in voting to fight fraud and corruption. However, there may also be a situation where Blockchain may stop being anonymous and voter records get exposed publicly. Many election security experts are concerned that the use

of Blockchain may in fact introduce new security vulnerabilities as records would not be protected while being transmitted to the voting Blockchain. Therefore, there is no guarantee that the data would be accurate or was not modified in transit. Also, in the future, there may be a situation when voter information becomes publicly accessible. It may be difficult to assure citizens that their private votes would not be made public. Thus, if governments are interested in implementing Blockchain technology for usage in voting, they must take significant steps to ensure that the privacy of the votes remains intact and they preserve the freedoms of free and democratic votes.

4. **Blockchain for Transparent Trade Tracing:** Blockchain technology may be used for ethical business practices such as in the implementation of the Blockchain in tracing diamonds or cobalt mining. Blockchain technology implementation in the diamond industry will help fight against the blood diamond trade as well as prevent the ability of synthetic stones to be claimed as natural. However, Blockchain companies face growing suspicion of their own business ethics as initial coin offering (ICO) fraud continues to make headlines. The Blockchain community may be used to identify fraudsters since as long as fraudulent ICOs continue to hit the headlines, the Blockchain industry as a whole will be facing dejection.

5. **Environmental Impact of Blockchain Technology:** It is critical to consider the environmental impact of Blockchain technology. Blockchain transactions rely on complex algorithms that require large amounts of computing (hash) power. The amount and speed of calculations directly influence the increase in the amount of resources required. Since blockchain mining is an energy-intensive task, it will require more processing power and more energy consumption. While efficient and green computing may be considered at some point of time, Blockchain will probably until then rely on traditional energy supplies in many areas. Many areas have begun to stop allowing companies to employ Blockchain technology, especially crypto-currency mining operations. This is due to the large increase in power consumption of these entities. Moreover, the issue is that electrical organizations probably will not have the option to flexibly adequate power. While Blockchain

technology benefits several industries, in no way should it have a bad impact on the environment.

6. **Decentralized or Majority Owned:** 51% attack is one of the limitations of the blockchain technology and must be considered when employing public Blockchain technologies. It occurs when a single entity possesses 51% of the Blockchains computing (hash) power and may be manipulated. In the past, the attack has been successfully executed against several crypto-currency Blockchains like Bitcoin Gold to Litecoin Cash. If a majority of a Blockchains hash is obtained, it may lead to an entity being able to block transactions and enable double-spending, effectively. This makes the decentralized nature of the Blockchain futile since now the majority is owned by the entity. This may significantly affect the control of accounting, exchange, and even social media transactions when there is no decentralization and control is exercised by a single entity. The controlling entity may be capable of creating havoc on financial exchanges by delaying transactions or assigning preferential transaction processing over other parties. Consider blockchain technology for voting, if the controlling party of the Blockchain wants a specific candidate to win 51% attack may be launched. They could slow down transactions, deny transactions, and completely destroy the democratic nature of voting through control of the Blockchain. Although it is the worst-case scenario for the ethics of the Blockchain, it is a potential consideration that must be taken seriously.

7. **Ethically More Benefits or Dangers:** There are several ethical considerations that must be taken into account when implementing Blockchain technology. The benefits may outweigh the potential negative impacts on privacy and environment. The privacy of the information on the Blockchain may not be protected at all times. Further, the information of the citizens whose data is stored may not be kept safe at all times. The Blockchain may not be secured against impedance of outside gatherings and kept from expected corruption of 51% entities. It may not be right to store the information for a lifetime or it may cause harm to an individual in the future. Blockchain is a fascinating technology that has many use cases and applications, but it may also support some never to be erased consequences (Figure 3.6).

```
                            ┌─────────────────────────────────────────┐
                            │ How Ethical is Blockchain Anonymity and │
                            │                Privacy                  │
                            └─────────────────────────────────────────┘
                            ┌─────────────────────────────────────────┐
                            │ There is no possibility to be forgotten │
                            │               or erased                 │
                            └─────────────────────────────────────────┘
                            ┌─────────────────────────────────────────┐
                            │        Blockchain Ethics in Voting      │
                            └─────────────────────────────────────────┘
┌────────────────────┐      ┌─────────────────────────────────────────┐
│ Ethical Concerns in├──────┤ Blockchain for transparent trade tracing│
│     Blockchain     │      └─────────────────────────────────────────┘
└────────────────────┘      ┌─────────────────────────────────────────┐
                            │    Environmental Impact of Blockchain   │
                            │                Technology               │
                            └─────────────────────────────────────────┘
                            ┌─────────────────────────────────────────┐
                            │     Decentralized or Majority Owned     │
                            └─────────────────────────────────────────┘
                            ┌─────────────────────────────────────────┐
                            │    Ethically More benefits or dangers   │
                            └─────────────────────────────────────────┘
```

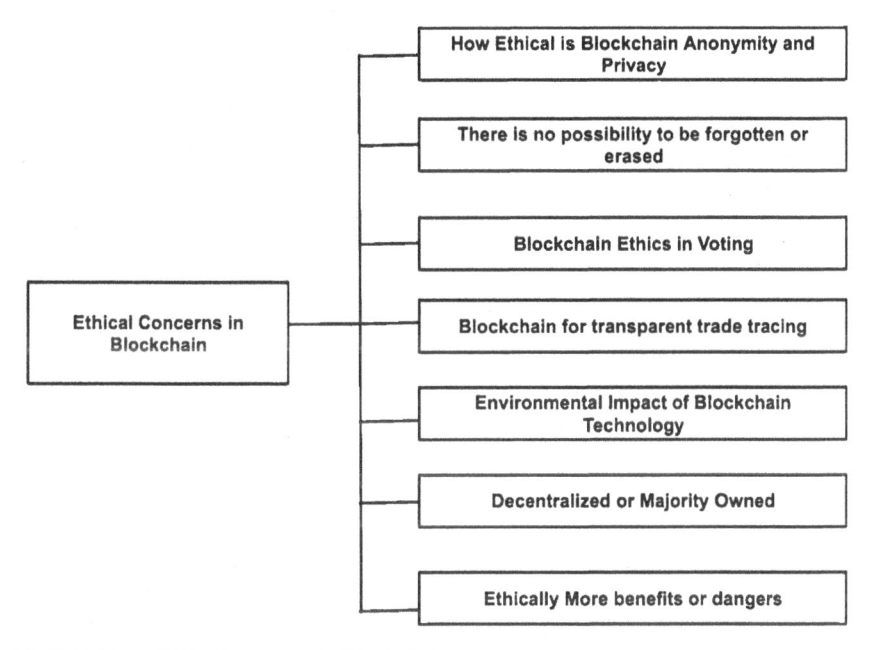

FIGURE 3.6 Ethical concerns in Blockchain.

Based on the security issues, several best practices may contribute towards ethical usage of blockchains:

- If an organization's Blockchain utilizes a proof-of-work (PoW) consensus mechanism, it needs to have security measures set up to forestall a 51% attack. This can be accomplished by being careful of mining pools, actualizing consolidated mining on a blockchain with a higher hash rate, or changing to an alternate consensus mechanism.
- The safest methods of storing funds are through hardware or paper wallets. Storing cryptocurrencies requires monitoring since these methods have minimal online touchpoints. These keep cryptocoins far from malevolent online hackers. If the business model involves regular trading, using a decentralized exchange (DEX) allows trading directly from the cryptocurrency wallet.
- Forestalling social engineering is moderately simple and helpful. Login credentials or private keys must never be sent out to anyone. Social engineering tactics should be looked out for. Setting up training programs for employees that shows them how to recognize

the various kinds of social engineering tricks and abstain from succumbing to them is very useful.
- When using any blockchain-based software, it is beneficial to check to see that it has gone through a third-party security audit. Since, the code behind it is also open-source, anyone can go in and review it for flaws or loopholes. Even with those precautions, security practices must be followed since many software-level blockchain security issues go undiscovered for years. Monitoring and vigilance are essential for dodging malware or stopping it as soon as possible after downloading it. If there are any performance issues on the system, checking task managers for mysterious programs running should assist in identifying malicious operations. Employees must avoid clicking on suspicious links. If a website is being operated, regular security checks must be run to ensure that it does not have injected malware.

3.2.5 CYBERSECURITY ETHICS IN CLOUD

Cloud computing may be defined as the delivery of computing services which include storage, software, servers, analytics, networking databases, and intelligence over the Internet, which is alluded to as the cloud. It offers quicker advancement, adaptable assets, and economies of scale. The domain of cloud computing is expanding rapidly and used by a great many individuals and organizations. In any case, moral issues identified with this innovation are not being thought of. The most prominent features of cloud computing are resource/storage virtualization, scalability, and elasticity, efficiency of resource sharing, usage optimization/optimized by usage, ease of usage, fast information sharing, delivery, and control, accessibility, and anonymity. Also, cloud computing relates to three advancements that are pertinent to an ethical analysis. To begin with, the moving of control from technology users to the third parties servicing the cloud due to re-appropriating and offshoring of information and communication technology (ICT) usefulness to the cloud. Second, the capacity of information in various physical locations across numerous servers around the globe potentially claimed and controlled by a wide range of associations. Third, the interconnection of different services over the cloud. Since there are different levels of functionality for different providers, it is connected to provide a specific service to an end-user. Some ethical issues in cloud computing are as follows [32]:

1. **Control:** Cloud computing creates the demand for outsourcing or offshoring of ICT tasks to third-party service providers. The information stored locally is stored in the cloud, which means that the users can place their computation and data on machines, but they cannot directly control them. Hence to ensure scalability and extend customers or users of a cloud, computer service relinquishes control over computation and data. In case there is an issue it can be difficult to detect who caused the problem. Further, without any solid evidence, it may not be possible for the parties involved to hold each other responsible for the problem in case of a dispute. For a networked organizational and technological structure, it may be further difficult to attribute consequences of actions to a specific individual or organization.

2. **Problem of Many Hands:** In a typical cloud computing scenario, a specific service delivered to a user depends on another system. This system in turn may depend on other systems as well. Thus, a cloud service to the end-clients can be based on a framework adjusted in the cloud by another organization. Cloud computing relies on the service-oriented architecture (SOA) such that all functionalities include services that can be aggregated into larger applications. These applications may be used to perform functions for end-users. However, if something undesirable happens, the complex structure of cloud services can make it extremely difficult to determine who is responsible for the disruption.

3. **Self Determination:** Granting control to the Cloud provider raises the question of information self-determination. Informational self-determination may be defined as the capacity of people to practice personal control over the collection, use, and divulgence of their own information by others. The Internet incorporates ubiquitous and unlimited data where sharing and storage among organizations self-determination is challenged. This raises questions related to privacy issues and also puts the confidence and trust of information society at stake. Organizations that offer cloud computing services must accommodate the interests of their users/customers. This may be done by being open and accountable about their data

management practices. The cloud service providers may seek informed consent from individuals, and also provide them with credible access and redress mechanisms.

4. **Accountability:** Personal data that is stored in the cloud must be managed properly. Accountability ensures a promising approach or responsibility to empower users to ensure what is being done. Users of an accountable cloud can check whether the cloud is performing as agreed. To ensure accountability in the cloud, there must be transparency regarding adequate information about how data is handled within the cloud and a clear allocation of responsibility. Along with recorded evidence, these key elements may be useful in deciding who is responsible at whatever point an issue happens or question emerges. Since responsibility requires point-by-point records of activities by its users in the cloud, it might prompt tension among privacy and responsibility. Consequently, it is critical to consider what is being recorded, and who the record is made accessible to.

5. **Ownership:** Although data is actively stored in the cloud by users, it is also capable of generating data itself for different purposes. This data may be used to provide accountability, to improve the service given, or different reasons, for example, execution or security. The digital interactions and tracks are gathered together using unique identifiers and sophisticated matching algorithms. This may also leave a trail of extraordinarily detailed personal information. If this information is not protected properly, it may be exploited and abused. There are a few limitations regarding how the information is used. When this data is located in at least one database of the cloud, it may be accessed easily and may also be used in ways that individuals never envisioned or intended. There may also be questions about ownership with respect to infringements on copyrights. It is essential to give clients access to practically boundless storage and computing power. This is one of the ways how cloud services could make it much simpler to share copyrighted material over the Internet.

6. **Function Creep:** It is yet another threat to data stored in the cloud. Function creep may be thought of as data collected for a specific purpose, over time which has become used for other (unanticipated, unwanted) purposes. Consider a circumstance wherein a database

with biometric information of citizens might be intended for validation purposes yet may then end up being exceptionally useful for crime investigations. In a world of cloud computing, due to relinquished control and reduced sight on what data is being used for, function creep may become a serious danger.
7. **Privacy:** Many companies that provide cloud services collect terabytes of data. A large portion of this data is sensitive personal information, which may further be stored in data centers across the globe. In a circumstance where individual information is stored in the cloud, ambiguity about protection can be possibly unsafe. Since information is no longer stored locally, control over the information is moved to the service providers. In such a circumstance, consumers need to confide in the cloud supplier that their own data is protected and will not be uncovered. However, diverse server providers have varying opinions on privacy, and it may never be clear to customers as to which services providers they are dealing with. Cloud computing witnesses' different services becoming intertwined in such a way that a hosted application of one company, for instance, maybe built on a development/deployment framework of another. Therefore there is a greater chance that to consumers, it will not generally be clear what they can anticipate from service providers in the cloud concerning privacy (Figure 3.7).

In order to address the ethical concerns in cloud computing, one can adopt several approaches:

- The precautionary principle may be used to prevent harm from unknown consequences while still not hampering progress and innovation. The principle expresses that one should avoid activities notwithstanding logical vulnerabilities about genuine or irreversible harm. Moreover, the weight of confirmation for guaranteeing the security of an activity falls on the individuals who propose it [31]. Many effects and undesirable consequences of cloud computing are yet to be identified. Although it does not mean its development and execution should be terminated altogether, the precautionary principle still urges the parties involved to anticipate consequences that are not certain. Since the burden of proof is placed on them, they may never utilize unpredictability to abstain from structuring and offering types of services that welcome moral and sound use of innate undesirable and contentious actions.

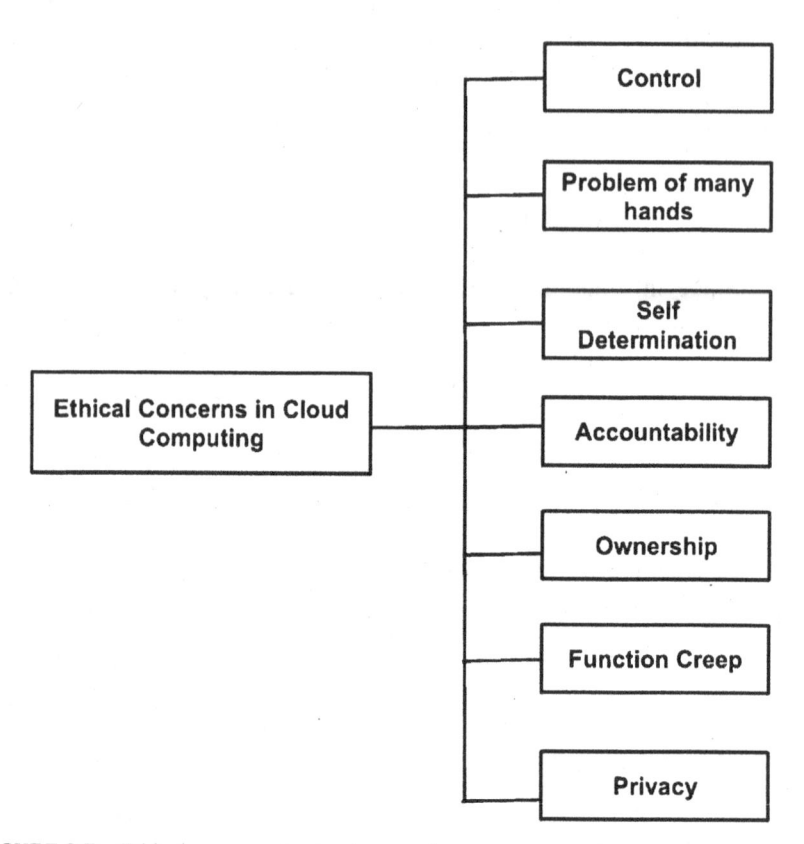

FIGURE 3.7 Ethical concerns in cloud computing.

- Since all innovation is dependent upon a few legislative strategies and casual understandings as much as national and international law and guidelines, due to the concept of governance, there will always be a question about how ethics can be integrated into technology. With respect to the type of technology and the context of its use, most governance arrangements are more favorable to the inclusion of ethics than others. However, in a regulated funding regime, explicit attention to ethics can be required. The European research funding in the current 7th Framework Program requires all proposals to pay explicit attention to ethical issues. All the while, different settings, for example, those of privately owned business development are substantially less dependent upon morals related oversight. The administration arrangements there are more helpful for authoritative objectives, for example, profit generation.

3.2.6 CYBERSECURITY ETHICS IN INTERNET OF THINGS (IOT)

The Internet of things (IoT) may be defined as the ever-growing network of physical objects over the Internet, identified by an internet protocol (IP) address. IoT guarantees web network, and the correspondence that happens between these items and other Internet-empowered devices and frameworks. Over the network, the physical devices connect to other physical devices, using wireless communication and offering contextual service. The combination of these technologies raises a number of ethical questions. since more and more activities may be accomplished unrelated to the surrounding objects, which may further enable communication. This gives the possibility of making new businesses, inconceivable today. Technology offers a variety of services, and these services may be tailored depending on the actions of an individual, the device, the infrastructure or nature at that particular moment. IoT finds its use in several other applications and services related to households, smart cities, and health monitoring devices [34, 35]. Taking into consideration these aspects, several ethical problems may be caused [33]:

1. **Ubiquity, Omnipresence:** Once the user is attracted to IoT, he/she is devoured by it, there is no clear way out. A way to give up is by using the artifacts (which may no longer be possible at some point, due to the producers which will equip them with Internet connection devices).
2. **Miniaturization, invisibility:** Computers components may start shrinking, the devices may become smaller and transparent. This may avoid any kind of inspections, audit, quality control and accounting procedures.
3. **Ambiguity:** It is difficult to tell apart natural objects, artifacts, and beings due to the easy transformation from one category into another. This transformation is based on tags, advanced design, and absorption in new networks of artifacts. This may lead to serious issues in identity and system boundaries.

4. **Difficult Identification:** For being connected to IoT, the objects must have an identity. The access to a group of objects, the management of these identities might raise great interest may also lead to problems of security and control in a globalized world.
5. **Ultra-Connectivity:** With IoT expanding, the connections will increase in number and reach unprecedented scales of objects and people. This may have an effect on the quantities of transferred data and products, which may also increase greatly (Big Data). Moreover, they could be used maliciously.
6. **Autonomous and Unpredictable Behavior:** There is a high probability of the interconnected objects interfering spontaneously in human events, in many unexpected ways for the users as well as designers. Since people are part of the IoT environments along with artifacts and devices, it will lead to the creation of hybrid systems with unexpected behavior. The incremental development of IoT may further lead to emerging behaviors without the users fully understanding the kind of environment they might be exposed to.
7. **Incorporated Intelligence:** It makes the objects be seen as substitutes in social life. In such a situation, the objects will be intelligent and dynamic, and will also manifest emerging behavior. They may carry on as expansions of the human brain and body. Being deprived of these devices may lead to problems. The best example of such a situation is users who consider themselves cognitively or socially handicapped without Internet, smartphones or social media.
8. **Difficult Control:** Due to the high numbers of hubs, switches, and data, IoT control and governance may not be centralized. This means that the information flows will be eased and while the transfers will be quicker and cheaper, it may not be easily controlled. There may appear certain properties and phenomena that require monitoring and governance adequately. This will have an influence on accountancy and control activities.

As we know, ethical behavior is concerned with questions that arise with respect to:

1. **Enforcing the Property Rights on Information:** With respect to the property right on data and information, the issues usually appear from the correct identification of the authors, mostly about who is the owner of specific data retrieved by an object connected

to IoT. When the information is sensitive, it is a serious issue. The IoT omnipresence ensures that the boundaries between the public and private space are invisible, and people tend to know where their information lies. In a situation where monitoring the individuals without them being aware of it will also raise serious privacy and ethical questions. The objects may be equipped with sensors that may enable them to see, hear, and even smell. The data acquired by the sensors might be sent in incredible amounts and in different ways through networks, which may again raise questions related to privacy. Using technologies like: radio-frequency identification (RFID), global positioning system (GPS) and near-field communication (NFC), the geographic place where a person is and his activities from one place to another can be easily tracked without his knowledge. This information may be easily collected from a chip implanted with the person's consent, and may later be used maliciously. It is also possible to create individual profiles depending on their consumption habits. Malicious users in this case may make decisions related to them.

2. **Ensuring the Access to Information:** If there is a contemporary attack on a system which may lead to information loss or spreading, there is a possibility that a virus or a hack may directly impact users. This may be related to accessibility of information. Consider a situation where a car is connected to IoT and there is some sort of interference in the control system. Such an attack may jeopardize the life of the passengers and is very much possible. Stuxnet attack is one of the most popular attacks on control systems and may be used to attack any physical target related to computers. The list of vulnerable systems includes the electric heating systems, food distribution networks, hospitals, traffic lights systems, transport networks, and even wires.

3. **Ensuring the Integrity of the Information:** Ensuring data integrity refers to ensuring the fact that data is complete, original, consistent, attributable, and accurate. It is important to protect data at all stages of its lifecycle, whether it is the time when data is created, transmitted, in use or at rest. Otherwise, there is no affirmation that the integrity of current information is kept up. Since IoT gadgets are associated remotely to a network and can send information, keeping up trustworthiness is explicitly significant for IoT devices.

IoT devices may frequently be associated with sensitive information and are used across a variety of industries. Consider an IoT device connected to the life sciences industry where they are regularly utilized in the control of medication item assembling or hardware monitoring. The IoT device might be familiar with delicate data like sensor checking temperature, light intensity, humidity, and so on.

4. **Enforcing the Right to Private Life:** It is possible that IoT may intensify the debate between individual privacy and the use of personal information for promoting effectiveness, safety, and security. Several questions may be raised as to who should control information, who should access it, who can use it, etc., although the answer may not always be clear-cut. In the event that we think about medical monitoring devices and the data they acquire, there may be questions pertaining to whether the personal health information (PHI) should be shared with the Centers for Disease Control in case they want to track a potential epidemic. For biomedical researchers to model potential treatment strategies, richer datasets might be needed, so should they be provided with the health information? In some cases, individuals must have a basic right to opt-out, delete, or mask their information from systems in the IoT. However, it may not be feasible or possible for an individual to control all the data generated about them by IoT systems. Also, strong individual privacy rights may also mean less social benefit, and therefore too many opt-outs may lead to eroding the public and private value of IoT datasets. This may likewise have a general negative effect on the social advantages.

5. **Accountability for Decisions Made by Autonomous Systems:** Many autonomous systems are capable of replacing human activities. There is a challenge of when and how these systems should be deployed, and who might be responsible and accountable for their behavior. If a smart system fails, is hacked, or acts with negative or unintended consequences, who is accountable, how, and to whom? Consider the situation of autonomous vehicles, which make multiple decisions without human intervention. As of now, we expect automobile companies to be accountable in case automotive systems, like anti-lock brakes, fail. In future if cars begin to drive themselves and accidents occur, who should be responsible for the same? With systems taking on more decisions made by people, it

will be progressively testing to make a plausible framework that also answers aspects related to responsibility and accountability.
6. **Promoting Ethical Use of IoT Technologies:** Since systems can be used for both good and bad, we might as well say that technologies have no ethics. For example, video surveillance may be tremendously helpful in ensuring that senior citizens stay in their homes longer or for parents to monitor their kids. Also, it can likewise expose private behavior to conniving viewers and may further invite unwanted intrusion (Figures 3.8 and 3.9).

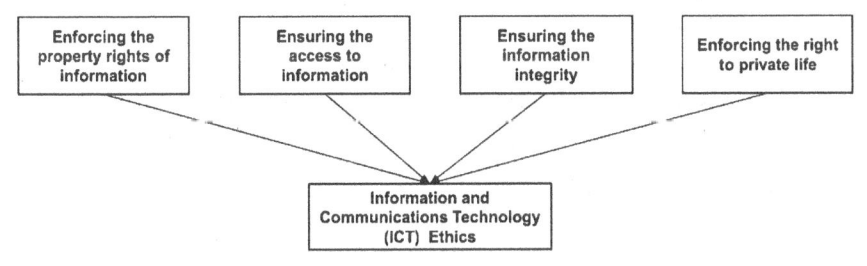

FIGURE 3.8 Information and communication technology (ICT) ethics.

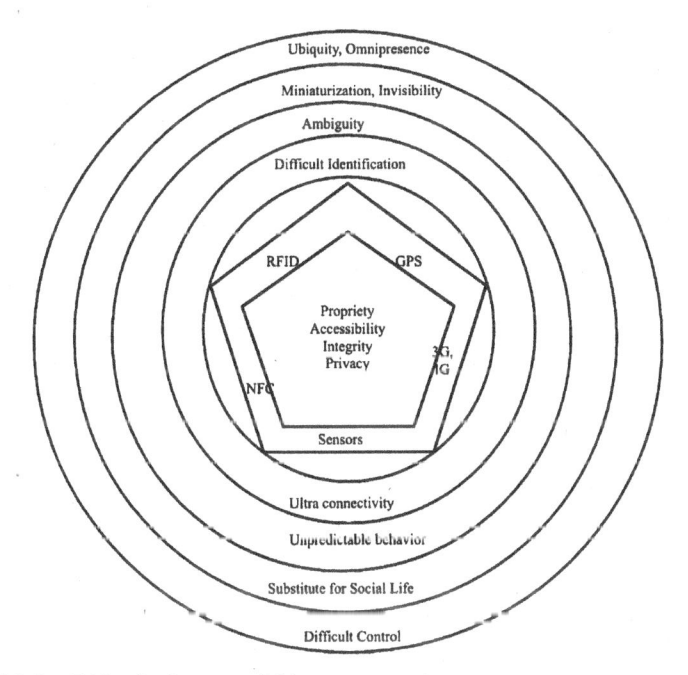

FIGURE 3.9 Ethics for Internet of things.

Based on the ethical challenges that we face for IoT; few steps might be considered [33]:

1. **Policy for IoT Safety, Security, and Privacy:** This may require developing a viable approach for promoting individual rights, data security, and trust. Further, inappropriate behavior, corruption, and crime may result in disincentives and penalties.
2. **A Legal Framework for Determining Appropriate Behavior of Autonomous IoT Entities:** This framework would be responsible and accountable for identifying parties for that behavior, and determining who can enforce compliance, how, and on what grounds.
3. **Focus on Human Rights and Ethical Behavior in the IoT:** This refers to a sense of how such behavior would be enforced. This is significant for IoT to advance human prosperity and add to the progression of society.
4. **Sustainable Development of the IoT as Part of a Larger Societal and Technological Ecosystem:** This alludes to the effect of IoT economical advancement on natural systems(for instance, 3D-printed organs, inserts), ecological frameworks, and natural resources).

3.2.7 CYBERSECURITY ETHICS IN ARTIFICIAL GENERAL INTELLIGENCE

AI may be defined as the ability of a computer program or a machine to think and learn [36, 37]. This field of study is responsible for trying to make computers smart. AI machines can learn from experience, conform to new information sources and perform some tasks like humans. Popular AI applications include expert systems, speech recognition, and machine vision [122, 123]. Some ethical issues in AI are as follows:

1. **Unemployment:** Simulated intelligence movement has made some experts accept that it will consistently and unavoidably assume control over enormous divisions of the workforce. This may bring mass-scale unemployment and social unrest. Therefore, several jobs that are currently being done by people, may become obsolete or automated sooner. Computerized reasoning-based frameworks and chatbots are overwhelming each industry [15].

2. **Inequality:** The economic system relies on compensation for contribution to the economy, which is often assessed using an hourly wage. There is always a question raised about how to distribute wealth created by machines. Numerous organizations are as yet reliant on hourly work with regards to services and products. The use of AI may make a company drastically cut down on relying on the human workforce. In such a situation, revenues go to fewer people. Therefore, individuals who have ownership in AI-driven companies are the ones who may make all the money.
3. **Humanity:** This aspect raises the question as to how machines Affect our behavior and interaction. Artificially intelligent bots are increasingly getting better and better at modeling human conversation and relationships [38]. Although humans are confined to attention and kindness that they can expend on another person, artificial bots are known to channel virtually unlimited resources into building relationships.
4. **Artificial Stupidity:** Machine Intelligence is the output of extensive learning, for humans or machines. For systems, there is a preparation stage in which they figure out how to identify the correct patterns and act as indicated by their information. When a system is completely trained, it moves to the testing phase. In this phase, it is hit with more examples and observed to see how it performs. While the training phase alone cannot cover all possible examples that a system may deal with in the real world the real question is how mistakes made by machines can be guarded? AI systems may be fooled in ways that humans might not be. Therefore, there is a need to ensure that the machine performs as designed, and that people may not overpower it to use it for their own ends.
5. **Racist Robots:** Although AI is capable of speed and capacity when it comes to processing that is far greater than that of humans, may not always be neutral. Therefore, it is hard to determine if machines may be trusted or fair. Although AI systems are created by humans, they can still be biased and judgmental. There is a need to eliminate AI bias. However, if used positively, or if used by those who aspire for progress in society, AI can become a catalyst for positive change.
6. **Security:** At the point when an innovation turns out to be increasingly amazing, it might be utilized for terrible reasons just as

well. This is not only confined to robots that are mass produced to replace human soldiers, or autonomous weapons, but also extends to AI systems that may cause damage if used maliciously. Since the enemies would be virtual and the combat would involve machines and networks, the field of cybersecurity will grow immensely. Hence, there is a need to keep AI safe from adversaries since such situations mean dealing with a system that is faster and more capable than humans by orders of magnitude.

7. **Evil Geniuses: Protection Against Unintended Consequences:** Along with adversaries, we may also need to worry about AI, if it turns against us. Consider an advanced AI system with appalling unexpected consequences. For machines, malevolence at play is exceptionally far-fetched. However, there may be a lack of understanding. Consider an example in which an AI system has been asked to eradicate cancer from the world. While computing, it comes up with a formula that indeed finds a way to eradicate cancer-by killing everyone on the planet. In this situation, the computer would have achieved its goal of eradicating cancer very efficiently, but not in the way humans intended it.

8. **Singularity:** Humans are one of the most astute species on earth. Human strength is the aftereffect of resourcefulness and insight. The motivation behind why humans show signs of improvement of greater, quicker, more grounded creatures is on the grounds that they can make and use tools to control different animals. People depend on physical tools like confines and weapons, and intellectual devices like training and conditioning. All these pose serious questions about AI. There are speculations about AI having the same advantage over humans. Could humans stay being dominated by sophisticated intelligent systems? While pulling the plug may seem like a good option for now, what if a sufficiently advanced machine anticipates this move and defends itself. Some researchers define this situation as singularity, i.e., the point in time when human beings may no longer be the most intelligent beings on earth [39].

9. **Robot Rights:** Once machines are advanced enough to perceive, feel, and act, there may arise questions on their legal status. One of the basic questions that may need attention is whether they should be treated like animals of comparable intelligence or if the suffering

of feeling machines would be taken into consideration. By what means can the human treatment of AI be characterized? There may also be several ethical questions about mitigating suffering, others about risking negative outcomes. While these risks are being considered, it is also important to remember that this technological progress means better lives for everyone. While AI has vast potential for a better future and for making lives easy, its responsible implementation is completely up to humans (Figure 3.10).

```
Cybersecurity Ethics in          ├── Unemployment
Artificial General Intelligence  ├── Inequality
                                 ├── Humanity
                                 ├── Artificial Stupidity
                                 ├── Racist Robots
                                 ├── Security
                                 ├── Evil Geniuses
                                 ├── Singularity
                                 └── Robot Rights
```

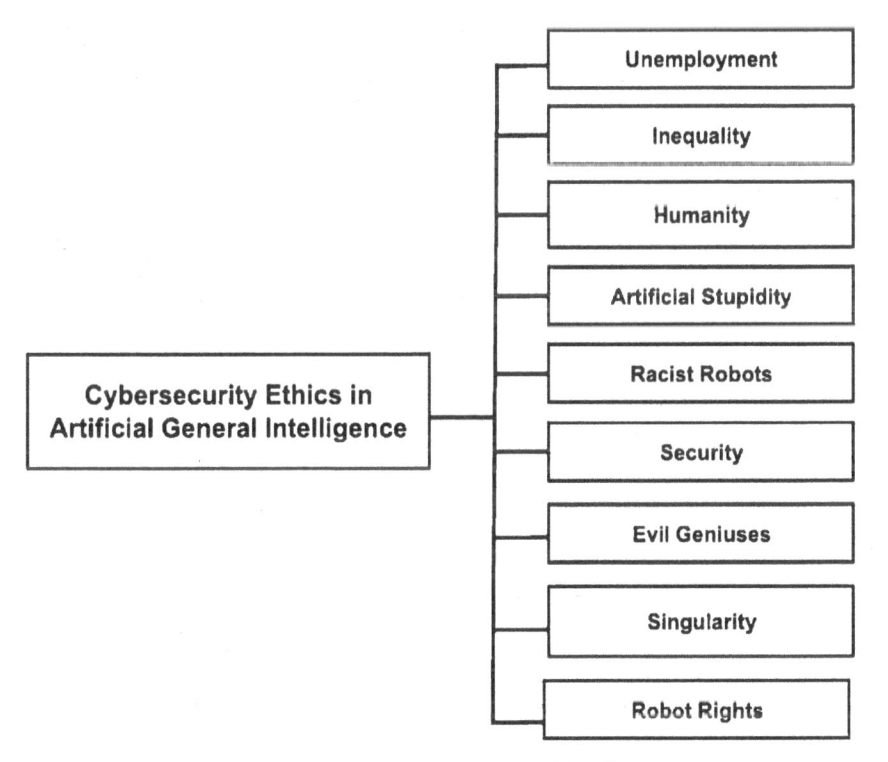

FIGURE 3.10 Cybersecurity ethics for artificial general intelligence.

An ethical approach towards AI could be ensured considering the following:

1. **AI Must be Thought of as a Force for Good:** A shared ethical AI framework may provide clarity against how this technology can best be used to benefit individuals and society. There is a

need to ensure that partialities of the past must not be accidentally incorporated with automated systems. Also, such systems must be carefully designed from the beginning. This means that the input should come from a diverse group of people.

2. **Intelligibility and Fairness:** It is equally important to ensure that an AI operates within parameters of intelligibility and fairness. This may be done if companies and organizations improve the intelligibility of their AI systems.
3. **Data Protection:** AI must not be used to disparage the data rights or privacy of individuals, communities or organizations. The process of gathering data and accessing it must be revised. The underlying idea is to ensure that companies have fair and reasonable access to data. Citizens and consumers can also protect their privacy simultaneously.
4. **Flourishing Alongside AI:** Since humans created AI, humans and AI working together may contribute immensely towards the development of society. People have the right to be educated. This can lead humans to flourish mentally, emotionally, and economically alongside AI.
5. **Confronting the Power to Destroy:** Due to several speculations that hint at the possibility of singularity, the autonomous power to hurt, destroy or deceive human beings must not ever be vested in AI. It is believed that well-intended AI research may be misused in ways that could turn against humanity. AI researchers and developers need to consider the ethical implications of their work as well as the field. It is crucial for governments, the scholarly community and industry to build up universal standards for the design, improvement, guideline, and deployment of AI.

3.2.8 CYBERSECURITY ETHICS IN INTELLIGENCE, SPYING, AND SECRET SERVICES

Intelligence has always been a significant aspect of maintaining state security. However, the era of globalization has metamorphosed the environment in which intelligence services operate. With the expansion in the quantity of threats that intelligence agencies are attempting to react to, intelligent work progressively happens in the public circle. Policymakers highlight the crucial role of intelligence in justifying policy decisions. While the discussion that follows may not be exhaustive, it outlines some of the main trends that may have emerged over the last few decades and the challenges they pose for ethical intelligence gathering. Some of the challenges are as follows:

1. **Changing Nature of Threats:** Many threats are interconnected, borderless, state-based, and their number keeps on increasing. Simultaneously, most security challenges are borderless and are regularly directed by non-state attackers or isolated cells are more difficult to track and could possibly be associated with terrorist organizations, which again are extremely difficult to identify. Also, intelligence agencies have seen a noteworthy increment in their responsibilities. They are responsible for handling issues ranging from potential pandemics to the effects of climate change. Not only that, but intelligence agencies are struggling to be effective, hone their ability to perform tasks, while simultaneously focusing on the ethical aspects of their responsibilities, which is a challenging task. In this way, it is imperative to consider how ethical intelligence must advance.

2. **Technology:** It has been the driving force for transforming the strategic environment in which intelligence agencies operate. Because of progression in innovation, organizations have a far more extensive reach and now have access to more information than ever before. However, this has led to an exacerbation of threats faced by the agencies. This is a direct result of the fundamentally expanding connection among state entities and non-state entities that threaten state security. Moreover, technology has also led to incredible pressure on agencies because executive decision-makers now expect immediate, real-time intelligence. This puts intelligence producers under pressure because now they need to compete with the wide assortment of online, unsubstantiated data accessible to their buyers. While intelligence collaboration has benefits of its own and may also become a necessity to protect the state, the reaction of civil rights groups raises many ethical questions about the appropriateness of technical means of intelligence gathering as well as its regulation. Advances in innovation must be viewed as giving chances to intelligence agencies. The real question is how they adapt while maintaining civil liberties.

3. **Outsourcing:** Due to the pressures of current increasing threats and the demand for real-time reporting, agencies come up with remarkable responses to outsource intelligence responsibilities. Offices have been known to redistribute tasks and cross examination exercises to private military organizations. These private companies are not subject to the same regulations as state agencies. This makes it even harder to track abuses and prosecute crimes. Opportunities are created by technology and they have led to a dramatic increase in the outsourcing of intelligence work to private sector entities. It is believed that states will be continuing their dependency on private sector support and expertise for responding to the challenges and technological environment that may face. It is necessary to consider Improved oversight and regulation.

4. **Cooperation:** The world is burdened with global challenges which need global response. The 21st century has witnessed considerable explosion of intelligence sharing, which is not only limited to military, defense, and diplomacy, but is also leaving its mark on foreign policy. Given the expanse of such a situation, cooperation is a must since it allows benefits of burden-sharing and increasing access to equipment, expertise, and technology. Cooperation lies at the heart of the tension existing between ethics and intelligence. Many international partners have similar rules, values, and ethics, while others choose not to follow them. A few organizations have likewise been blamed for purposely subcontracting intelligence activities to unified offices. This has led to technically adhering to national ethical and legal obligations Finally, agencies may also provide information to partners that may or may not be used in ways that are contrary to national law and democratic values. The majority of intelligence relationships fall under non-treaty arrangements in international law. Therefore, they are flexible and not legally binding. This makes it easier to work under the radar and avoid regulations. Also, public enquiries related to extraordinary performance and surveillance have exhibited national laws and ethical values with the knowledge of their executive. Therefore, cooperation is necessary for responding to current threats. It has led to the creation of an ethical vacuum due to which intelligence agencies have lost oversight and accountability.

5. **Norms and Values:** The technologically complex world of the 21st century has seen dramatic increase in numbers and types of threats. There is also an increase in new forms of intelligence production. There is lack of clarity in deciding about what is ethically acceptable and under what circumstances is both a reality and a responsibility. Like ethics, norms, and values can be expected to change over time. As of now, the normative dimension of democracy is arguably stronger. This is because of the increased transparency and the multitude of stakeholders involved in governance. Consequently, the arrival to a correspondingly dubious circumstance where intelligence services acquire data through morally flawed or unlawful strategies presents genuine difficulties for democracies. The reaction of the public to recent scandals has been quite nuanced. In the past, there has been clear outrage among common society gatherings and certain fragments of people in general. Further, terrorist threats seem to have also led to a surprisingly muted reaction, while in many countries' polls indicate that members of the public support mass surveillance as an acceptable method of intelligence collection. Additional surveys indicate acceptance of enhanced interrogation techniques. There is a sense that societal norms may be unpredicted, and due to this, it may be extremely difficult for intelligence agencies to adhere to unclear ethical practices. Unfolding of this situation in the future will determine the limits of what is ethically acceptable and the context in which intelligence agencies will operate (Figure 3.11).

The undeniably straightforward condition comprising intelligent gathering has not only led to challenges but has also created opportunities. It empowered the opening up of the intelligence black box and has immense potential to change ideas, cultures, conduct, and the frameworks within which intelligence agencies operate. The never-ending debate on the ethical practice of intelligence gathering has been strengthened by the involvement of multiple actors on the national, regional, and international levels. Moral choices might be hard to make but an approach to do the same while involving all relevant stakeholders would facilitate the creation of appropriate guidelines. This may allow intelligence agencies and policymakers to rebuild trust and operate more effectively in today's public environment. While numerous recommendations and strategies are recent and doubtful, they show the opening of more channels than any time in recent memory to accomplish a superior harmony between individual rights and national security, among ethics and effectiveness in democratic societies.

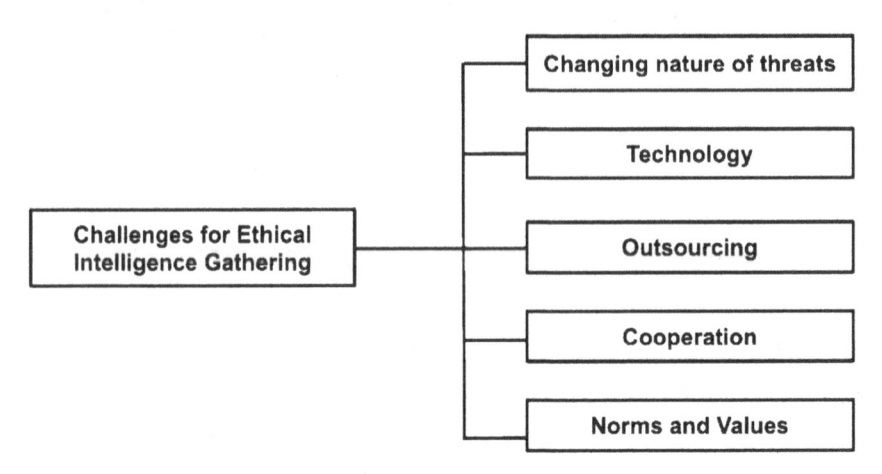

FIGURE 3.11 Challenges for ethical intelligence gathering.

3.2.9 CYBERSECURITY ETHICS IN BIOMETRICS

Over the last few decades, continuous development of computer technology has led to reduction of prices and the improvement of performance, and increase in practical and social applications like biometrics technology. The field of biometrics innovation is fundamentally utilized in the fields of work punching, exchange installment, visa, border access control, and access control [40]. Be that as it may, technological advancements require a ceaseless examination son the moral issues that may emerge, and biometrics is no special case. Therefore, it is necessary to consider the ethical issues that may crop up from the use cases and applications of biometrics. There is a need to explore ways to deal with these ethical issues:

1. **Privacy:** It is one of the biggest challenges to deal with for Biometric technology. Biometric information is usually collected by observing individuals and is used to identify individuals on the basis of fingerprints, faces, hand shapes, etc. All these information are sensitive and personal. The biggest controversy that arises is whether the biometric information that is stored in the biometric

system is still personal information. Some researchers believe that the biometric information stored in the biometric system is not personal information. Their claim is mainly based on the following two arguments:

i. The stored biometric data does not have any meaning since it may not be classified as personally identifiable;
ii. The biometric image simply cannot rebuild from the template.

These stored biometric information numbers may be easily extracted from individuals and are unique and can be used to identify individuals. The overall purpose of collecting biometric information and turning it into numbers is to identify and/or authenticate a person's identity. Templates may also be used to identify individuals. As far as the second argument is considered, it has been reported in the literature that it is possible to reconstruct biometric images from the template. Henceforth there is no uncertainty that biometric data is stored in a biometric system and may still be classified as personal information. Since biometrics are perpetual, hard to change, and by and large noticeable to other people, when they are forged or leaked, they cannot be reset. This poses additional security risks and aggravated privacy issues. Therefore, biometric information must be treated as sensitive data. The two most prominent issues are function creep and informatization of the body. Functional transformation may be defined as the use of biometric information beyond the original purpose. Functional changes may occur in situations where an individual's informed or uninformed circumstances are inevitable. According to some researcher's functional metamorphosis incorporates policy vacuum or missing, not satisfied with a given purpose or function, landslide effect, or secret application. The analysis shows that the information contained in the biometric system may be superfluous. Further, the biometric framework cannot neglect repetitive data. Informatization of the body is one more significant perspective and is additionally an extraordinary kind of functional transformation. The term informatization of the body, may be defined as the ability to retrieve a significant amount of information about individuals from biometric systems. This mined information may prove to be very rich since it incorporates some sensitive data, like the medical information, transaction records, and so on. The medical effects of biometrics may be divided into two categories: direct effect, which cause harm to the body itself, for example, radiation to the body and the spread of an infection and, indirect effect that pertains to the disclosure of medical information, including the current state of mind and body and the potential risk of illness. While direct medical influence is unreasonable, indirect medical effects deserve discussion. The indirect medical influence may be used to obtain medical information from biometric systems. If this medical information is leaked, there is a greater risk.

2. **Autonomy:** When collecting biometric information, there are several ethical questions that must be considered. The questions may be in the form of what personal information should be collected and if any other personal information apart from the biometric information needs to be included. There may be questions related to whether it is ethical to inform the collector of the potential risks, or how the recipients should be informed about how their information is stored. The purpose of the information used, the individuals who can get the information, how long the information may be stored, and if the consent of the recipient is needed repeatedly when using the biometric information. These series of questions are not limited to privacy issues, but deal with autonomy. A huge piece of practicing autonomy is informed consent. Anton Alterman recommended that biometrics ought to have informed consent in the application, and people who willfully submit biometric data ought to [41]:

   i. Be completely educated regarding possible risks;
   ii. Be capable of understanding the possible effects of their actions;
   iii. Make such conduct with no threat.

   In order to ensure the individual's informed consent, it is mandatory for the individual to understand the purpose and meaning of the biometric system. While grown-ups are considered to have adequate capacity to get data, the issue is for the most part the kid's informed consent when utilizing biometrics. Consent issues similar to these also come from vulnerable populations like the elderly, mentally ill, and poorly understood people. The present personal information is confidentially mined without the knowledge of individuals. This might be ascribed to progresses in surveillance technology and the potential for remote sensing of certain biometric advancements. Monitoring probes is one of the most common examples for the same. The observing test is utilized to record a person's picture and whereabouts without the individual's knowledge. While some are for security, crime avoidance, and investigation considerations, the Irish Bioethics Committee states that the assortment of biometric data should be shielded under a couple of preconditions::

i. Effectiveness, which refers to secret collection of biometric information that is capable of achieving social security, prevention, and even reduction of crime;
ii. Proportionality, referring to the level of checking of secret collection of biometric data, measures, and personal freedom. The degree of restriction must be equivalent.
iii. Need, which alludes to checking measures for covertly gathering biometric data. This is necessary to ensure public safety and for achieving national well-being goals.
iv. The least infringement, which refers to secretly collecting biometric information for individual rights. This also promotes minimization of violations of interests.
v. Transparency, which alludes to arrangements, measures, and activities identified with protecting social security ought to be made known to the general population.
vi. Compensatory, referring to the fact that if errors are found in monitoring, people following ethics must correct mistakes in time, and give compensation.

The secret collection of biometric data which fulfills the above conditions can be morally safeguarded, else it will not be guarded.

3. **Social Exclusion:** Many biometric technologies are yet to undergo development and innovation, there are still challenges related to deployment as well as scalability. The identification system relies on specific scenarios. Presently, biometric acquisition devices are incapable of handling values of individuals other than normal values, and some individuals may not be easy to identify and are thus excluded. When such systems are linked to social welfare, the unidentifiable individuals may be excluded from social welfare. This may lead to injustice. These groups of individuals could be individuals with disabilities or poor understanding, individuals

with psychological disorders, the old, individuals of specific races, and homeless people. Thus, it is important to have a similar ethical responsibility to guarantee that these people do not bring about disproportionate damage (Figure 3.12).

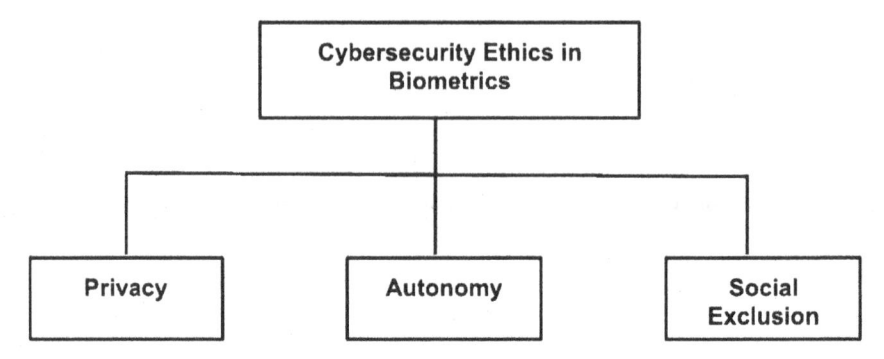

FIGURE 3.12 Cybersecurity ethics (challenges) in biometrics.

3.2.10 CYBERSECURITY ISSUES IN BUSINESS

Business world has been impacted majorly over the last few decades by globalization and digitization. This has not only altered the ethical issues but also their gravity, as evidenced by the widening of problems and complaints. As business management is becoming more complex, a multitude of ethical issues appears simultaneously. Therefore, an in-depth understanding of ethical problems is required. This would ensure identification of mitigating options is required, for which these review efforts have been undertaken [42]. Although the business world is largely dominated by technological advancement, all recent developments in business management, call for considering ethical issues. Addressing these issues may lead to strengthened business perspectives. Corporate responsibility, ethics, and accountability are prominent issues in the business world. Organizations do not harbor a perfectly ethical climate and many ethical issues like legal liability, workplace safety, child labor, bribery, cybercrimes, overbilling, privacy threats and disclosures due to social networking, frauds, misleading, fake reimbursements, etc., can be observed in different businesses and management. There are issues related to cyberethics that have profoundly strengthened after the development and popularization of social media. Managers and management personnel are always under ethical pressure and stress from different stakeholders like owners, government entities, employees, customers, suppliers, competitors, and other managers. The trust, integrity, and honesty of management and companies are under permanent stress. There are several levels on which business managers may face ethical issues. Some of the levels are personnel, organizations, trade, society, and the globe. Ethical issues may also arise with managers, customers, suppliers, employees, companies, and government entities. Therefore, it is possible that many layers of ethical issues can crop up, however, ethics in the business world is based on integrity and trust. If integrity and honesty of a management/company become doubtful or its trust is shaken, the ethical issues are strengthened and need attention. In situations where deals are to be made with customers trust and integrity are highly essential and also are referred to as the backbone of a business. Ethical issues may also arise due to conflicts of interests, poor management of employees and people, diversity, and cross-culture composition of working teams, ineffective communication, ethical conduct on social media, workplace safety issues and ignoring legal liabilities. Hence, the moral issues in the business world and current administration are rising as a huge problem. Business ethical issues and their practices in different organizations have gained very high importance currently because these are now readily exposed on social media. Not agreeing with business ethics may affect the businesses negatively, and may also lead to losing reputation and popularity of companies. This, in turn, could lead to a decrease in customers' number, loss of business, and reductions in revenues. Some of the cybersecurity ethical issues faced by the businesses are as follows:

1. **Ethical Issues in Accounting and Finance:** Ever since account matters were put online, ethical issues in accounting and finance increased significantly. Falsifying online records is very common and while it was confined to accounts previously now, it is prevalent for online statements, reports, purchase, and payments. The major motive is to save tax and to pay the minimum to shareholders. There are a few exploitative account matters like dressing and misdirecting financial investigation, manipulation of accounts, bribery, money frauds, overbilling of expenses and purchase, fake reimbursements, compensation to executives, and an indication of lesser revenues, etc. Some other ethical issues of accounting and finance include Fraudulent Financial Reporting, Misappropriation

of Assets, Disclosure, and Penalties. The impact of these types of ethical issues is very deeply affecting companies, shareholders, states, and governments, shareholders, and last customers through high priced products and services.

2. **Social Media Ethical Issues:** Social media is used as a platform for personal and work-related activities by most of the employees. Hence, it is a difficult task to separate private and official use. As the use of networking and social media websites like Facebook and Twitter is increasing and popularizing, ethical issues are also escalating. Many ethical issues are known to arise in situations where business employees' access and use of social websites incorporates disclosing confidential secrets, conflicts, private information and potential use of child labor and workplace discrimination. This may prompt irreversible harm of the organization's reputation and credibility. In addition, utilization of social media during duty hours might be regarded as abuse of organization time and resources. Companies are capable of imposing restrictions that may be considered as a ban on private rights by the employees. Such situations have created serious issues for the employers, and most of them label it as employees' misconduct online since these activities by employees are thought of as disloyalty and breach of employment rules. Therefore, many organizations recommend changing rules of business ethics regarding the use of social media by employees to pull out of this situation.

3. **Technology/Privacy:** Advancement of innovative technology has made it possible to observe, monitor, and record employees' movements, performance, presence at their seats and working activities. Employees' computers, communications, emails, and visited internet sites can be checked by managers and employers. Although employers are capable of legally checking official Emails of employees, it is unethical on the part of electronic surveillance to become spies. Although observing workers through video cameras and recording their conduct and development can keep the working environment safe on one side, it also hinders them mentally because they are thinking all the time that their movements are being observed by others. Therefore, it is the ethical duty of employers to create a balance between employee privacy concerns and preventing unethical behavior of employees that may potentially hurt business.

There is a need to frame internal policies in this regard. Further, all workers must be educated ahead of time about the degree to which the computer and internet devices provided to employees will be monitored and checked (Figure 3.13).

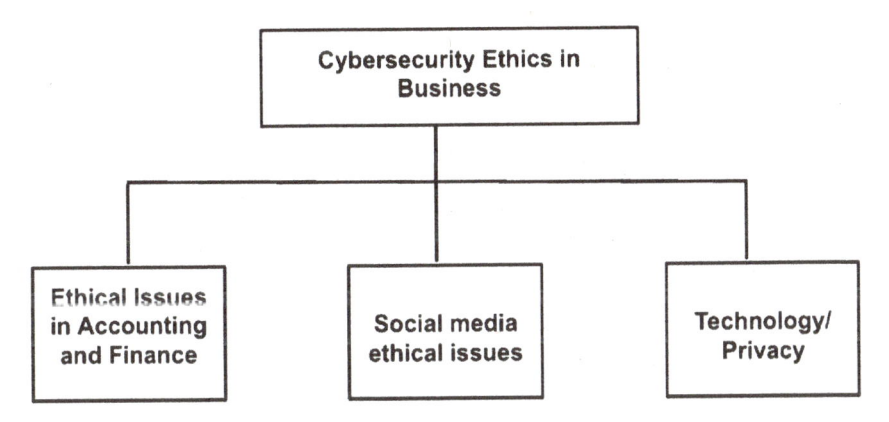

FIGURE 3.13 Cybersecurity ethics in business.

After reviewing important ethical issues, it is highly important that management options and strategies should be discussed which can be employed to rectify the prevailing situation of ethical issues in business management. Business managers must put in all efforts for understanding ethical issues of their organization. They may likewise incorporate moral wisdom and administrative wisdom to comprehend such issues. The following approaches may be adopted to ensure some level of ethics in the business environment:

- While an ethical behavior might not be profitable for the company, unethical behavior frequently generates substantial losses to the business, especially on a long-term basis. Hence, it is necessary that officials try to follow ethics as much as possible.
- The managers may create an ethical and moral culture in the workplace.
- Organizations may ensure ethics by routine assessment and evaluation of ethical issues in a company, identifying their quantum and main causes, grouping the issues according to their types, planning code of conduct for all individuals, framing policies for execution of organizations' techniques and observing the moral and security laws in letter and spirit.

- An ethical office must be established for every organization. Its representatives must be placed at all big workplaces to ensure no law and regulation breaches take place.
- Training programs could be launched for managers periodically, for managing ethical issues.
- Continuous monitoring of ethical issues may be required to complete and make programs successful.
- The cooperation of government entities and international organizations may prove highly useful in this regard.

3.3 SUMMARY AND REVIEW

- Cyberspace comprises the Internet, telecommunication systems, computer systems, embedded processors and control systems. Each of these have their own ethics and standards.
- Healthcare cybersecurity is one of the significant threats in the healthcare industry. Electronic health records (EHRs), likewise alluded to as EHRs, contain a large group of sensitive data about patients' clinical chronicles, making medical network security an essential IT concern.
- Digital Certificates provide identity and trust. Digital Certificates empower numerous security use cases that must be tended to in the HIPAA specialized protections.
- Autonomous weapons (AWS) could play a major role in cyber warfare. The AWS suggests the possibility of eliminating the human operator from the battlefield which raises several ethical questions.
- Financial institutions incorporate wealth management firms, investment brokers, and credit unions. These are very attractive to cyber-attackers as they host a variety of sensitive data like bank account numbers, social security numbers, etc.
- A blockchain is a digital record of exchanges (transactions). Blockchains can be used for recording cryptocurrency-based transactions, like Bitcoin.
- Blockchain technology relies on cryptographic proofs and game theory consensus mechanisms, hence is why it is impossible to hack While there are many safety features in blockchain technology it does not mean that security issues are non-existent. Blockchains have exposure to their own particular arrangement of security issues.

- Cloud computing has several features like resource/storage virtualization, scalability, and elasticity, efficiency of resource sharing, usage optimization/optimized by usage, ease of usage, fast information sharing, delivery, and control, accessibility, and anonymity.
- IoT ensures internet connectivity, and the communication that occurs between these objects and other Internet-enabled devices and systems. In IoT networks, the physical devices work by connecting to other physical devices, by means of wireless communication and offering contextual service.The combination of these technologies raises a number of ethical questions.
- Artificial intelligence (AI) machines can learn from experience, change in accordance with new information sources and perform human-like tasks. Popular AI applications include expert systems, speech recognition and machine vision.
- With the increase in the number of threats that intelligence agencies are struggling to respond to, intelligence work increasingly occurs in the public sphere. Policymakers highlight the crucial role of intelligence in justifying policy decisions.
- The field of biometrics technology is mainly used in the fields of work punching, transaction payment, visa, border access control and access control. Privacy, Autonomy, and Social Exclusion are some cybersecurity ethics challenges faced by biometrics.
- The business world is largely dominated by technological advancement and globalization. Corporate responsibility, ethics, and accountability are prominent issues in the business world. Associations do not harbor an entirely moral atmosphere and numerous moral issues like lawful risk, working environment security, child labor, bribery, cybercrimes, overbilling, privacy threats, and exposures because of social networking, frauds, deluding, counterfeit repayments, and so forth can be seen in various organizations and the board.

QUESTIONS TO PONDER

1. Consider autonomous vehicles. Should a driverless vehicle kill its passenger to spare five strangers?
2. You are a medical student. At some point, you see one of your fellow students placing clinical hardware from the stock room into

their sack. When you ask them about it, they state they just need to rehearse their clinical abilities and not to tell anybody. What might you do?
3. Do you agree that organ donation should be an opt-out system rather than an opt-in system? How ethical or unethical do you think it is?
4. Would it be unethical to breach patient confidentiality?
5. Consider you are a doctor. You observed that a colleague made a mistake with a patient's medication? Would you inform anyone of this incident? Why/Why not?
6. IoT-based smart home monitoring systems could be immensely useful to monitor the elderly. What are the issues associated with such a system?
7. List some morals that should be programmed into intelligent machines?

KEYWORDS

- **computer systems**
- **cybersecurity**
- **cyberspace**
- **embedded system**
- **internet etiquette**
- **telecommunication systems**

CHAPTER 4

Introduction to Cyber laws

In the last three chapters, we discussed ethics and ethical issues related to cybersecurity. We highlighted various types of ethical issues as well as ethics in several domains that underpin the idea of cyberspace as well as cybersecurity. We know that an unethical approach in cyberspace has severe repercussions, which may take a lot of time and money to resolve. Sometimes these issues might be impossible to fix. Therefore, there is the concept of cyber laws. In this chapter, we will introduce cyber law in the realm of cyberspace and cybersecurity, the need for these laws, and we will draw out the differences between ethics and laws. Further, we will delve into some typical examples of cybercrimes and discuss why they are illegal and unethical. Finally, we will read about some challenges in cyber law.

4.1 CYBER LAWS

As we know, ethics is a code of responsible behavior, while law is an outcome of ethics. Since cyberspace operates on a global level, the number of cybercrimes increases rampantly. Several activities over cyberspace like illegal access, illegal interception, data interference, misuse of devices, computer-related forgery, computer-related offenses, computer-related fraud, offenses related to child pornography, and offenses related to copyright and neighboring rights smay be termed as cybercrimes, which may have a drastic impact over the society [24, 43]. Therefore, there is a need to introduce some kind of legislation that mandates acceptable behavioral use of technology over cyberspace.

> **Definition 4.1: Cyber law**

> Cyber law may be defined as legislation that underpins acceptable behavioral use of technology like computer hardware and software, the Internet, and networks. It guarantees that users are protected from any harm by empowering the investigation and indictment of online crime. Further, it applies to the activities of people, gatherings, general society, government, and private associations.

Cyber law has a role in society and business. Cybercrimes like fraud, forgery, money laundering, theft, and other illegal activities are performed via computers, the Internet, and networks. Cyber law is responsible for investigating crimes perpetrated in the physical world but enabled in cyberspace. In a situation where organized crime is associated with using the Internet to distribute illegal substances may face prosecution under cyber laws. From the business perspective, cyber law is used to protect companies from unlawful access and theft of their intellectual property (IP).

Although cyberspace is global, cyber law is different in different countries. While cybercrime impacts the global community, the adoption of cybercrime legislation varies among countries. Most of the countries have cyber laws, few have draft legislation, and some even have no cyber laws. Numerous states are involved in developing new cyber laws as addenda to their present codes. Many countries revise their already existing national codes with legislative language on cybercrime. In the past several nations have instituted their own national cyber laws remembering their own national prerequisites. Countries vary widely in their ability to investigate and punish cybercrime. Further, changing levels of risks across different countries make it difficult to pass common cyber laws all over the globe.

4.2 THE NEED FOR CYBER LAWS

Cyber law is worried about the law governing the Internet. It addresses several issues like cybercrimes, electronic commerce, IP and data protection, and privacy. Cybercrime predominantly deals with computer crimes, computer-related crimes, and computer abuse. E-commerce involves electronic data used in commercial transactions. Electronic commerce laws are responsible for addressing issues of data authentication by electronic and/or digital signatures. IP deals with copyright law pertaining to computer software, computer source code, websites, etc., software, and source code licenses, trademark law with respect to domain names, semiconductor law which highlights the protection of semiconductor integrated circuits design and layouts, and patent law concerned with computer hardware and software. Data protection and privacy laws are responsible for addressing legal issues that emerge while gathering, putting away and communicating sensitive personal information. This is usually done by data controllers like banks, hospitals, email service providers, etc.

Due to the advancement in technology, especially cybersecurity, it may be said that current criminals can take more with a computer than with a firearm, and tomorrow's terrorists might have the option to accomplish more harm with a keyboard than with a bomb [44]. Cybercrimes that aim at stealing valuable data and vandalizing the economy are a threat not only to cyberspace but various other domains too that harbor sensitive information. Thus, enforcing cyber laws in cyberspace is mandatory:

- Cyber laws are vital because they touch almost all aspects of transactions and behavior on and concerning the Internet, the World Wide Web and Cyberspace.
- Cyber laws ensure the right to enter into legally enforceable digital contracts.
- Cybercrimes, for example, malware attacks, computer frauds, phishing, hacking, unauthorized access, email hijacking, denial of service and pornography, and becoming frequent, and may potentially harm sensitive data [15, 16, 45], the economy as well as the reputation of an individual or an organization, hence cyber laws may help prevent them.
- Almost all transactions are in electronic form and accountability is significant.
- In numerous non-cyber crime cases as well, significant proof might be found in systems, mobile phones, for example, in instances of separation, murder, abducting, tax avoidance, and so on.
- Cyber law is an important part of Digital Forensics. Laws related to the 4th amendment and chain of custody are significant in handling devices and cases in case of such crimes.

4.3 CYBERETHICS AND CYBER LAW

Let us recall the triad that we mentioned in Chapter 1. Figure 4.1 mentions how ethics are a code of responsible behavior and should be followed. Law being an outcome of ethics is something that must be followed. Policies are guidelines that organizations usually follow to promote cybersecurity in their environments. We see how these three are connected to each other. Although they all aim to ensure better cybersecurity in a given environment, they are very different from each other.

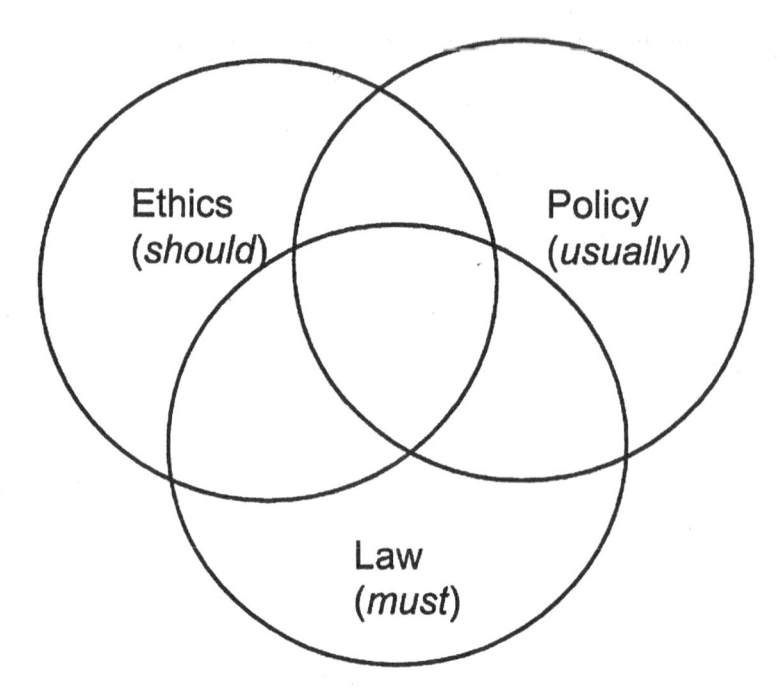

FIGURE 4.1 Ethics, law, and policy triad.

As we know, ethics are rules that control an individual or society, created to decide what is good or bad, right or wrong, in a given circumstance. Ethics are also responsible for regulating an individual's behavior or conduct and helps them live a good life. This may be done by applying the moral rules and guidelines. Law, on the other hand, maybe thought of as the systematic set of universally accepted rules and regulations.

These principles and guidelines might be made by a suitable authority, for example, government, which might be territorial, national, international, and so forth. It might be utilized to oversee the activity and conduct of the individuals and can be upheld, by imposing penalties. Further, there is a difference in scope and application of ethics and laws.

> **Ethics versus Legal:**

> WhileLegal' can apply to a more widespread scope, 'Ethical' applies on an individual basis. Legal has its premise in ethics, while Ethics has its premise in morals. The two of them judge a specific conduct or activity either right or wrong in their respective opinions. 'Legal' may be said to have more of an objective view, while 'Ethics' has a personal as well as varied view depending on the individual.

Due to the significant increase in the number of cyber-criminal activities in cyberspace that are not just unethical but also illegal, the role of ethics becomes even more significant. Therefore, there is a need to combine ethics and Law in regulating the activities of the cyber world. Moreover, it is crucial for curbing the menace of cybercrime in society. Information technology (IT) has made the world a global village and has enhanced every sphere and sector of society like the economy, commerce, social, and educational sectors. However, this is one space where a lot of work is required to be done. It is high time now for careful inspection of the legal and ethical aspects since not enough guidelines are available in this field as compared to those available in conventional branches of science and technology. There is a significant need for cyberethics to be strengthened by cyber laws. Cyber laws across the world are varied and different laws in different parts of the world have sought to reiterate and reinforce ethical principles concerning ethical behavior in cyberspace. No wonder, it has been opined that an information governance framework should contain measurable and strategic goals that will be beneficial for the provider and citizens and promote ethical standards. Many countries have come up with new legal national frameworks to govern and regulate cybersecurity. Hence, it is imperative that cyber legal frameworks and cybersecurity legal frameworks must now be evolved keeping in mind the evolved principles of cyberethics.

Consider the following example: obtaining data from networks. Here, there are two separate concerns:

- First, collecting network measurement data (e.g., packet traces), and
- Second, it is publishing the data.

Clearly, the ethical issues, in this case, are respecting users' privacy and respectful uses of published traces, while the legal issue is communication privacy laws. Most cybercrimes have some components of Illegal as well as unethical behavior.

Keeping in mind the growing number of cyber laws in the world, including cybersecurity laws, it is increasingly clear that the said legal frameworks incorporate cyber ethical principles and standards, e.g., a demonstration of not hacking a system is a moral guideline. If an ethical principle is violated, then such an act is prescribed as a cybercrime which is punishable with imprisonment and fines. If a person does any cybercrime activity, then it becomes an offense under the existing cyber laws. Further, the increased focus of people on protection and preservation of personal privacy as also data privacy has brought forward the need for effectively codifying ethical considerations pertaining to the protection of privacy. For example, it is ethical not to invade anyone's privacy, however, in case if someone does invade someone else's privacy, that becomes an actionable wrong for which various remedies are provided under different cyber legal frameworks. Further, extraordinary cyber laws have specified that an individual, who does not follow anticipated cultural moral standards and behavior concerning activities on cyberspace, would be held to legal consequences. These incorporate different remedies including damages by way of compensation as additionally criminal liability imprisonment and fines. Similarly, it is ethical to respect another person's personal data and sensitive personal data. These ethical principles form the basis for data protection legal frameworks. These frameworks come within the broad umbrella of cyber law jurisprudence.

4.4 NEED FOR CYBER REGULATION BASED ON CYBERETHICS

The domain of cyber law is concerned with providing adequate, strong, validity, and sanction for cyber ethical principles and cyberethics as an evolving discipline. For ensuring proper cyber laws, governments in every country, public policymakers, computer professionals, organizations, and private citizens must all take an interest and contribute. There is a need to exploit this global information in a socially and ethically sensitive way for our future benefit and applications.

The requirement for guidelines in the cyber world cannot be overemphasized since technological advancement has conveniently transformed the world into a global village [124]. Cyber law calls for the safe and lawful collection, retention, processing, transmission, and use of personal data of individuals. Cyberethics, therefore, provides the foundation for cyber legal principles concerning the protection of data and other aspects of human endeavor in cyberspace. The eventual fate of the crossing point of cyber law and Cyberethics expect a great deal of criticalness. Cyberethics as a discipline is substantially evolving. Cyber law will play a very important role in the evolution of cyberethics [46]. The obligation will be on officials and administrators to begin perceiving the convergence between cyber law and cyberethics and increasingly grant legal recognition, validity, and sanction to cyberethics principles. This becomes although more significant, as we are coming across times where cybercrimes are going to be far more difficult and dangerous. Cybercrime as a paradigm is constantly evolving [14].

The global cost of cybercrime is constantly increasing. The cybersecurity community, major media, researchers, and industry experts have collectively agreed to the prediction that cybercrime damages will cost the world in trillions annually by 2021. There may be a significant transfer of economic wealth in history involving incentives for innovation and investment. It may also be more profitable than any other global trade. The emergence of darknet presents completely different challenges of cyberethics. Darknet encapsulates countless ethical standards concerning cyberethics. It is also important to examine the role of cyberethics in the darknet as the same can be instrumental in the development and crystallization of various cyber ethical principles.

4.5 CYBERCRIME CATEGORIES FROM THE LEGAL PERSPECTIVE

The Internet is not a safe place, as it witnesses cybercrime growing at a rapid rate since people are connected to devices as well as each other digitally. Cybercrimes influence associations, organizations, and governments billions of dollars every year. These crimes not only target the organizations' economy but also hamper their reputation. Illegal activities on the Internet keep on increasing rampantly and show no signs of slowing down. In fact, cybercrime is growing more and more. From the Legal perspective, cybercrime can be classified into the following types (Figure 4.2).

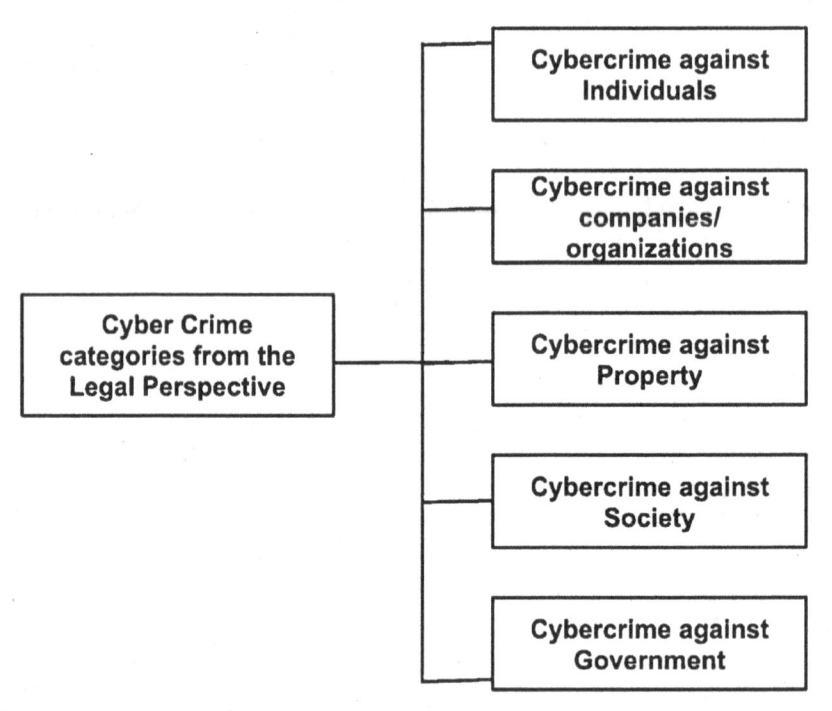

FIGURE 4.2 Cybercrime categories from legal perspective.

1. **Cybercrime Against Individuals:** These are cybercrimes that directly have an impact on any person or their properties. Examples of such cybercrime include social engineering, phishing, email harassment, cyberstalking, and spreading illegal adult materials. Credit Card extortion, human trafficking, social engineering, data fraud, and defamation are some different instances of cybercrimes against people.
2. **Cybercrime Against Companies/Organizations:** Cybercrimes against organizations are very common. If a company website is hacked, it becomes a serious problem that may have dire consequences for the company, as well as its employees, associates,

and customers. It may cost the company a lot of money as well as its reputation. Some examples of cybercrime against companies/organizations include data breaches, cyber extortion, malware distribution, etc.
3. **Cybercrime Against Property:** It could be destroyed property or stolen property. Some online crimes against property may affect computers or servers. Some online violations against property may influence systems and servers. A portion of these violations incorporate DDOS attacks, hacking, malware transmission, cyber, and typosquatting, system vandalism and copyright infringement.
4. **Cybercrime Against Society:** When cybercrime is committed against various individuals, it is known as cybercrime against society. It affects society as a whole. Financial crimes against open associations, selling unlawful products, trafficking, web-based betting, fraud, etc., are some examples of cybercrime against society.
5. **Cybercrime Against Government:** When cybercrimes are committed against the government, it may be considered as an attack on the nation's sovereignty. It can result in prosecution by federal cybersecurity and law enforcement agencies. Such activities may also be classified under cyber terrorism and include breaking into government systems and networks, destroying, and closing down military sites, and spreading publicity. Cybercrimes against the government may likewise incorporate hacking, unauthorized accesses, digital warfare, cyber terrorism, and software piracy.

4.6 CYBER LAW AND INTELLECTUAL PROPERTY (IP)

Cyberspace is prone to crimes related to IP.

> **Definition 4.2: Intellectual Property (IP)**

Intellectual property may be defined as the domain of the Law that manages ensuring the privileges of the individuals who make unique works. It considers everything from unique plays and books to creations and company identification marks. IP laws are capable of encouraging new technologies, artistic expressions, and inventions for promoting economic growth.

Individuals who know that their creative work will be safeguarded and that they can benefit from their work may feel encouraged to produce novel things so as to enhance technology, make operations more proficient, and make magnificence in our general surroundings. IP is a field of legal practice and cyber law. It is an extremely new territory of Law that amalgamates inventive lawful speculations with new takes on traditional doctrine. Both Legal theories and traditional doctrine have been fundamentally affected by the improvement of new innovation and the development of the Internet. Since Internet users can promptly duplicate material on the web, copyright questions have gotten complicated. Cyber law has evolved over the last few decades for responding to the wide range of offenses that can be committed online. Both IP and cyber law, subsequently, are currently growing fields of expanding significance. IP rights identified with cyber law for the most part fall into the accompanying classifications (Figure 4.3).

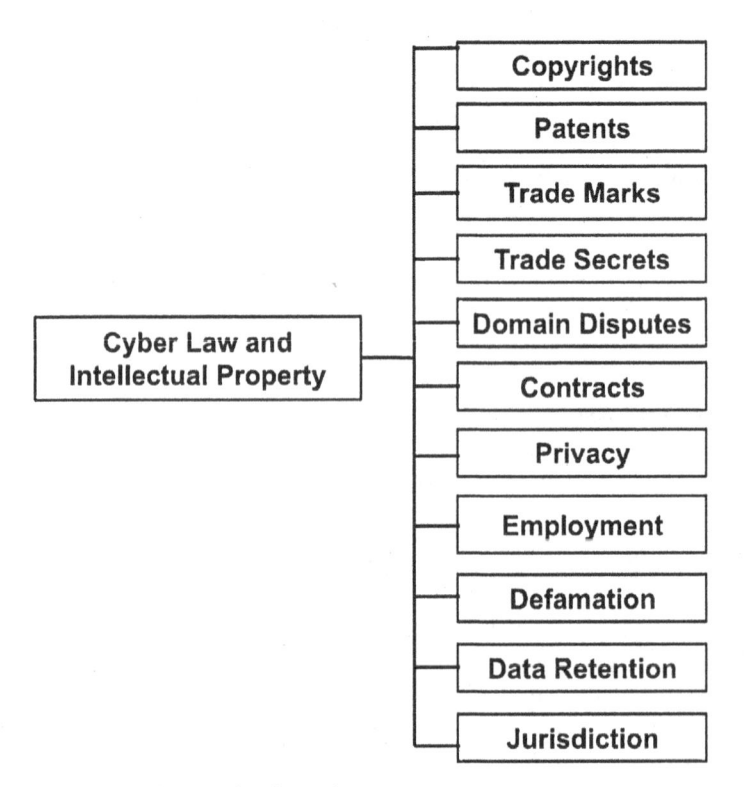

FIGURE 4.3 Cyber law and intellectual property.

4.6.1 COPYRIGHTS

Copyright grants rights to the creator of original works, for example, musical, dramatic, literary, and artistic works, and other scholarly works, for example, programming codes. Copyrights provide protection to almost any piece of IP one can transmit over the Internet. It may include books, music, movies, games, etc. Copyright laws are mainly for encouraging the creation and distribution of such works. The basic thought is to protect a person's work, whether it is published or unpublished. Since thoughts and ideas are not recorded or expressed, they cannot be copyrighted. Also, copyrighted expressions must be tangible. Moreover, copyrights are responsible for protecting a form of expression, rather than the subject matter of a work. The 1976 Copyright Act governs copyrights and gives creators selective rights to replicate their work, prepare subsidiary works, convey duplicates of the work, or perform or show the work in public. Copyrights are automatically granted to original works, irrespective of whether the copyright is registered. Copyright attorneys may typically be litigators who are responsible for assisting their clients enforce their copyrights. They may also be transactional attorneys who can guide their clients, manage, and license their copyrightable assets.

4.6.2 PATENTS

Patents are used for the protection of the invention. These are utilized on the web for two principle reasons, new software and new online business strategies. There are three unique sorts of patents in the United States:

1. **Utility Patents:** These are the patents that secure creations with a particular capacity. They may include things like chemicals, machines, and technology.
2. **Design Patents:** These are the patents responsible for protecting the unique way of how a manufactured object appears.
3. **Plant Patents:** These are the patents that protect plant varieties that are asexually reproduced, including hybrids.

While inventors may be solely responsible for the creation of their inventions, they may not conclude that their creation is patented before they apply and are approved for a patent. The US Patent and Trademark Office is responsible for approving patents in the United States. The process can be complex and time-consuming, and therefore hiring an IP attorney may prove to be beneficial. Intellectual attorneys would guarantee that innovators have sufficient paperwork to get the patent for protecting their invention and making it profitable.

4.6.3 TRADEMARKS

A trademark is essentially a word, expression, image, or structure, or a combination that recognizes and differentiates the source of the products of one party from those of others. Trademarks intend to protect the names and identifying marks of products and companies for avoiding confusion between brands. Trademarks are responsible for making it easy for consumers to differentiate competitors from each other. Trademarks may also be defined as exclusive rights for using a certain design in commerce. While registering a trademark may not prevent others from producing a similar good, but will definitely forbid them from showcasing the good with an imprint that is comparative enough to befuddle consumers about the good's origin. Trademark lawyers are capable of assisting in the registration process. They can provide advice on the development and use of trademarks, and may even represent their clients in litigation. In a situation where there is a likelihood of confusion between two marks, a party with a registered trademark may sue the other party for infringement. Common defenses against charges of infringement or dilution include fair use. In fair use, a mark may be used in good faith for its primary meaning and nominative use. A trademark is automatically assumed as soon as a business begins using a certain mark to identify its company. A trademark may likewise utilize the symbol TM without filing their symbol or name with the government.

4.6.4 TRADE SECRETS

A trade secret refers to any practice or process of a company which is generally not known outside of the company. It may also incorporate formulas, patterns, and processes. Online organizations are known to utilize proprietary innovation protections for a few reasons, although it does not prevent reverse engineering. Trade secret laws are responsible for protecting multiple forms of IP. Trade secrets provide companies authorization to maintain the confidentiality of economically beneficial information. Consider the food industry, it has many trade secrets. These trade secrets range from the recipe for soft drinks to spices in various food chains. A few organizations depend on proprietary innovations rather than licenses, which are more standardized. This is due to the fact that obtaining a patent requires full disclosure. Further, patents expire after 20 years, but trade secrets can be kept indefinitely. There are no formal ways of protecting trade secrets. Also, there is no legal recourse for preventing someone from using a trade secret once it has been made public. This is the reason why lawyers craft non disclosure and non-compete employment contracts that not only protect trade secrets but also comply with employment law.

4.6.5 DOMAIN DISPUTES

Domain disputes are very similar to trademarks. They are usually about who owns a web address. By and large, the individual who runs a site may not be the proprietor of the site. Moreover, since domains are cheap, many people buy multiple domains hoping for a big payday. Although the IP right does not protect domain names, the name that makes up the domain name may itself be protected by copyright, trademark right or a geographical name. It may also consist of a surname, a trading name or a corporate name.

4.6.6 CONTRACTS

While many people think that contracts do apply online, that is not always the case. Given a situation where an individual is registering for a website, the individual has to agree to terms of service. This is a contract. Contracts and IP complement each other. No contract that is signed by an individual or a company is unimportant. Contracts must be reviewed to make sure that the individual or the company is maximizing and not damaging IP assets. This is on the grounds that utilizing contracts, IP rights might be sold or licensed or even given away. Unpleasant agreements may prompt prosecution and unnecessary charges. Some risky areas when dealing with contracts are employees and contractors, development agreements, web design agreements, etc. Agreements to license a product or IP to another company, agreements to license a product or IP from another company, distribution agreements, domain name and trademark license agreements, and patent licenses, cross-licenses, and pools are some other risky areas of contracts. If there is a situation where employees, contractors, consultants, or other companies are involved in developing an IP, there must be a contract with that person or entity even before work is started. The earliest start of work may also raise important rights, and it may be possible for the contractor to become the author or owner of its work, or possibly a joint owner. Therefore, it is mandatory that contracts specify who owns the IP that is created and how the IP will be treated in the future.

4.6.7 PRIVACY

Online organizations must consider the user's security and privacy. Since increasingly more data is sent over the web, there is a greater risk of privacy being compromised. Hence cyber laws with respect to privacy are very important. Since governments, different associations, or people utilize the Internet to assemble data about people or gatherings, privacy issues prevail. People transmit private information over the Internet and store private data on computers. This data is not stored under their physical control, thus making privacy concerns increasingly important. Internet privacy faces challenges on a daily basis. Ad companies equip cookies to track browser history. Law enforcement professionals scan social media profiles for evidence of criminal activity, thus harming individual privacy. Many organizations work towards protecting privacy online and therefore litigation, policy, and research opportunities for lawyers exist. With government surveillance of online communications, extensive political and legal implications are raised. Several organizations work against efforts to broaden government surveillance, and these concerns may rise to Constitutional levels. The 4th amendment is responsible for protecting people from unreasonable government searches and seizures. This may regularly be ensnared by government online surveillance and intelligence gathering. The development of jurisprudence in this area is still in progress. While not involving the 4th amendment, privacy concerns are likewise raised by the activities of huge companies. These corporations are capable of managing and manipulating a huge chunk of online personal data. Many nonprofit organizations working on Internet privacy may also be engaged with these issues.

4.6.8 EMPLOYMENT

Frequently worker contract terms are connected to cyber law in view of non-exposure and non-content statements. The two clauses must also take into account the Internet. It might likewise incorporate how representatives utilize their organization email or other computerized assets.

4.6.9 DEFAMATION

The web is a well-known stage for carrying out crimes. Crimes like cyber-bullying often lead to defamation of individuals. While proving defamation was not considered from the viewpoint of the Internet for a long time, now it includes the Internet. Defamation describes both libel, for writing, and slander, for spoken. Whether it is libel or slander, defamation is capable of harming an individual's reputation. It may also lead to a personal injury lawsuit. Although defamation is not a criminal act, it is significant because it may harm an individual's emotional, physical, and financial well-being. As social media platforms grow and online news sources, defaming has become very easy. This also increases the threat of various defamation suits. It is difficult for courts to strike a balance between an individual's right to free speech and another is right to not be subject to defamation. Hence for proving defamation, some key elements must exist. The key elements may be in the form of a statement of fact, a published statement, a statement that caused an injury, a statement that is false or a statement that is not privileged.

4.6.10 DATA RETENTION

Since data has value, handling data is a primary concern, especially because it is transmitted through the Internet and is stored in multiple locations. Data retention is a big issue when it comes to litigation. In lawsuits, requesting electronic records and physical records has become common. However, while there are no current laws that require keeping electronic records forever, the same cannot be said for physical records.

4.6.11 JURISDICTION

Jurisdiction is a significant attribute of court cases. Cyber Crimes have complicated jurisprudence. Consider a circumstance where a cybercriminal is situated in one particular location of the United States and their victim is situated in another. The question raised could be related to which state will follow jurisdiction. Since different states have different rules, it may complicate the issue. Another point that must be considered depends on what court, federal or state, a case was filed in.

4.7 CHALLENGES IN CYBER LAW

One of the biggest challenges in developing cyber laws is the concept of jurisdiction. Jurisdiction might be characterized as the official capacity to settle on lawful choices and decisions. Jurisdiction seems challenging for developing cyber laws because the Internet is virtual. Further, it does not have any territories. Thus, while mediating the legitimate network regularly faces a contention of Law. Henceforth, what makes an act unlawful or criminal are the laws appropriate to the jurisdiction to which it was committed. In a situation where two jurisdictions agree that an individual's action is illegal, it is hard to foresee which Law will apply if both jurisdictions have different severity of punishment or fines. Conflict may likewise fall back on who has authority over a case in a digital world that has no physical boundaries. With the issue of jurisdiction it makes the activities of law enforcement officers and agencies daunting when pursuing a cybercrime, every law enforcement staff or agency has their area of authority. Law enforcement agencies are capable of authorizing and enforcing the Law within their jurisdictions. Consider a situation where a police officer is commissioned in Seattle. He may have no authority to arrest someone in Delaware. Even in cases involving deportation, countries are not obliged to hand over a criminal to another. Communication-related to understanding and deportation may not hold good at any time without serious consequences. This is the reason a few nations have settlements whereby they consent to do as such, in spite of the fact that in those cases as well, it is typically a costly and protracted procedure. Since jurisdictional issues frequently slow down or totally hinder the authorization of cybercrime laws, the process of drafting an extradition treaty with another nation seems tough. Some cyber law Issues requiring discussions are mentioned as follows:

1. **Are all the crimes recognized that we see every day in cyberspace?:** Cyber Crime is an evolving field, and therefore, as technology moves, new types of crimes surface. Therefore, it is practically incomprehensible for the Law to precisely distinguish various kinds of violations and propose remedies accordingly. Hence cyber law needs to depict offenses just as a general term. Thus, corresponding incidents must be interpreted and mapped to different offenses mentioned in the Act. Making offenses with not more than a specific number of years for imprisonment as bailable has been considered as one of the weaknesses in many countries. However, this may also be interpreted as a proportion of shielding innocent victims from being hassled. In the event that this arrangement can be abused by guilty parties to control evidences, it would be vital for the Police to guarantee that evidences are secured rapidly. Judiciary must also cooperate in certain cases to sanction E-Discovery for securing evidence before they are erased. Cyber Forensic capability can be relied on for the same.

2. **Are our police responsive to the public when a complaint is made?:** It is important to train the Police force when it comes to Cyber Crimes so that it yields results. One of the greatest underlying issues that many citizens still face is the inability to get their complaints registered with all Police Stations as well as the presence of Cyber Crime Police Stations. Sometimes citizens in some places are encouraged by some officers to avoid registration of cases. Cyber Crimes must be reported by the public without hassles. Another issue that crops up is that in many countries, the Police face a lack of support from Intermediaries Internet Providers, Phone service providers and Web site owners. This is particularly when some information is required for investigation. Numerous large associations secure the criminals by their privacy protection strategies and hinder the quick investigation of crimes. This requires a national-level policy formulation and further discussion concerning how to get sensitive data without relinquishing privacy protection.

3. **Are intermediaries and corporates co-operative with law enforcement?:** Intermediaries and Corporates avoid working

with law enforcement as it seems like an intrusion to their work. Therefore, there may not be enough cooperation even if the corporate interest is involved. More often than not, Crimes perpetrated inside the corporate system are not announced. Even if some crimes are reported and identified, the criminal is just backed out of the activity and not handed over to the law enforcement. Intermediaries have tons of investigative information and therefore must undergo periodical compliance audits. This may also ensure that they take reasonable precautions to prevent the occurrence of Cyber Crimes within their domain. Commercially minded business entities must look at the crimes from the human perspective and set aside some investment for information security and consumer education.

4. **Are the expertise in cyber law and cyber forensics being amplified?:** For creating the right pool of talents in cyber law and cyber forensics, the educational plan in training needs to be geared up:
   i. Creating awareness of cybercrimes at the high school level;
   ii. Presenting cyber laws in the curriculum in graduation level;
   iii. Presenting information security in the educational program at the technical and management instruction.

5. **Are cyber laws being misused for internet censorship and privacy invasion?:** It is important to ensure that reasonable steps are being taken by an individual in order to satisfy a legal requirement. However, the notification has come for severe criticism. This is because it can also be misused as a means of Internet Censorship. The fear is a result of a few cases in the past where the intensity of the executive to give directions to block websites has been utilized without legitimate governing rules. There is a need to guarantee citizen participation in the usage of delicate controls, for example, hindering of sites to evade the provisions being misapplied (Figure 4.4).

With this, we acquaint ourselves with introduction to cyber laws in general and the relationship between ethics and laws in cyberspace. In this chapter, we explored the need for cyber laws. The need for cyber regulation based on cyberethics and understood cybercrime categories from the legal perspective. We also familiarize ourselves with IP in cyberspace as well as the challenges faced in cyber laws.

Introduction to Cyber laws 153

```
                    ┌─────────────────────────────┐
                    │ Are all the crimes recognized│
                    │ that we see every day in    │
                    │ Cyberspace?                  │
                    └─────────────────────────────┘
                    ┌─────────────────────────────┐
                    │ Are our Police responsive to │
                    │ the public when a complaint  │
                    │ is made?                     │
                    └─────────────────────────────┘
┌──────────────────┐┌─────────────────────────────┐
│ Challenges in    ││ Are Intermediaries and      │
│ Cyber Law        ├┤ Corporates co-operative with│
│                  ││ law enforcement?            │
└──────────────────┘└─────────────────────────────┘
                    ┌─────────────────────────────┐
                    │ Are the expertise in cyber  │
                    │ law and cyber forensics     │
                    │ being amplified?            │
                    └─────────────────────────────┘
                    ┌─────────────────────────────┐
                    │ Are Cyber Laws being misused│
                    │ for Internet Censorship and │
                    │ Privacy Invasion?           │
                    └─────────────────────────────┘
```

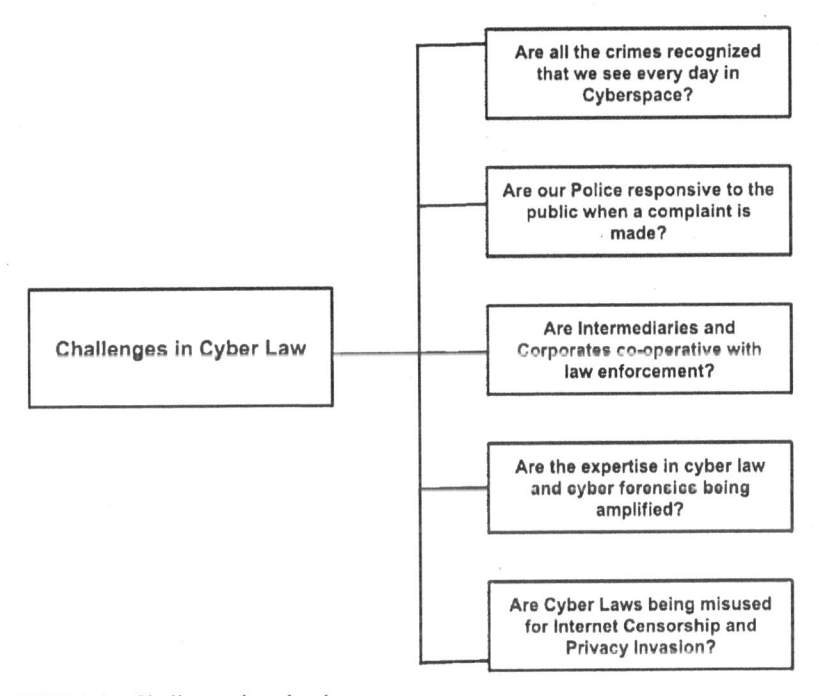

FIGURE 4.4 Challenges in cyber law.

4.8 SUMMARY AND REVIEW

- Ethics is a code of responsible behavior, while law is an outcome of Ethics.
- Cyber law may be defined as legislation that underpins acceptable behavioral use of technology like computer hardware and software, the Internet, and networks.
- Cyber law is responsible for investigating crimes perpetrated in the physical world but enabled in cyberspace.
- Over the last few decades, cybercrimes like stealing valuable data and vandalizing the economy have become a threat to cyberspace and other domains too that harbor sensitive information. Thus, enforcing cyber laws in cyberspace is mandatory.
- While 'Legal' can apply to a more widespread scope, 'Ethical' applies on an individual basis. Legal has its premise in ethics, while Ethics has its premise in morals.

- Intellectual property (IP) is a field of legal practice and cyber law. Intellectual property laws are meant to encourage new technologies, artistic expressions, and inventions for promoting economic growth.
- Copyright laws are mainly for encouraging the creation and distribution of such works. The fundamental thought is to ensure a person's work, whether it is distributed or unpublished.
- Patents are used for the protection of the invention. These are used on the Internet for two main reasons, i.e., new software and new online business methods.
- A trademark is basically a word, phrase, symbol, or design, or a combination that identifies and distinguishes the source of the goods of one party from those of others. Trademarks protect the names and identifying marks of products and companies to prevent confusion between brands.
- A trade secret is any training or procedure of an organization that is commonly not known outside of the organization. It may also incorporate formulas, patterns, and processes.
- Domain disputes are very similar to trademarks. They are usually about who owns a web address. In many cases, the person who runs a website may not be the owner of the website.
- Contracts must be reviewed to make sure that the individual or the company is maximizing and not damaging intellectual property assets.
- The 4th amendment is responsible for protecting people from unreasonable government searches and seizures.

QUESTIONS TO PONDER:

1. Eric and Su hacked the same website from Spain and Japan, respectively, which led to the disclosure of nearly 10,000 user accounts. The user account incorporated information like name, age, address, healthcare provider, and last four digits of some personal identification number. When the company got to know about the leaked accounts, they tracked down both the hackers. While Eric faced imprisonment for 7-years, Sue was charged 10,000 dollars for the felony. While both of them committed the same crime, they were charged differently. What could have been the reason? Do

you think it is fair? Is it just because Sue is a woman and Eric is a man?

2. What is meant by "a modern thief can steal more with a computer than with a gun, and tomorrow's terrorist may be able to do more damage with a keyboard than with a bomb"? Explain with an example.

3. As we know, cybercrimes have both illegal as well as unethical components. Consider the cybercrimes listed below and discuss the illegal and unethical components that may be incorporated into these:
   i. Child pornography;
   ii. Software piracy;
   iii. Spear phishing;
   iv. Physically breaking one's computer;
   v. Identity theft;
   vi. Forgery;
   vii. Gaining illegal access to company's data;
   viii. Ransomware attacks.

KEYWORDS

- **computer hardware and software**
- **cyberethics**
- **cyber law**
- **cybercrimes**
- **e-commerce**
- **intellectual-property**

CHAPTER 5

Cyber laws in the United States

We moved to the legal aspects of cybersecurity once we got acquainted with ethics related to cyberspace and cybersecurity. In the previous chapter, we introduced the concept of cyber laws and discussed their importance in cyberspace as well as cybersecurity. We drew out the similarity and differences between cyberethics and cyber laws, and we found that they are related to each other. As we mentioned before, laws are the outcomes of ethics. Cybercrimes from the legal perspective and cyber law with intellectual property (IP) manifest the need to discuss cyber laws, which brings us to the last chapter of the first part of the book, i.e., cybersecurity with respect to ethics and law. In this chapter, we will discuss the two types of cybercrime laws, the current legislative framework of cyber laws in the United States, discussions, proposed revisions, and current statuses of some laws followed by famous case studies.

5.1 TYPES OF CYBERCRIME LAWS

As cyberspace spans over the network, cybersecurity is of utmost importance, since cybercrimes increased day-by-day [11, 14, 16, 24]. There is a variety in the crimes committed, and therefore, there are several cyber laws. Overall, cybercrime laws can be classified into two categories. Each category has a list of cybercrimes that fall under the laws which we will discuss in this section. The cybercrime laws are discussed in subsections.

5.1.1 SUBSTANTIVE CYBERCRIME LAWS

Substantive cybercrime laws are the laws that are concerned about forbidding internet fraud, hacking, intrusions into services and systems, child pornography, online gambling, child pornography, and so on. They may be of the following forms:

1. **Misrepresentation and Related Action as for Data, Identification Records, and Authentication Features:** Fraud-related activities may be defined in many ways. For example: Intentionally fabricating an identification report, validation features, or a flimsy identification archive; Knowingly transferring a false identification, identification document or an authentication feature, that was that was taken or created without legal authority; Knowingly possessing with the intent to use unlawfully in excess of five identification documents, authentication features, or flimsy identification proofs; Knowingly producing, transferring, or possessing a document-making implementation or authentication feature with the intent to use it in the production of a false identification document. The punishment for such an offense is usually imprisonment or fine, or both.

2. **Aggravated Identity Theft:** This alludes to intentionally moving, having, or using, an individual's methods for identification illicitly. The punishment for such a felony is usually imprisonment. The offense may be conducted in the form of theft of public money, property, and rewards, false personation of citizenship, relating to bank, mail or wire fraud, relating to fraud and false statements, etc. Aggravated identity theft may also be a terrorism offense.

3. **Fraud and Related Activity with Respect to Access Devices:** These can be of many forms. Purposely fabricating, utilizing, or dealing at least one fake access device, using one or more unauthorized access devices, knowingly having a telecommunications instrument that has been adjusted or modified to get unapproved utilization of telecommunications services may be deemed as fraud. Further, knowingly using, producing, trafficking or possessing a telecommunications instrument that has been modified or altered to obtain unauthorized use of telecommunications services may be termed as crime. People committing the offenses are subject to penalties. A person who is outside the United States will be subject to fines, punishments, detainment, and relinquishment if the offense concerns an access device issued, that is being owned, managed, and controlled by an account user, financial institution, credit card system member, or other entity within the jurisdiction

of the United States. The law also takes into account whether the person is transporting, delivering, conveying, transferring to or through, or otherwise storing secrets, or holdings within the jurisdiction of the United States, any article that has been or may be used to aid the commission of the offense.

4. **Fraud and Related Activity with Respect to Computers:** Such fraudulent activities involve knowingly accessing a computer without authorization or exceeding authorized access. Intentionally accessing a computer without authorization in order to access sensitive information, i.e., data identified with financial establishments, interstate or foreign correspondences, information that is owned by government agencies are considered a felony. Consciously transferring of program, information, code, or command, and as a result of such conduct, deliberately may cause damage without authorization, to a protected computer. Fraud related activities may cause physical injuries, may be a threat to public health or safety, and damage computer systems owned by the government.

5. **Fraud and Related Activity with Respect to Electronic Mail:** Fraud or Extortion as for electronic mail may include accessing a computer system without authorization in order to initiate the transmission of multiple commercial electronic mail messages, using a computer to retransmit multiple electronic mail messages to deceive or mislead recipients, Falsifying header information in electronic mail messages, etc.

6. **Fraud by Radio, Television or Wire:** Such fraudulent activities are performed with the aim of acquiring property and money and are typically upheld by flimsy or deceitful acts, portrayals, or guarantees, by transmissions using wire, radio, or television communication. The communication may be interstate or foreign commerce. Moreover, the offences may also consolidate works, signs, signals, pictures, or sounds to execute such a plan or guile, and the convicts are either fined or imprisoned.

7. **Malicious Activities Related to Communications Lines, Stations, or Systems:** Deliberately or maliciously injuring or destroying property, material, works, telegraphs, stations, systems or different methods for correspondence, worked or constrained by the United States is viewed as a lawful offense. In the event that these are

utilized for military or civil resistance elements of the United States, that is in the process of constriction or has already been constructed, for malicious activities that may lead to some kind of obstruction, hindrance, or delay in the transmission of any communication the convict may be either or imprisoned or both.

8. **Transportation or Importation of Obscene Matters:** Transporting obscene matters that use common carrier or interactive computer service to the United Systems is considered a felony. These may include anything that is obscene, lewd, lascivious, or filthy books, pamphlets, pictures, motion-picture films, papers, letters, writings, prints. Importation and distribution of any matter of indecent character, or lewd, filthy phonograph recording, indecent, electrical transcription, lewd, or other article or thing fit for creating sound may be considered felony.

9. **Transportation of Obscene Matters for Trading and Distribution Purposes:** Selling or distributing obscene, lewd, lascivious, or filthy books, pamphlets, pictures, films, papers, letters, writings, prints, silhouettes, drawings, figures, images, casts, phonograph recordings, electrical transcriptions or other articles equipped for creating sound or some other matter of profane or improper character is considered a felony. Transportation of such obscene matters is considered a felony and the convict may face either imprisonment or fine or both.

10. **Indecent Visual Representation of the Sexual Abuse of Minors:** An individual who knowingly produces, distributes, receives, or possesses an obscene visual depiction of a minor with the objective of distributing it may be convicted and may have to face penalties or imprisonment. The visual depiction may be in the form of paintings, sculptures, cartoons, and drawings that portrays a minor engaging in sexually explicit conduct.

11. **Sexual Exploitation of Children:** People who utilize, use, convince, prompt, tempt, or force any minor to take part in any sexually explicit conduct with the purpose of producing any visual depiction of such conduct, may be convicted. Parents, legal guardians, or people having custody or control of a minor who consciously allow the minor to take part in, or to help any person to participate in, sexually explicit conduct with the target of delivering any producing any visual depiction of such conduct shall be convicted.

12. **A Few Exercises Relating to Material Including the Sexual Abuse of Minors:** Exploitation of minors can be conducted in many ways. Deliberately transporting visual depiction of offensive and lewd pictures of minors using any and all means including by computer systems or sends, any visual illustration is a lawful offense. Consciously receiving such visual depictions in mail is also a felony.
13. **A Few Exercises Relating to Material Comprising or Containing Child Pornography:** Many activities contribute to the perpetuation of child pornography. Consciously transmitting and receiving pornographic content over the internet, reproducing child pornography for distribution purposes, and advertising and promoting child pornography through mails are considered felony.
14. **Deceiving Domain Names on the Internet:** Using a misleading domain name deliberately on the Internet for misleading a person into viewing material constituting obscenity is a felony. Further, knowingly using a misleading domain name on the Internet with the objective of deceiving a minor into viewing material that is subversive to children on the Internet is also a felony.
15. **Misdirecting Words or Digital Pictures on the Internet:** There are numerous methods of deluding words or computerized pictures on the Internet. Using a misleading domain name deliberately on the Internet to mislead a person into viewing obscenity, or with an intent to deceive a minor into viewing materials that might be harmful for minors is considered a felony, and the convict may face a fine or imprisonment.
16. **Utilizing Interstate Services for Transmission of Data About a Minor:** Intentionally starting the transmission of data like name, social security number, electronic mail address, address, and telephone number of another individual, being aware that the individual is below the age of 16 years, with the purpose of enticing, encouraging, offering, or soliciting any person for engaging in any sexual activity is considered a felony and the convict may face fines or imprisonment.
17. **Criminal Offenses Related to Copyright:** This could be in the form of reproduction or distribution of copyrighted material. It is performed with the end goal of business advantage or private monetary profit.

18. **Unauthorized Publication or Use of Communications:** Individuals involved in receiving, transmitting, or assisting in the transmission of any foreign or interstate communication by means of wire or radio must not disclose or publish the existence, contents, to anyone except the authorized channels of transmission or reception. The authorized persons could be agents, or attorneys, people utilized or approved to advance such correspondence to its goal, appropriate accounting or conveying officials of the different communicating centers over which the correspondence is transmitted, etc.

5.1.2 PROCEDURAL CYBER LAWS

Procedural Cybercrime laws might be characterized as laws concerned about the power to protect and get electronic information from outsiders, including internet service providers (ISPs), position to block electronic correspondences or, the power to look and hold onto electronic proof. They may be of the following forms:

1. **Interception of Wire, Oral, or Electronic Communication:** Interception of wire, oral, and electronic communication may involve disclosure and interception of oral, wire or electronic communications, manufacture, distribution, and possession. It might likewise incorporate promoting of wire, oral, or electronic correspondence intercepting device, approval for block attempt of wire, oral, or electronic interchanges and reports concerning caught wire, oral, or electronic interchanges.
2. **Conservation and Divulgence of Stored Wire and Electronic Correspondence:** Cybercrimes identified with this might be as unlawful access to stored interchanges, deliberate revelation of customer interchanges or records, delayed notices, counterintelligence access to telephone tolls and transactional records, wrongful disclosure of videotape rentals or sale records, etc.
3. **Pen Registers and Trap and Trace Devices:** Cybercrime laws may witness general prohibition on pen register and trap and trace device use. Applications that include order and issuance for a pen register, and trace and trap device are considered felonies and must be allowed only after certain legal steps and documentation procedures. Pen register or a trap and trace device installation assistance

must be performed by individuals who have the right to access the devices. Correspondingly emergency pen register and trap and trace device establishment must be only performed by authorized individuals.

5.2 CYBER LAWS IN THE UNITED STATES

The role of the government addressing cybersecurity is a complicated process. It is necessary to secure federal systems and fulfill the appropriate role in protecting nonfederal systems. Despite the fact that there is no legitimate system enactment set up, many affirmed resolutions consider different parts of cybersecurity. Some notable provisions are in the following acts that form the current legislative framework of the United States are as follows [47, 48]:

1. **The Counterfeit Access Device and Computer Fraud and Abuse Act of 1984:** It is answerable for forbidding a few assaults on computer system frameworks identified with the administration and on those utilized by banks and in trade.
2. **The Electronic Communications Privacy Act of 1986 (ECPA):** It is answerable for forbidding unapproved snooping or electronic eavesdropping.
3. **The Computer Security Act of 1987:** This made the National Institute of Standards and Technology (NIST) reinforce security guidelines for government computer systems, with the exception of the national security systems that are utilized for defense and insight missions. The Secretary of Commerce is responsible for ensuring security standards.
4. **The Paperwork Reduction Act of 1995:** This made the office of management and Budget (OMB) responsible for developing cybersecurity policies.
5. **The Clinger-Cohen Act of 1996:** This gave agency heads the responsibility to ensure the adequacy of agency information security policies and procedures. It also introduced the position of chief information officer (CIO). Under this act, the Secretary of Commerce authority must make declared security principles necessary.

6. **The Homeland Security Act of 2002 (HSA):** This made the department of homeland security (DHS) responsible for cybersecurity as well as for the general responsibilities for homeland security and critical infrastructure (CI).
7. **The Cyber Security Research and Development (R&D) Act:** It was enacted in 2002. This act inaugurated research responsibilities in cybersecurity for NIST and the National Science Foundation (NSF).
8. **The E-Government Act of 2002:** This is the primary legislative vehicle that guarantees federal management and activities for making data and services accessible on the web. Besides it also incorporates different cybersecurity necessities.
9. **The Federal Information Security Management Act of 2002 (FISMA):** This refined and reinforced NIST and agency cybersecurity responsibilities. It also established a central federal incident center, and made OMB, instead of the Secretary of Commerce, accountable for implementing federal cybersecurity standards (Figure 5.1).

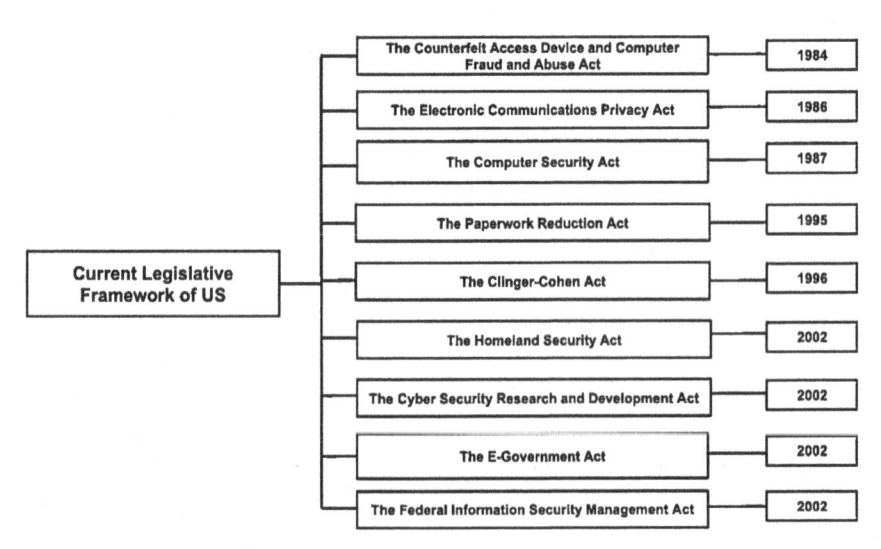

FIGURE 5.1 Current legislative framework of the United States.

We have already seen the provisional acts that are a part of the current legislative framework. However, there are several other acts that are related to cyberspace as well as cybersecurity. Further, inclusion of cybersecurity makes it mandatory for the acts to be revised so that the issues related to cybersecurity and cyberspace are well taken care of. In this section, we will be discussing some of these acts [49].

5.2.1 POSSE COMITATUS ACT OF 1879

Posse Comitatus Act is a United States government law that was marked on June 18, 1878. The reason for this act was to confine the forces of the federal government in utilizing federal military facilities for enforcing domestic policies within the United States. In this manner, it limits the utilization of military powers in regular citizen law implementation inside the United States. However, it does not remain constant for federal government facilities. The possible updates of this act reflect the intersection of cybersecurity with the federal government. One could argue the law keeps the military from collaborating on cybersecurity with common organizations, which may potentially be deficient of appropriate skills and capacities of the military and Department of Defense (DoD). Also, it is not easy to tell apart a criminal cyber-attack and an attack involving national defense. The act may also be amended to clarify that the military can operate domestically regarding cyber threats to privately owned infrastructure.

5.2.2 ANTITRUST LAWS AND SECTION 5 OF THE FEDERAL TRADE COMMISSION (FTC) ACT

This act prohibits unfair methods of competition (UMC), especially affecting commerce. The deceptive acts could be in form of a representation, omission, or practice misleads or is likely to mislead the consumer, a consumer interpreting the representation, omission, or practice that is considered reasonable under the circumstances, or then again a material that has misdirecting portrayal, oversight, or practice. Data sharing understandings between private companies might be dependent upon antitrust investigation, since the sharing of data among contenders could open doors for cooperation with the objective of controlling exchange. Be that as it may, data sharing understandings to battle cybersecurity might be in compliance with antitrust standards as long as their goals are to combat cyber threats instead of limiting rivalry. Also, so as to create successful and productive data sharing understandings to battle cybersecurity threats, an explicit exclusion from the antitrust laws for the agreements may be required.

5.2.3 NATIONAL INSTITUTE OF STANDARDS AND TECHNOLOGY (NIST) ACT

This act gives agencies responsibilities related to technical standards. Once amendments were added, the act was responsible for identifying relevant research topics like computer and telecommunication systems, information security and control systems. The act also contributed towards setting up a system guidelines program at the NIST. In any case, regardless of NIST's ebb and flow position to lead research on computers and data security, a few concerns have been raised about whether those exercises ought to be upgraded considering the advancing dangerous condition for cybersecurity [23]. For instance, leading exploration on identity management and the security of data frameworks [119], networks [120], and industrial control systems (ICS).

5.2.4 FEDERAL POWER ACT

The Federal Power Act established the Federal Energy Regulatory Commission. It is responsible for the interstate sale and transmission of electric power. Since the electric grid is vulnerable to cyber-attacks, several concerns have been raised for this act. The Energy Policy Act of 2005 made the Federal Energy Regulatory Commission responsible for creating unwavering quality guidelines for power frameworks and for reacting to quickly emerging cybersecurity threats.

5.2.5 COMMUNICATIONS ACT OF 1934

This act was responsible for combining and organizing federal regulation of telephone, telegraph, and radio communications. It also created the Federal Communications Commission (FCC) to oversee and regulate these industries. The FCC was responsible for domestic and international commercial wired and wireless communications. The President is given the authority in a national emergency to control stations or devices capable of emitting electromagnetic radiation in the situation of a war or threat of war. The growth of information and communications technology has an impact over the economy. Thus the act should incorporate a cybersecurity aspect to it. The power given to the President, may lead to shutting down of Internet communications during a war or national emergency. This is also known as the internet kill switch and is a debatable issue.

5.2.6 NATIONAL SECURITY ACT OF 1947

This act reorganized military and intelligence functions in the federal government to ensure defense and national security. It also led to the creation of the National Security Council, the Central Intelligence Agency, and the position for Secretary of Defense. Further, it led to the establishment of procedures for accessing classified information. Along with cyber-attacks, other limitations that exist for cybersecurity-related practices involve limitations on sharing of information. There ought to be an arrangement for assurance of data from entities and parties sharing data.

5.2.7 U.S. INFORMATION AND EDUCATIONAL EXCHANGE ACT OF 1948 (SMITH-MUNDT ACT)

This act approves the domestic dissemination of data and material about the United States planned basically for unfamiliar crowds, and for different purposes. The law might have been interpreted for prohibiting the military from conducting information operations in cyberspace. This is because some of those activities might be considered as propaganda that may reach citizens of the United States, since the United States does not confine Internet access as per regional limits.

5.2.8 STATE DEPARTMENT BASIC AUTHORITIES ACT OF 1956

The act provides certain basic authority to the Department of State. It attests that the association of the Department of State, including the places of facilitator for counterterrorism and for HIV/AIDS (human immunodeficiency virus/acquired immunodeficiency syndrome) reaction. With the Internet getting worldwide, a few concerns have been raised about international endeavors in cybersecurity by the United States. One approach to do so would be by setting up an organizer for the internet and cybersecurity issues inside the Department of State.

5.2.9 FREEDOM OF INFORMATION ACT (FOIA)

This act enables any person to access existing, identifiable, unpublished executive-branch agency records without any explanation or justification. However, the accessed material should fall within any of FOIA's categories of exemption from disclosure. Sharing of cybersecurity data between the federal government and nonfederal entities is essential. However, there may be questions about the records being subject to public release under FOIA, resulting in potential economic or other harm to the source. Some of the exemptions that may particularly apply to cybersecurity information are as follows:

- **Exemption 1:** Information appropriately sorted for national defense or international policy purposes as confidential under criteria built up by executive request.
- **Exemption 2:** Information explicitly absolved from divulgence by a statute other than FOIA if that rule meets criteria spread out in FOIA.
- **Exemption 3:** Trade secrets and business or money related data acquired from an individual that is favored or prohibited to disclosure.

While the existing protections are still in place, many private sector entities may have questions with respect to the public release of sensitive records that may not be specific enough to protect, or they may be too narrow to protect all records of concern. This may require some kind of amendment.

5.2.10 OMNIBUS CRIME CONTROL AND SAFE STREETS ACT OF 1968

This act is competent enough to assist State and local governments in decreasing the incidence of crime, increasing the effectiveness, fairness, and ensuring coordination between law enforcement and criminal justice systems at all levels of government. While Title I is responsible for establishing federal grant programs and other forms of assistance to state and local law enforcement, Title III is concerned with comprehensive wiretapping and electronic eavesdropping statute. This regulation outlawed both activities in general terms and allowed government and state law implementation officials to utilize them under severe impediments. As cybercrime incidents continue to increase, the abilities of the State and local law enforcement agencies to invest sufficient resources in enforcement activities is questioned.

5.2.11 RACKETEER INFLUENCED AND CORRUPT ORGANIZATIONS ACT (RICO)

This is a United States government law that is answerable for giving extended criminal punishments and a civil cause of action for acts proceeded as a major aspect of a continuous criminal association. In addition, it augments the common and criminal outcomes of a list of state and federal felonies when carried out in a manner characteristic of the conduct of organized crime. It has been recommended to include computer fraud within the definition of racketeering.

5.2.12 FEDERAL ADVISORY COMMITTEE ACT (FACA)

This act specifies the situations under which a federal advisory committee can be established as well as its responsibilities and limitations. The act commands the gatherings of such boards of trustees be available to people in general and that records be accessible for open assessment. This act has been condemned as conceivably obstructing the full advancement of open/private associations in cybersecurity, especially as for blocking private-sector correspondences and contribution on policies and strategy.

5.2.13 PRIVACY ACT OF 1974

The Privacy Act of 1974 limits the divulgence of personally identifiable information (PII) held by government offices. Further, it expects organizations to furnish access to people with office records containing data on them. It likewise settled a code of fair information practices for assortment, management, and spread of records by organizations, including prerequisites for security and secrecy of records. The act may be considered for specific modifications with regards to cybersecurity, thinking about PII, as well as how it can be used. Examples include explicitly permitting the sharing among federal agencies or with appropriate third parties.

5.2.14 COUNTERFEIT ACCESS DEVICE AND COMPUTER FRAUD AND ABUSE ACT OF 1984

This act is responsible for providing criminal penalties, like asset forfeiture, unauthorized access, and wrongful use of computers and networks of the federal government or financial institutions. The act is not confined to a country but also takes into account foreign commerce or communication. The act depicts getting secured data, harming or threatening harm to a system, utilizing the computer system to submit misrepresentation or commit fraud, dealing with stolen passwords, and spying as wrong practices. It also criminalizes electronic trespassing and exceeding authorized access to federal government computers. This act created a statutory exemption for intelligence and law enforcement activities. The act may be updated by increasing penalties for most violations of the act like damaging computers or password trafficking. It also asserts that provisions should be focused narrowly enough for avoiding creation of unintended liability for legitimate activities. Punishments for credit card theft could be expanded. The act may wrongly permit prosecution for some acts, such as Internet scanning for reducing vulnerabilities to cyberattacks. Hence, there is a need to narrow the applicability of some provisions of the act for reducing the risk of enforcement which some researchers and professionals deem overzealous.

5.2.15 ELECTRONIC COMMUNICATIONS PRIVACY ACT OF 1986 (ECPA)

The act is responsible for taking into account the key security privileges of citizens and the legitimate needs of law imposition as for information shared or stored in different sorts of electronic and telecommunications services. From the time when the act was passed, the Internet and associated technologies have amplified exponentially. The act consists of three parts:

- The wiretap act is responsible for prohibiting interception of wire, oral, or electronic communications. It basically targets wiretapping and electronic eavesdropping, possessing wiretapping or electronic eavesdropping and snooping devices, use or exposure of data acquired through illicit wiretapping or electronic snooping, and revelation of data made protected through court-requested wiretapping or electronic listening in, so as to impede justice.
- The stored communications act (SCA) prohibits unlawful access to stored communications.
- The pen register and trap and trace institution oversees the establishment and utilization of pen devices, and trap and trace devices, prohibiting unlawful utilization of a pen register or a trap and trace device. This act is concerned about setting up rules that law implementation must follow before they can get to information stored by service providers. It depends on the sort of customer data included and the kind of services being presented. To enforce disclosure by a third party, the authorization law enforcement must obtain an order. The order could range from a simple subpoena to a search warrant based on probable cause. Further, there is a need to follow a legal structure that can be effectively applied to modern technology. This legal structure must be responsible for protecting users' reasonable expectations of privacy. There can be extra changes to laws administering the security of electronic interchanges to encourage sharing of proper cybersecurity data, including the advancement of an unknown reporting mechanism.

5.2.16 DEPARTMENT OF DEFENSE (DOD) APPROPRIATIONS ACT, 1987

This act is liable for giving explicit power to the U.S. Unique Operations Command (USSOCOM) as for direct activity, strategic surveillance, experimental warfare, foreign internal defense, common issues, and psychological operations. It is likewise answerable for dissecting acts that relate to counterterrorism, theater search and rescue, humanitarian assistance, and other such exercises as might be determined by the President or the Secretary of Defense. Since some military activities are performed in a clandestine manner, for concealing the nature of the operation and also in order to collect intelligence, these actions are termed as covert actions. It is accepted that in the cyber domain recognizing whether an activity is or ought to be viewed as undercover or covert is tricky. It is often difficult to identify an attacking adversary's intent and location. It might be required to reassess the authorities of the DoD in light of the intelligence capabilities that assist in responding to and conducting offensive cyberattacks.

5.2.17　HIGH-PERFORMANCE COMPUTING ACT OF 1991

This act is responsible for building up a government elite registering program and necessitates that it address security needs. Moreover, it ensures that NIST establishes security and privacy in high-performance computing for federal systems. The act is additionally liable for setting up the Networking and Information Technology Research and Development (NITRD) Program, which delivers the necessary yearly report. Throughout the most recent couple of years, a few concerns have been raised that the program does not yield adequate strategic planning and does not adequately encourage cybersecurity innovative work and R&D. It highlights the necessity to understand the scientific principles of cyber-physical systems as well as some improving strategies for structuring, creating, and operating such frameworks with safety, high reliability, and security.

5.2.18　COMMUNICATIONS ASSISTANCE FOR LAW ENFORCEMENT ACT OF 1994 (CALEA)

This act is responsible for ensuring that telecommunications carriers help law implementation in performing electronic monitoring on their computerized systems. Further, it guides the telecommunications industry to configure, create, and convey arrangements that meet necessities for carriers to help approved electronic monitoring like call-identifying information for a target such that the privacy and security of other communications is not compromised. The act might be reexamined to improve its viability intending to cybersecurity concerns. This may not be the best component for gathering data sent by means of the Internet, therefore reassessment might be needed. Some recommendations include making changes to laws that govern the securing electronic interchanges to encourage sharing of suitable cybersecurity data, including the advancement of an anonymous reporting mechanism.

5.2.19 COMMUNICATIONS DECENCY ACT OF 1996

Since the internet witness's indecency and obscenity via telecommunications systems, this law was introduced. The law targets lascivious or explicit material, especially when it appeared to children that are not yet 18 years old. Additionally, the profanity and harassment provisions could likewise be deciphered as applying to visuals, violent terrorist propaganda or combustible language. As per the act, no client or provider interactive computer service will be treated as the speaker or distributor of any data. This has been translated to clear ISPs and certain online administrations of obligation regarding third-party content on those systems or sites. A great deal of content over the Internet, for example, terrorist chat rooms or propaganda websites might be a national security or operational threat that is not spoken to inside the Communications Decency Act. Further, if any material is obscene, the law does not give government organizations the power to necessitate that the ISPs facilitating the content take it offline, which ought to be considered. It might likewise be contended that the intelligence value picked up by protecting and checking the sites exceeds the potential threat risk.

5.2.20 CLINGER-COHEN ACT (INFORMATION TECHNOLOGY MANAGEMENT REFORM ACT) OF 1996

This act is responsible for giving agency heads authority to acquire information technology (IT) and requires them to ensure the adequacy of agency information security policies. It also established the position of agency CIO, concerned with assisting agency heads in IT acquisition and management. It requires the OMB to regulate significant IT acquisitions and furthermore expects OMB to declare, in conference with the Secretary of Homeland Security, mandatory government computer principles. These standards are usually based on those developed by the NIST. This law also exempts national security systems from most provisions. Increase in technology has prompted worries among cybersecurity specialists about possible weaknesses at different points along the supply chain for IT items. There have been debates on limiting the authority and jurisdiction of CIOs since their establishment. Further, policymaking, and infrastructure maintenance-related aspects also need revisions.

5.2.21 IDENTITY THEFT AND ASSUMPTION DETERRENCE ACT OF 1998

According to this act, identity theft is a federal crime. Individuals who commit or attempt to commit identity theft may be convicted under this act. Forfeiting property or intending to use it for fraud purposes also constitutes identity theft. This act guided the Federal Trade Commission (FTC) to record protests of fraud, give victims instructive materials, and allude grievances to the appropriate consumer reporting and law enforcement agencies. A potential update on the act might be approving compensation to data fraud victims for their time spent recouping from the damage brought about by the genuine or expected identity theft or fraud.

5.2.22 HOMELAND SECURITY ACT OF 2002 (HSA)

The essential strategy of the Homeland Security Act is to forestall terrorist assaults inside the United States. It is also responsible for reducing the vulnerability of the United States to terrorism, and minimizing damage and assisting in recovery for terrorist attacks that may occur in the United States. The DHS must give state and local governments and private entities with threat and vulnerability information, emergency management support, and specialized assistance with recuperation plans for critical data frameworks. It allows the Secretary of Homeland Security to assign qualified advances as subject to specific protections from liability in claims identifying with their utilization because of an act of terrorism. This act likewise established techniques and mechanisms to encourage data sharing among bureaucratic offices and proper nonfederal government and critical infrastructure personnel. It authorized DHS to establish a system of volunteer experts for assisting local communities in responding to attacks on information and communications systems. It has also been instrumental in strengthening some criminal penalties pertaining to cybercrime. It was liable for making the Directorate of Science and Technology inside DHS and appointed it expansive R&D duties. A few adjustments to the act might be in type of foundation of center for cybersecurity and interchanges inside DHS, required coordination with the DHS Office of Infrastructure Protection and division explicit organizations, foundation of the United States computer emergency readiness team (US-CERT) inside the center, specified data-sharing methods for government offices and other entities and foundation of a program inside the center to offer help to the private sector and a risk management strategy for security.

5.2.23 FEDERAL INFORMATION SECURITY MANAGEMENT ACT OF 2002 (FISMA)

This act created a security framework for federal information systems. This framework emphasized on risk management, and provided explicit duties to the OMB, the NIST, and the heads, CIOs, chief information security officers (CISOs), and inspectors general (IGs) of government organizations. It guaranteed that significant computer systems distinguish and give proper security protections, and create, record, and actualize agency-wide data security programs. It additionally gave OMB an obligation regarding regulating government data security strategy and assessing organization data security programs, however, excluded national security frameworks, aside from concerning implementation of responsibility for meeting necessities and answering to Congress. The act empowered update of obligations regarding the Secretary of Commerce and NIST for information system norms and furthermore moved duty regarding proclamation of those guidelines from the Secretary of Commerce to OMB. It orders the NIST cybersecurity principles to be integral with those produced for national security frameworks, to the degree attainable. It supports intermittent risk assessments, to decide and execute vital security controls in a cost-effective way, and to assess those controls. It indicated data security obligations regarding organizations' CISOs, including office-wide information security programs, guidelines, standards, strategies, and preparation of security and other workforce. It likewise supports security awareness training, remedial actions addressing insufficiencies, and techniques for taking care of security incidence and guaranteeing coherence of tasks. It is also liable for the foundation of a central federal incident center, supervised by OMB, to examine incidents and ensure technical assistance identifying them, to educate organizations. Due to insufficient efficiency in providing adequate cybersecurity to government IT systems, a focus on procedure for reporting operational security must be appended. There is a lack of widely accepted cybersecurity metrics, and insufficient means to enforce compliance both within and across agencies, which must be taken care of. Utilization of proper measurements, constant checking, and prioritizing risk-based as opposed to least safety efforts and modifications in reporting requirements might be some possible updates.

5.2.24 TERRORISM RISK INSURANCE ACT OF 2002

This act is responsible for providing federal cost-sharing subsidies for insured losses resulting from acts of terrorism. Thus, it is responsible for giving impetuses to the advancement of insurance coverage for misfortunes from acts of terrorism. Since losses from cyberattacks are not specifically included, the act may require some modifications.

5.2.25 CYBER SECURITY RESEARCH AND DEVELOPMENT ACT, 2002

This act is concerned with the NSF for conceding awards for essential exploration to upgrade computer security and for improving academic research and faculty development programs in computer and network security. It also encourages the establishment of multidisciplinary centers for research on computer and network security. The act orders that NIST build up programs to grant postdoctoral and senior research fellowships in cybersecurity and help foundations of higher learning that join forces with revenue-driven entities to perform cybersecurity research. Research related to cybersecurity is insufficiently coordinated and prioritized, therefore the act requires modification. There is a need to revise the methodology using which NIST creates checklists and drafts guidance and technical standards for federal IT systems.

5.2.26 E-GOVERNMENT ACT OF 2002

This act is in charge of guiding federal IT management. It also oversees the initiatives that make information and services available online. It was instrumental in establishing the Office of Electronic Government within OMB, which was led by an administrator with a range of IT management duties, including cybersecurity. It additionally established the interagency CIO Council and determined working with the NIST on security norms as one of its responsibilities. This act additionally doled out organization CIOs responsibility for checking usage of government cybersecurity principles in their offices. There are several requirements for security and protection of confidential information. This includes electronic authentication and privacy guidelines. It also takes into account the Federal Information Security Management Act of 2002 (FISMA). Because of specific holes in cybersecurity aptitude, the act may observe some potential updates. There is a need to address the increase in commercial availability of PII, so as to increase privacy.

5.2.27 IDENTITY THEFT PENALTY ENHANCEMENT ACT

The act is liable for instituting penalties for aggravated identity theft. It is possible for convicts to get extra penalties (2 to 5-years' detainment) for the theft committed pertaining to other federal crimes. Since identity theft is exceptionally common in the United States, improving the identity theft laws is an unquestionable requirement. It has been recommended to alter the identity theft and aggravated identity theft rules with the goal that criminals who abuse the identities of corporations and associations can be prosecuted. Further, adding new crimes as predicate offenses may amend the aggravated identity theft statute. Since the range of potential victims includes not only individuals but organizations as well, it is necessary to clarify the identity theft and aggravated identity theft statutes.

5.2.28 INTELLIGENCE REFORM AND TERRORISM PREVENTION ACT OF 2004 (IRTPA)

This act is responsible for establishing the position of the Director of National Intelligence. Further, it set up mission duties for a few entities in the intelligence, homeland security, and national security networks. It is additionally liable for discussing issues identified with the assortment, examination, and sharing of security-related information. This act is likewise concerned with building a Privacy and Civil Liberties Board inside the Executive Office of the President. Since this act does not contain a solitary reference to cyber, cybersecurity, or related exercises, it is mandatory that it is updated so as to reform the intelligence community and the intelligence as well as the intelligence-related activities of the United States Government. Many organizations, programs, and activities in the act are responsible for addressing cybersecurity-related issues. This act is also responsible for addressing many types of risks to the nation and threats that are a result of manmade and naturally occurring events. The act has classifications based on how the federal government distinguishes, evaluates, routs, reacts to, and recoups from current and developing threats. Several updates with respect to cybersecurity-related issues may be incorporated to the act. However, updates may have an effect on numerous organizations and activities.

5.2.29 CYBERSECURITY ENHANCEMENT ACT OF 2014

This act allows the Director of the NIST to facilitate the development of a set of cybersecurity standards and best practices for CI. The act mandates that the Director of NIST facilitate intimately with the private sector in creating these guidelines in the most ideal manner. In any case, the federal, state, and local governments are prohibited from utilizing data shared by a private entity to develop such standards to control that entity. The Cybersecurity Enhancement Act, commands the Director of NIST to fill in as a coordinator for the federal government's association in the advancement of global cybersecurity standards. The Director is also responsible for consulting with federal agencies and private sector stakeholders, and also for developing a strategy for increased use of cloud computing technology by the government. This will include support for private sector efforts for enhancing standardization and interoperability of cloud computing services. The government organizations and offices, working through the National Science and Technology Council and the NITRD Program, must build up a bureaucratic cybersecurity R&D strategic plan that needs to be updated once in every 4 years. The strategic plan must be developed keeping in mind the industry and academic stakeholders. This is done to guarantee that the arrangement is not duplicative of private-sector R&D endeavors. The act is also in charge of creating scholarships and grants for federal cybersecurity workers, as well as a cybersecurity education and awareness programs.

5.2.30 NATIONAL CYBERSECURITY PROTECTION ADVANCEMENT ACT (NCPAA) OF 2015

The act is responsible for providing liability protections to companies that share cyberthreat information with the DHS's National Cybersecurity and Communications Integration Center (NCCIC). The protecting cyber networks act (PCNA) is closely related to this act and is designed to encourage companies to share information related to cyber threat indicators along with defensive measures together and also with NCCIC. This might be done by giving liability protections and risk insurances to organizations that take part in such sharing for cybersecurity purposes, just as for organizations that neglect to go about because of such sharing. The bill gives comparable protections to organizations that participate in scanning, identifying, acquiring, monitoring, logging, or analyzing information that is stored on, processed by, or is being transmitted across systems. NCPAA permits NCCIC to share data in regards to cybersecurity risks with privately owned businesses, notwithstanding other non-federal entities. The act additionally joins a few arrangements intended to constrain the privacy impact of data sharing, remembering a denial for government utilization of shared data to take part in surveillance for tracking individual's PII. The demonstration guarantees responsibility by commanding that the DHS set up and every year audit privacy and civil liberties policies and procedures that affect the receipt, maintenance, use, and exposure of data imparted to NCCIC compliant with the bill.

5.2.31 FTC: FEDERAL TRADE COMMISSION ACT SECTION 5

FTC Act Section 5 is both a data security guideline (which requires proper cybersecurity measures) and a privacy law. The law might be applied to all foundations in the US aside from banks and other carriers. Under this act, uncalled for or beguiling acts or practices that influence trade are named lawful offenses. Consider a circumstance wherein an organization tells its consumers that their own data will be protected. FTC is responsible for taking law enforcement actions to make sure that the company secures the data. If an organization violates consumers' privacy rights, or misleads consumers by failing to maintain security for sensitive consumer information, or causes substantial consumer injury, it may be charged by the FTC. These practices are deemed unfair and deceptive that have potential to affect commerce.

5.2.32　DATA SECURITY AND BREACH NOTIFICATION ACT OF 2015

This act requires certain commercial entities that are taken into account by the FTC, common carriers subject to the Communications Act of 1934, and nonprofit organizations that are capable of using, accessing, transmitting, storing, discarding, or gathering decoded nonpublic individual data for actualizing safety efforts to secure electronic data against unapproved access and acquisition. It is compulsory that these reestablish the integrity, security, and privacy of their information systems following the disclosure of a security breach. It is critical to decide if there is a risk that a breach will bring about identity theft, monetary misfortune or harm or money related fraud individuals' personal information. In the situation of a data breach, notice of a breach must be sent to influence U.S. residents, the FTC and the U.S. Secret Service or the Federal Bureau of Investigation (FBI) if an unapproved individual gets to and obtains the individual data of in excess of 10,000 people, and consumer reporting agencies if notice must be given to in excess of 10,000 people. These notifications should be given if a breached entity processes individual information on behalf of a non-breached entity or when a supplier of electronic information transmission, storage, or network connection services gets mindful of a break. The act issues explicit punishments that the FTC and states may force to uphold against infringement of the act.

5.2.33　DEFENSE FEDERAL ACQUISITION REGULATION (DFAR)

Defense federal acquisition regulation (DFAR) is a cybersecurity regulation that primarily takes into account US DoD contractors. It may also be stated as a DoD-specific supplement to the FAR (federal acquisition regulation). It is responsible for providing acquisition regulations which may be specific to the DoD. It is mandatory for the DoD government acquisition officials, contractors, and subcontractors to cling to the guidelines in the DFARS. The DFARS fuses prerequisites of law DoD-wide approaches, assignments of FAR specialists, deviations from FAR necessities, and strategies and systems that significantly affect general society. All contractors and subcontractors processing, storing, or transmitting Controlled Unclassified Information need to satisfy minimum security guidelines determined in the DFARS, failing which may prompt loss of agreements with the DoD.

5.2.34 FDA: REGULATIONS FOR THE USE OF ELECTRONIC RECORDS IN CLINICAL INVESTIGATIONS

The Food and Drug Administration (FDA) Regulations for the Use of Electronic Records in Clinical Investigations is a cybersecurity law. The act holds true for organizations pertaining to clinical investigations of medical products, including sponsors, clinical investigators, institutional review boards (IRBs), and contract research organizations (CROs). A large portion of these individuals and associations are likewise social insurance suppliers, so their tasks would in all probability fall under the health insurance portability and accountability act (HIPAA) rules as well. The act likewise considers electronic systems inside the associations that might be utilized to make, alter, maintain, archive, recover, or communicate records utilized in clinical examinations. The guidelines command that systems guarantee exactness, reliability, and consistent performance, access to authorized individuals is limited, and that trails are audited. It is important to adhere to written policies that hold individuals accountable and also promotes cybersecurity training.

5.2.35 CFTC: COMMODITY FUTURES TRADING COMMISSION DERIVATIVES CLEARING ORGANIZATIONS REGULATION

This act applies to derivatives clearing associations. These associations give a stage to clearing transactions in commodities for future delivery or commodity option transactions. To protect themselves, derivatives clearing associations must build up a broad and strong data security program that incorporates annual compliance report that must be sent to the board and CFTC, vulnerability testing of independent contractors twice every quarter Internal and external penetration testing at least annually, control testing once every 3-years, yearly security incident response (IR) plan testing and annual enterprise technology risk assessment (ETRA).

5.2.36 FPA: PRIVACY ACT OF 1974

The FPA applies just to organizations of the US Federal Government. The reason for FPA is to guarantee evaluating, assortment, maintenance, use, and dissemination of PII. This data is generally kept up in systems of records by government organizations. Under this act, if there is any sort of revelation of data from an arrangement of records constrained by the government organization without the consent of the subject, the individual is denied except if the divulgence is allowed under one of 12 legal special cases. All US government organizations must not unveil any record that is contained in an arrangement of records using any and all means of correspondence to any individual, or to another office, without a composed solicitation from, or the earlier written consent of, the person to whom the record relates. The federal agencies must permit people to access their records or to any data relating to them that is contained in the system, and grant them and, if they demand, an individual based on their very own preference to go with them, to audit the record and have a duplicate made. They should likewise keep up any record concerning any individual, putting forth sensible attempts to guarantee such records are exact, significant, convenient, and complete, and guarantee reasonableness in any determination pertaining to capabilities, character, rights, or opportunities of, or advantages to, the person.

5.2.37 CYBERSECURITY INFORMATION SHARING ACT OF 2015

The Cybersecurity Information Sharing Act (CISA) of 2015 was signed on December 18, 2015. The two main components of the law are authorizing institutions to monitor and execute cautious measures on their own information systems to counter cyber threats, and to give certain assurances to urging establishments to deliberately share data about cyber threat indicators and protective measures with the state, local, and federal governments, along with private entities and companies. These protections take into account non-waiver of privilege, protection from liability, and protections from freedom of information act (FOIA) disclosure, although, importantly, some of these protections apply only when sharing with certain entities. The following points reflect a non-exhaustive summary of some key provisions made by CISA:

1. **Monitoring and Defending Information Systems (Protection from Liability for Monitoring):** CISA states that no reason for activity will lie or be kept up in any court against any private entity. Its further states that such actions may be promptly dismissed if monitoring is conducted in accordance with CISA.
2. **Sharing or Receiving Cyber Threat Indicators or Defensive Measures:** Cyber threat indicators might be characterized as data that is important to portray or recognize an assortment of listed threats. They may likewise be techniques for abusing a security weakness or making a legitimate client accidentally empower such misuse. The data on the real or potential harm brought about by an incident, may incorporate a portrayal of the data exfiltrated. This might be related with explicit cybersecurity threats.

    Defensive measures applied to an information system (or to information on that system) assist in detecting, preventing, and mitigating known or suspected cybersecurity threats or security vulnerabilities. Defensive measures are capable of destroying, rendering unusable, providing unauthorized access to, or substantially harming third-party information systems.

    Cybersecurity purpose refers to the purpose of protecting an information system or information from cybersecurity threats or security vulnerabilities.
3. **Scrubbing Personal Information before Sharing:** An organization meaning to share a cyber threat indicator must expel or execute specialized capabilities arranged to evacuate any data that is not legitimately identified with a cybersecurity threat. The organization must not know about the data at the hour of sharing to be personal information of a particular individual or data that recognizes a particular person.
4. **Protections for Sharing and Receiving Information:** In order to protect information, information sharing must be done with respect to CISA's requirements. This is comprehensive of the necessity with respect to evacuation of individual data.

    Security from liability. CISA states that no reason for activity will lie or be kept up in any court against any private entity. Further, any such activity will be speedily excused for the sharing or receipt of a cyber threat indicator or defensive measure led as per this title.

Federal sharing must use the DHS process to obtain liability protection. This liability protection holds true when the information is shared by the DHS process as detailed by the CISA. In any case, there is a special case that covers interchanges by a managed non-Federal entity with such an entity's Federal administrative authority in regards to a cybersecurity threat.

5. **Antitrust Exemption:** CISA states that it is anything but a federal or state antitrust infringement for organizations to share cyber threat indicators or defensive measures for forestalling, investigating, or mitigating threats.
6. **Non-Waiver of Privilege:** Sharing information with the federal government does not postpone benefits and other legitimate securities. This also includes trade secret protection. There is no similar arrangement for offering with state and local governments or different organizations.
7. **Proprietary Information:** The defensive measures and cyber threat indicators shared will be considered as the sharing entity's money-related business, and restrictive data by the federal government.
8. **Exemption from Federal and State FOIA Law:** Data shared under CISA is absolved from divulgence under the FOIA (5 U.S.C. 552), just as under any State or local provisions requiring divulgence of data or records.
9. **Data Cannot Be Used to Regulate or Take Enforcement Actions Against Lawful Activities:** Information appropriated under CISA will not be utilized by any local, tribal, State, and Federal, government for guidelines. This incorporates enforcement actions, legitimate exercises of any non-Federal element or any exercises taken by a non-Federal entity as per compulsory standards. These may be activities related to sharing cyber threat indicators, monitoring, and operating defensive measures. The information may be additionally used to develop or implement new cybersecurity regulations.

There has been an ongoing debate on whether CISA should lead it to expand its intentional data sharing over its gauge levels:

- In order to determine what information is capable of being shared, companies need to investigate whether a cyber threat indicator or defensive measure embroils sensitive business data and exercise specific consideration in assessing the expenses and advantages of sharing the data. Since CISA forces no necessity to cyber information, consequently regardless of whether an organization decides to share data, it is allowed to recognize various sorts of data. Disclosing information about an institution's specific cyber vulnerabilities and incidents may lead to legal, competitive, and reputational risks, which may be greater if that information is known to the competitors and customers. If the information is very sensitive, companies may want to focus on the ability to share on an anonymous basis. Public companies may need to perform a risk assessment and survey how their choice to share data under CISA may cooperate with their disclosure decisions under the securities laws, furnished that sharing cyber information with the government might be viewed as an indicator of materiality requiring divulgence in a public filing.

- In order to determine whether information must be shared under CISA, organizations must make a sensible evaluation of the qualities and constraints of the insurances offered by the law. The liability protection is not pertinent in circumstances where there is imparting to government organizations other than DHS, aside from when an organization communicates with its federal regulatory authority in regards to a cybersecurity threat. And keeping in mind that there is a non-waiver of privilege protection that applies to data sharing to the federal government, there is no such arrangement that applies to sharing with state or local governments or different organizations. CISA's securities may likewise have issues that request further investigation. One such issue might be investigating the importance of the insurances against federal and state administrative and authorization activity.
- An organization that decides to share information under CISA must establish or adapt certain techniques and systems to gather, screen, and report the sorts of data. The organization must consider this data proper for sharing, remembering that the assurances apply just when sharing is led by the CISA's necessities. For example, sharing information that conforms to the description of cyber threat indicators and defensive measures. Another example is complying with the requirements for removal of personal information. An organization's legal, IT, and compliance functions must be responsible for coordinating in order to ensure that information shared complies with CISA. It is also necessary to consider that information sharing must manage the cost of maximum protection, and that consistency with CISA is all around archived.

5.2.38 CALIFORNIA'S SB-327 BILL FOR IOT SECURITY

California's SB-327 Bill for IoT Security is a bill administering the law for internet of things (IoT) and is the first-of-its-kind. Likewise, signing this SB 327 bill makes California the first state in the USA to receive such enactment. Under this law, any and all makers of Internet-connected or smart devices must ensure that the gadgets involved have reasonable security features which secure the devices and any data contained in that from a wide range of unauthorized access, destruction, use, adjustment, or divulgence. The regulation holds true as of January 01, 2020. While the law has been passed keeping in mind the security threats that IoT devices are prone to, this law does not clearly state what constitutes a reasonable security feature. However, it does specify how any such feature must be:

- Appropriate to the function and nature of the device;
- Appropriate with respect to the information the device may contain, transmit, and collect;
- Designed for protection of the device and any information contained with respect to unauthorized access, destruction, use, modification, or disclosure.

In many cases, the passwords assigned by the manufacturers are unique to devices and may satisfy the reasonable security requirements. Manufacturers providing devices with shared default passwords is a common practice. This implies that after installation, if the end-user does not change the passwords, the device may be easily accessible by anyone. Many U.S. federal agencies present detailed specifications on how the security of such devices can be managed. This can enhance the transparency for consumers. Others like the state governments are yet to engage in IoT's cybersecurity.

5.3 SOME IMPORTANT CYBER LAWS: HISTORY AND THEIR CASE STUDIES

In the previous section, we acquainted ourselves with the current legislative framework of the United States and some laws along with their possible updates and loopholes with respect to cybersecurity. While most of the laws discussed are general cybersecurity laws, there are some laws that have specific case studies associated with them. The concept of introducing case studies to the laws is indicative of the familiar situations that people might go through every day while interacting in cyberspace. It makes people realize the legal and illegal things that go on in cyberspace from a practical point of view. In this section, we will familiarize ourselves with some very popular cybersecurity laws, along with their history as well as some interesting case studies related to them.

5.3.1 COMPUTER FRAUD AND ABUSE ACT (CFAA)

The Computer Fraud and Abuse Act (CFAA) is a United States cybersecurity law that was sanctioned in 1986. The law forbids access of a computer without authorization, or in excess of authorization. It is also the federal anti-hacking statute that prohibits unauthorized access to computers and networks. The CFAA has its own definition for 'protected computers:'

- These computers are reserved for financial institutions or the US Government, or on any system, when the lead comprising the offense influences the system's utilization by or for the money-related association or the government.
- They may even be used in or influencing interstate or international business or correspondence, including a system situated outside the US that is used in a manner that influences interstate or international trade or correspondence of us.

Because of the interstate nature of many Internet communication, any ordinary computer (or device) can come under the jurisdiction of the law, including cell phones. According to this law, anyone who knowingly accessed a computer without authorization or, on the other hand, surpassed authorized access, and by methods for such direct acquired data that has been controlled by the United States Government as confidential, may be convicted of fraud. It is against the law to willfully communicate, deliver, transmit, or cause information regarded as confidential by the United States Government to be communicated, delivered, or transmitted. One must not even attempt to communicate, deliver or transmit such information. This law also prohibits intentional access to a computer without authorization or intentionally exceeding the authorized access. One must not acquire data contained in a monetary record of a budgetary foundation, or of a card guarantor, or data contained in a document of a consumer reporting agency on a consumer. It is illegal to obtain data from any division or office of the United States, or data from any secured computer. Breaking into a computer deliberately, without approval to get to any nonpublic computer of a division or organization of the US, accessing such a computer of that department or agency that is exclusively for the utilization of the government of the United States or, within the instance of a system not only for such use is furthermore viewed as misrepresentation or fraud. Also, if the system is employed by or for the government of the US and such conduct affects that use by or for the government, accessing that system is taken into account as a felony under this law. Further purposely and with intent to cheat, get to a secured system without approval, or surpassing authorized access, and by methods for such conduct assisting the proposed fraud and acquiring anything useful, unless the thing of the fraud and therefore the thing obtained consists only of the utilization of the system could also be considered fraud. A convict may purposely cause the transmission of a program, data, code, or order, and because of such lead, deliberately cause harm without authorization, to a secured computer. He/she may purposefully get to a secured system without authorization, and because of such lead, recklessly cause harm. It is possible to intentionally access a protected computer without authorization, and the result of such conduct, causes damage and loss. Likewise, purposely, and with intention to cheat deals with any password or comparable data through which a system might be accessed without authorization, in circumstances where such dealing influences interstate or international trade, or such computer is utilized by or for the Government of the United States, with the aim of extricating money or valuable thing from a person, transmitting in interstate or foreign commerce any communication may be charged with penalties. These activities may harm a secured system, acquire data from it without authorization or in abundance of authorization, or to influence the integrity and confidentiality of data acquired from a secured system without authorization or by exceeding authorized access. This regulation also holds true for demanding or requesting for money or other thing of value in relation to damage to a protected computer, where such damage was caused to facilitate the extortion. This is one of the cyber laws that has been revised over and over. Certain cases demanded that the law be updated. Three very famous cases pertaining to this law are mentioned as follows:

➢ Case Study 1: The United States vs. Robert Morris:

Robert Morris was the first person to be indicted under the 1986 Computer Fraud and Abuse Act (CFAA) for distributing a worm that harmed and undermined secured systems. He was a Harvard and Cornell graduate who built up the principal broadly spread Internet 'worm' or the Morris Worm named after him [50]. A worm could also be a self-repeating malware that conceals itself on system hard drives and spreads to different systems on its network without any human intervention. The worm was released on November 02, 1988, using MIT's (Massachusetts Institute of Technology) systems to disguise the actual fact that he was a Cornell student. The worm was not meant to be offensive when Morris wrote it. However, it did not turn out the way Morris had expected it to. Morris trusted that just one duplicate of the worm would taint every system, but in an attempt to bypass computers it already had a reproduction; he had modified the worm to copy itself each 7th time it got a yes' reaction. The Morris worm began replicating at a much greater rate than Morris had expected. This led to flooding of hard drives and caused extensive damage. Although a friend of Morris tried to send a warning to other users, many systems had already been compromised. Within a couple of days, the Morris worm traveled across Arpanet, the precursor to today's Internet, and infected quite around 6,000 computers at research centers, military installations, and universities.

The expense of expelling the worm from every system went from $200 to $53,000. An estimated $100,000 to $10 million was lost because of lack of access to the online. Morris was soon identified because of the origin of the worm, and authorities looked to arraign him under the 1986 CFAA, which outlawed acquiring unapproved access to government systems. It brought prosecutors 8 months handy down a prosecution in light of the fact that there was an inside discussion over whether it might be possible to prove the fees. This is because prosecutors had to prove that Morris intended to cripple the systems' network. On July 26, 1989, Morris was indicted for spreading the Internet's first worm virus, affecting every 6,000 universities, research facilities and military systems. Morris was found guilty in 1990.

Despite his claims that he did not want to vandalize other system networks or acknowledge how rapidly the worm would spread, Morris was fined and condemned to community service. He was given a light-weight sentence is sort of a $10,050 fine, 400 hours of community service, and a 3-year probation. Following his sentence, it had been observed that public opinion on Morris' case was split. While numerous people considered him to be a criminal who had malignantly harmed property, acquired tips, and led to national frenzy among computer users, others figured he helped computer users out by uncovering security blemishes. Within the 20 years since the discharge of the Morris worm, computer security has formed an industry of its own. The Morris worm served as a wake-up call to the online engineering community about the risk of programming bugs, and it set up a platform for network security to become a legitimate area of R&D.

➤ Case Study 2: The United States vs. Lori Drew:

The case United States v. Lori Drew encompasses multiple events on the social networking site MySpace that led to the suicide of a 13-year-old girl named Megan Meier [51]. Meghan was a secondary school student living together with her family in Missouri and had been under the care of a psychiatrist for a period of 5 years. She had been prescribed certain drugs following her diagnosis of Attention Deficit Disorder and depression. Megan's parents enrolled her during a private Catholic school for her 8th grade year with strict code rules to assist alleviate her anxiety surrounding her weight and private appearance. The Meier family lived inside a similar neighborhood of the litigant, Lori Drew, and Megan were colleagues with Lori Drew's daughter. This initiated the creation of the fictitious MySpace profile of 16-year-old 'Josh Evans.' Lori Drew admitted that she created the account together with her daughter and co-worker Ashley Grills following the rumor that Megan Meier had spread gossip regarding Drew's daughter. When Megan Meier created her own MySpace account of her own, she received a message and a 'friend invitation' from Lori Drew who was pretending to be a 16-year-old boy. A web correspondence began between Megan and the boy, who she believed to be Josh Evans, with regular exchange of messages over MySpace.

October 15, 2006, witnessed a change in the friendly rapport of the messages from the Evans account and the following message was sent to Megan Meier: "I don't know if I would like to be friends with you anymore because I've heard that you simply aren't very nice to your friends." After this message was sent, numerous open announcements were posted on MySpace through the Evans account. Megan Meier discussed with her mother that she was called fat and a whore over the MySpace platform. A final message sent on to Meier from the Evans account detailed "the world would be a far better place without you." Two days following the primary demeaning message sent to Meier from the Evans account, on October 17, 2006, Megan Meier hanged herself in her bedroom (Pokin). Although Megan's parents knew about her online involvement, it was only after six weeks following their daughter's suicide that Meier's parents discovered that Josh Evans did not exist. A neighbor who was aware of Evans' account from her daughter notified the Meiers of the fake account and how the Drew family had a role in it. Although the Josh Evans account was operated by several people, Lori Drew admitted to participating in the creation of the account with the goal of discovering data in regards to what Megan Meier was stating about her daughter. In spite of the fact that the messages sent to Megan Meier were from numerous individuals, including Lori Drew and her daughter, Ashley Grills and companions of Lori Drew's daughter, declaration disclosed that Lori Drew was conscious of everything of the account's activity. Lori Drew was charged with the CFAA, as she had knowingly conspired to access a computer utilized in interstate and foreign commerce to obtain information from that computer to further a sinuous act. She was also booked under intentional infliction of emotional distress.

The conspiracy was administered infringing upon MySpace Terms of Service understanding, where a flimsy account was made to get information from a juvenile account holder. This information was obtained to harass, torment, embarrass, and humiliate the juvenile member. After the girl committed suicide, Lori Drew and co-plotters erased the records by deleting the account so as to stop the invention of the group's actions. Upon fabricating the false MySpace account of Josh Evans,' Drew and co-plotters fictitiously expressed their way of life as a 16-year-old boy. This was done to make the account appear realistic. Further, Drew, and co-schemers utilized a photo of a boy as a profile picture without his assent or awareness. The profile was utilized to contact a juvenile MySpace account holder to encourage a coquettish relationship and the message "the world would be a far better place without you" was deleted upon the suicide of Megan Meier.

Lori Drew was prosecuted on four points by the jury of the us District Court for the Central District of California. Drew was seen as blameworthy on the essential count of conspiracy however was found consistently vindicated on the last three checks of disregarding MySpace's Terms of Service understanding. In November 2008 Drew petitioned for exoneration of her conviction. For conviction to be upheld under this section, it is required that the system accessed is involved in interstate or foreign communication, and therefore, the defendant must have intentionally accessed the system without authorization. With these two conditions met, data must have then been acquired through the unauthorized computer access. Any computer connected to the web qualifies as a computer used in interstate or international business, and perusing content on any site can qualify as social affair data with unauthorized access. The idea of Drew's conviction was her unauthorized access through violating MySpace's Terms of Service agreement by interacting on MySpace with a false identity. It had been taken under consideration that the Terms of Service agreement couldn't fulfill unauthorized access,' which is the reason Drew's conviction could not be maintained. The judge was notified that there was a scarcity of actual notice to the general public with respect to the CFAA and disregarding a Terms of Service Agreement. In the event that a Terms of Service understanding can direct what is thought of authorized this is able to make the statute vague because it does not explicitly list what parts of the Terms of Service are disregarding surpassing authorized use. The Judge conceded Lori Drew movement for absolution since it had been on the idea of a particularly broad definition of how a user can exceed unauthorized use of a computing system.

> **Case Study 3: The United States vs. Aaron Swartz:**

Aaron Swartz used to be an American computer programmer, author, political coordinator and Internet activist. He was once indicted for some infringement of the CFAA of 1986 [52], in the wake of downloading countless scholarly articles utilizing the MIT computer network from an archive (JSTOR) for which he had an account as a Harvard researcher. JSTOR is a digital repository that archives and disseminates online manuscripts, Geographic Information Systems, scanned plant specimens and content material from tutorial journal articles. Swartz was a researcher at Harvard University, and was therefore provided a JSTOR account. Visitors to MIT's open campus have been authorized to access JSTOR through its network.

According to authorities, Swartz downloaded a large range of educational journal articles from JSTOR using MIT's computer network, over the course of a few weeks in late 2010 and early 2011. They stated Swartz downloaded the documents to a laptop linked to a networking swap in a controlled-access wiring closet. Press reports asserted that the door to the closet was unlocked.

On January 6, 2011, Swartz was captured close to the Harvard grounds by two MIT cops and a U.S. Secret Service agent on the charges of breaking and entering with a purpose to commit a lawful offense. July 11, 2011 witnessed Swartz being indicted in Federal District Court on four felony counts: fraud and misrepresentation, wire fraud, unlawfully getting data from a secured system and wildly harming a secured system. On November 17, 2011, Swartz was prosecuted through a Middlesex County Superior Court jury on state charges of breaking and entering deliberately, theft, and

> unauthorized access to a computer network. Prosecutors charged Swartz with 4 felony counts, anyway later quickened this to 13 counts through outlining each date he downloaded files and transforming them into independent counts, accordingly expanding the greatest sentence he looked to 50 years and his achievable fines to $1 million. Swartz committed suicide, facing trial and the possibility of imprisonment, as a result of which the case was dismissed.
>
> It was neither physical nor economical. The leak was once found and plugged. JSTOR suffered no real monetary loss. It did no longer press charges. Like a pie in the face, Swartz's act was once disturbing to its victim, however, of no lasting consequence. MIT and JSTOR have sought to strike a balance on cybersecurity, imparting some human beings highly free use of their offerings whilst restricting get admission to others. Network owners, copyright holders, and publicly accessible databases should really have some recourse towards human beings who ruin their systems. But the regulation ought to also be written to keep away from miscarriages of justice, like the one that took place in the Swartz case.

Based on the case studies above, we can infer that the CFAA has several loopholes. The law can be reformed by taking into account certain points:

- First, the loss felony threshold is too high and might require some revisions. One of the biggest loopholes of the law is that a hacking misdemeanor turns into a felony in the event that the estimation of data acquired, or the money-related misfortune that the intrusion causes, surpasses $5,000. Practically, it is not a very big amount and can be easily met if the victim is a large corporation. This number could be increased to circumscribe the far too expansive criminal scope of the CFAA.
- Second, as there is a $5,000 threshold to bring a civil suit under the CFAA, this number must be raised as well. That is on the grounds that, on the off chance that we need to live with a dual civil statute as drafted by CPAA, a large portion of what is currently arraigned criminally ought to be civil.
- Third, the definition of damage should be confined to just incorporate real harm to information, or its technique for access, on a system. Innocuous occasions of a worker erasing their emails or the erasure of back up information for which there are various promptly accessible duplicates, ought to be dispensed with to the extent of

CFAA harm. Otherwise, CFAA criminal obligation can be applied for basic undertakings, for example, altering a Word record without consent, or turning off a system without authorization. While both the acts are innocuous, they can be perused as compromising CFAA damage under a broad reading of the statute.
- Finally, unauthorized access ought to be carefully kept to just incorporate occasions where a technical, code-based obstruction to access is circumvented. Unauthorized access must not be based on organization, authoritative, or any sort of connection between entities, people, or entities and people. This methodology would guarantee that courts can stay away from debates about whether conduct was approved from getting buried in the examination of connections created in the physical world.

5.3.2 GRAMM LEACH BLILEY ACT

Likewise alluded to as the Financial Modernization Act of 1999, the Gramm-Leach-Bliley Act (GLB Act or GLBA) is a United States government law that demands money-related foundations to explicate how they share and protect their customers' private information. It is mandatory that financial institutions communicate to their customers regarding how the customers' sensitive data is being shared and educate clients regarding their entitlement to quit on the off chance that they lean toward that their own information not be imparted to outsiders. The institutes may also need to apply specific protections to customers' private data along with a composed data security plan made by the foundation. Financial institutions who comply with the GLBA are at lower risk of penalties or reputational damage, which may be easily caused by unauthorized sharing or loss of private customer data. Further, there are a few protection and security benefits required by the GLBA safeguards rule for clients. Some of the rules are as follows:

- It is necessary to secure private information against unauthorized access;
- It is necessary to notify the customers regarding private information sharing between financial institutions and third parties. Customers must also be given the choice of opting out of private information sharing;

- It is necessary to track user activity must, which may also include any attempts to access protected records.

Since consistency with the GLBA secures consumer and customer records, it is instrumental in strengthening consumer reliability and trust. This guarantees clients gain affirmation in regards to the security of their own data. Safety and security of information can also lead to customer loyalty, which may enhance the reputation of the institution, repeat business, and other benefits for financial institutions. The GLBA makes it essential for financial institutions to ensure confidentiality and security of customers' nonpublic personal information (NPI), which may be in the form of names, addresses, Social security numbers, bank card account numbers, phone numbers, credit, and income histories, and any other personal information that financial institutions have that is not public. The Safeguards Rule asserts that financial institutions need to create a written information security plan which describes the program for protecting their customers' information. This arrangement must be custom-made explicitly to the establishment's tasks, sensitivity of information, size, and complexity. The Safeguards Rule is a set of statutes that must be followed by financial institutions. They include:

- Designating one or more employees to coordinate its information security program.
- Distinguishing and surveying the risks to client data in each applicable territory of the organization's activity, and assessing the adequacy of the current protections for controlling these risks.
- Planning and executing a protections program, and consistently checking and testing it.
- Selecting service providers for maintaining appropriate safeguards, making sure a contract requires them to look after safeguards, and directing their treatment of client data.
- Adjusting and Evaluating the program based on given circumstances, including changes in the firm's business or operations, or the results of security testing and monitoring.

For achieving GLBA compliance, the Safeguards Rule insists on money related foundations giving exceptional consideration to employee management and preparing, information systems, and security management in their data security plans and implementation. The GLBA's protection assurances just manage financial organizations, particularly organizations that are occupied with banking, insuring, stocks, and bonds, investing, and budgetary counsel:

- Financial institutions must develop precautions for ensuring the security and confidentiality of customer records and information. This must be done to ensure against any foreseen threats to the security or trustworthiness of such records, and to ensure against unauthorized access to or use such records or information which may potentially lead to substantial harm or inconvenience to any customer.
- It is essential that money-related organizations give individuals a notification of their information-sharing policies when they become a customer, and annually thereafter. This notice must incorporate information regarding policies on: uncovering NPI to affiliated and non-affiliated third parties, unveiling NPI after the client relationship is ended, and securing NPI.
- GLBA ensures that consumers have the right to opt-out from a limited amount of NPI sharing. Specifically, it may be possible for a consumer to address the financial institution not to share information with specific companies. Consumers have no privilege under the GLBA to stop sharing of NPI among affiliates. An associate might be characterized as an organization that controls, is constrained by, or is under basic control with another organization. Several exemptions under the GLBA permit information sharing over the consumer's objection. Moreover, financial institutions may disclose information to credit detailing organizations, finance-related corporations as a part of the sale of a trade, to abide by any other guidelines or as fundamental for a transaction demanded by the consumer.
- Financial institutions are proscribed from disclosing, other than to a consumer reporting agency, account numbers or access code to nonaffiliated third parties for using the information in direct mail marketing, telemarketing, or other marketing through means of electronic mail. This means that, even if a consumer fails to opt-out of a financial institutions' transfers, credit card numbers, pins, or other access codes cannot be sold, as they had been in some previous cases.
- Many other types of pretexting are prohibited by the GLBA. Pretexting might be characterized as the act of gathering individual data under misrepresentations. Pretexters act like authority figures for eliciting personal information about the victim. The GLBA forbids

the use of fraudulent, false, and fictitious proclamations or records to get client data from a budgetary foundation or legitimately from a client of a money-related establishment. It additionally disallows the utilization of forged, fake, lost or taken documents to get client data from a budgetary organization or legitimately from a client of a money-related foundation. Under the GLBA, asking a person in order to obtain customer information by means of flimsy, fraudulent, or fake reports or fake, lost or stolen records might be viewed as a lawful offense.

Although the GLBA tries to modernize the financial services, there are several issues associated with this law:

- To begin with, the GLBA does not secure consumers by unreasonably putting the weight on the person to ensure protection with an opt-out standard. Placing the burden on the customers to protect their data may weaken customer power to control their financial information. The agreements opt-out provisions do not infer that foundations will give a means of protection to customers irrespective of whether they opt-out of the agreement. This arrangement assumes that money-related organizations will share data except if explicitly advised not to by their clients. Further, if clients neglect to react, it gives establishments that opportunity to release client NPI.
- Second, the GLBA notices may seem confusing and may also limit the transparency of information practices. Numerous protection and opt-out policies are generally convoluted, confounded, and deluding as quite possibly they may have been created by entities whose interests are better off when there is no effective notice. GLBA does not deal with the lack of transparency completely, especially in the privacy notices. Many privacy notices are devoid of any specific information about how the data is actually used. GLDA notices notify consumers that their personal information will be shared, these do not inform consumers about who will get the data or the reasons for which it will be utilized.
- Third, the GLBA does not amplify consumers' control over affiliate information sharing. This is because consumers do not have the right to opt-out against affiliate information sharing.
- Fourth, it is very much possible for financial institutions to avoid opt-out requirements by misusing the special cases in the GLBA.

This is on the grounds that the service provider exemption permits money-related organizations to impart data to non-affiliated third parties regardless of consumer's opt-out.
- Fifth, the enforcement and compensation mechanisms for GLBA seem to be very weak. GLBA's enforcement mechanisms may not be sufficient for assuring compliance following even weaker privacy protections. Since authorization rests exclusively with government organizations, the individual may not have a private right of action.

> **Case Study 4: GLBA PayPal-Venmo Case:**

February 27, 2018, saw the FTC detailing an agreement with PayPal, Inc., to settle charges that its Venmo distributed installment administration deluded buyers in regards to security and the degree to which consumers' monetary accounts had been secured [53]. This turned out to be principally founded on the infringement of the GLBA's Privacy and Safeguards Rules. The complaint claimed that Venmo violated the Privacy Rule in 3 separate ways. First, Venmo did not offer a clean and conspicuous privacy notice that did not call interest to the nature and significance of the nature of the notice. Moreover, the privacy notice in Venmo's mobile application (the "Venmo App") was mentioned in gray text on a mild gray background that was no longer conspicuous to Venmo customers. Venmo failed to provide an accurate notice describing how it shares the user's personal information. Venmo's privacy notice mentioned that its shared users' personal data with members of their Venmo social web in the event that they distinct their account transactions as public. Instead, Venmo shared this data by default with everyone online, including those who did not even have a Venmo account. Finally, Venmo did not provide a privacy notice in a way that every consumer expects to receive one. A hyperlink in the Venmo App incorporated the privacy notice, and the users were not required to acknowledge its receipt as a necessary step for obtaining a financial product or service.

The FTC complaint additionally affirmed that Venmo distorted its data security practices by expressing that it uses bank-grade protection systems and data encryption to defend the transaction records. FTC claimed that Venmo violated the Safeguards Rule by failing to (1) to have a written information security program; (2) investigate the risks to the protection, confidentiality, and integrity of customer records; also, (3) actualize essential safeguards, for example, introducing security notices to clients that their passwords had been changed. In the settlement, Venmo is prohibited from distorting the degree of assurance provided by its privacy settings and the degree to which Venmo executes or adheres to a specific degree of security. Venmo is additionally prohibited from disregarding the Privacy Rule and the Safeguards Rule and is required to attain biennial third-party checks of its compliance with these policies for 10 years. It was taken into account that consumers suffered real harm from Venmo's misrepresentations and the case sent a solid message that money related establishments like Venmo need to concentrate on protection and security from the very beginning.

> **Case Study 5: GLBA TaxSlayer, LLC Case:**

In 2017, The FTC announced a consent order with TaxSlayer, LLC, a web tax guidance services provider, to settle claims that the business enterprise violated the GLBA Safeguards Rule and Privacy Rule [54]. As part of the online tax education process, TaxSlayer customers are asked to provide sufficient sensitive personal information, along with Social Security number, phone number, address, income, marital status, family size, financial institution names, and financial institution accounts. Between October and December 2015, hackers were capable of accessing account information for approximately 8,800 TaxSlayer customers. This led to a wide variety of counterfeit tax returns being filed. The FTC alleged that TaxSlayer violated the GLBA Safeguards Rule by means of failing to develop a written comprehensive protection program (till November 2015), conduct a risk assessment to identify reasonably foreseeable inner and external risks to security, and enforce information security safeguards that might assist in preventing cyber-attacks.

The FTC further claimed that TaxSlayer did not enforce appropriate risk-based authentication measures, along with requiring consumers to select strong passwords. The FTC also alleged that TaxSlayer violated the GLBA Privacy Rule by using failing to provide its customers with a clear and conspicuous preliminary privacy notice and deliver the notice in a way that guaranteed the buyers got it. Some Gramm-Leach-Bliley guidelines to take from the FTC's TaxSlayer case are assessing whether or not an enterprise is a financial institution challenges to the GLBA, delivering GLBA privacy notices in a manner that consumers are expected to receive it, using appropriate authentication procedures, which may encompass multi-factor authentication and fulfilling continuous commitments under the GLBA Safeguards Rule by means of driving forward with to assess and modify data security programs considering changes to business venture activities, the outcomes of tracking or testing, or any other applicable factors.

Given the limitations of the GLBA, several privacy advocates and industry groups have raised genuine concerns and requested some significant changes to the GLBA to guarantee more noteworthy insurance and consumer security. Some of these changes include:

- Financial institutions need to implement an opt-in approach for using personal information. This would ensure the minimization of any unwanted or unknowing disclosure of information. It would also place the weight of obligation on those actors who will gain from the divulgence of data.
- Financial institutions must provide and accept alternative opt-out methods in case an opt-out framework is maintained. They must also provide simple opt-out processes, which may incorporate simple access to protection strategies at branch offices and online through a solitary site with opt-out information.
- Financial institutions must include what information is going to be used in their privacy reports. This may ensure greater transparency and accountability.
- Money-related establishments must give clients a legal right of access to become familiar with industry practices so as to know how the data is gathered, who its affiliates are, and what the data gathered for is utilized. Monetary establishments must give essentially expressed and clear protection approaches. Also, money-related organizations must keep acceptable standards for clarity. This might be accomplished by showing more clear and more straightforward security and privacy reports.
- Money-related foundations may extend authorization to give states simultaneous purview to implement the provisions of GLBA. This may ensure a more efficient enforcement program.
- People must be given the right to protect their privacy and look for cures and remuneration under GLBA. As GLBA at present stands, there is no private right of action. People must be given the option to survey data that is unveiled or to address incorrect or inadequate information.

5.3.3 ELECTRONICS COMMUNICATIONS AND PRIVACY ACT

The Electronic Communications and Privacy Act 1986 was enacted by the United States for extending government restrictions on wiretaps from telephone calls for including transmissions of electronic information by computer system. The electronic communications privacy act (ECPA) was a correction to Title III of the Omnibus Crime Control and Safe Streets Act of 1968 (the Wiretap Statute). This act was initially designed to prevent unauthorized government access to private electronic communications. However, ECPA added new provisions that disallow access to stored electronic correspondences, including the SCA. The ECPA additionally included arrangements that grant the tracing of telephone interchanges.

Electronic correspondence alludes to any sort of transfer of signs, signals, writings, pictures, sounds, information, or insight of any nature sent in whole or to some extent by a wire, radio, electromagnetic, photo-electronic or photo-optical framework; It may affect interstate or foreign commerce. However, it does not include any of the following:

- A wire or oral communication;
- A correspondence made through a tone-only paging device;
- Correspondence from a tracking device;
- An electronic funds transfer information stored by a monetary organization in a correspondence or communication framework utilized for the electronic storage and transfer of assets.

Title I of the ECPA is responsible for protecting wire, oral, and electronic communications while still in transit. Title II of the ECPA is the SCA. It is responsible for protecting communications held in electronic storage, specifically messages stored on computers. Not only its protections are weaker than those of Title I, but they also not impose heightened standards for warrants. Title III is responsible for prohibiting the use of trace devices for record dialing, directing, tending to, and flagging data utilized during the time spent sending wire or electronic correspondences without a court request.

There have been issues raised over several faults existing for the ECPA, which has been criticized for failing to protect all communications and consumer records. This is because the law is obsolete and outdated. The law is also out of touch with how people currently use, share, and store information. Under this law, government agencies have the right to demand service providers any personal consumer data stored on the service provider's servers. Emails that are stored on a third-party server for more than 180 days are considered abandoned under the act and can be obtained easily by anyone. Just a composed proclamation ensuring that the data is pertinent to an examination, without legal audit is required for acquiring the email content. At the point when the law was initially passed, emails were stored on a third party's server for a constrained time frame. It was uniquely to encourage the transfer of email to the consumer's email client, which was commonly situated on their own or work computer. As online email services started becoming popular, users started storing emails online for an indefinite amount of time, rather than 180 days. For accessing emails that are stored on a personal computer, the police would require a warrant for seizing the contents of the email, irrespective of their age. If the emails are stored on an internet server, there is no need of obtaining a warranty, starting 180 days after receipt of the message. Members of the U.S. Congress suggested reforming this procedure in 2013.

ECPA extended the list of crimes that can legitimize the use of surveillance, and the quantity of legal individuals who can approve such observation. It is conceivable to procure information on traffic and calling patterns of an individual or a group without a warrant. This might allow an agency to obtain valuable intelligence and possibly invade privacy without any scrutiny, since the actual content of the communication is left untouched. Although workplace communications are believed to be secured, gaining access to reports and significant notifications is very easy. In order to gain access to reports, an employer must give notice or an administrator to report that the employee's activities are not to the organization's advantage. This means that an employer can monitor communications within the company with minimal assumptions. The ongoing debate is, where to limit the power of the government to see into civilian lives. This must be done with balancing the need to curb national threats. Some famous case studies related to this law are as follows:

> **Case Study 6: The United States vs. Katz:**

> Charles Katz lived in Los Angeles and was one of the main basketball handicappers in the nation in the 1960s [55]. He earned putting down bets for interstate gamblers and keeping a portion of the rewards. However, interstate betting was viewed as unlawful under the government guideline. In this way so as to keep away from detection and jail, he utilized open pay phones along Sunset Boulevard to direct his business. The Federal Bureau

of Investigation (FBI) began investigating his gambling activities, and in 1965 secretly recorded him reporting his handicaps to bookmakers by a covert listening tool the FBI had attached to a phone booth near his condo in Los Angeles. FBI dealers arrested Katz and charged him with transmitting wagers across U.S. state lines over the telephone, which is a criminal offense under U.S. Federal playing regulation. At trial, Katz's legal professional argued that the telephone booth he used should be considered a "constitutionally protected area" according to the 4th amendment. This meant that the FBI's recordings of his conversations must be excluded as evidence since the FBI had not acquired a court order permitting them to put the listening device on the telephone booth. The judge rejected this argument, and Katz was convicted based on recordings as evidence. He subsequently appealed his conviction to the U.S. Supreme Court.

A 7-1 decision was issued in Katz's favor by the Supreme Court on 18 December, 1967. This overturned his conviction. The choice surprisingly redefined the law administering the 4th amendment's assurances, which changed into previously confined to the searches and seizures of individuals' houses, papers, etc., spoken of within the amendment, and was traditionally understood to require an actual "trespass" or other physical intrusions by law enforcement officers. The court improved the amendment's protections beyond these conventional areas, pointing out that what an individual tries to safeguard as private, even in a zone available to the general population, might be intrinsically secured. Hence, the Supreme Court dominated in Katz v. United States, increasing the 4th amendment protection against unreasonable searches and seizures to cover electronic wiretaps.

➢ **Case Study 7: The United States vs. Warshak:**

When the government suspected Berkeley Premium Nutraceuticals, Inc. (Berkeley) of defrauding its customers, NuVox Communications (NuVox), an ISP, to warehouse the emails of Steven Warshak (defendant), who also happened to be the owner of Berkeley [56]. The government then coerced NuVox to turn over those warehoused emails without obtaining a warrant. In view of the contents of the emails, the preliminary court indicted Warshak. Anyway, Warshak appealed, charging that the administration's inquiry and seizure of his messages established an infringement of his 4th amendment rights. The United States Supreme Court had to concede him certiorari.

5.3.4 SARBANES OXLEY ACT OF 2002

The Sarbanes-Oxley (SOX) Act was passed on 30 July, 2002, with the plan to shield speculators from false money-related reporting by enterprises. It is otherwise called the SOX Act of 2002 and the Corporate Responsibility Act of 2002, as it involves publicly traded companies. High-profile frauds shook investor confidence and raised questions on the trustworthiness of corporate financial statements. This led to the general public demand a renovation of the already existing decades-old regulatory standards. The SOX Act of 2002 is responsible for making severe new standards for accountants, inspectors, and corporate officials and forcing more inflexible recordkeeping necessities. The changed act likewise included new criminal punishments for abusing security laws. The rules and implementation strategies laid out in the SOX Act of 2002 reworded and enhanced existing laws managing security guidelines. This additionally incorporates the Securities Exchange Act of 1934 and different laws upheld by the Securities and Exchange Commission (SEC). The new law appended reforms and improvements in four principal areas namely Corporate responsibility, Increased criminal punishment, Accounting regulation, and New protections. The SOX Act of 2002 a lengthy and complex piece of legislation. Its key provisions are referred to by their section numbers: Section 302, Section 404, and Section 802.

As per Section 302 of the SOX Act of 2002, senior corporate officials should actually affirm recorded as a hard copy that the organization's fiscal reports conform to SEC exposure prerequisites. All material aspects with respect to operations and financial condition of the issuer must be fairly presented. Officers who sign off on financial statements despite being aware of its inaccuracy and fallaciousness may be subject to criminal penalties. Section 404 of the SOX Act of 2002 asserts that management and auditors must manifest internal controls as well as reporting methods in order to validate the adequacy of those controls. It has been mentioned several times in the past that requirements in Section 404 may have a negative impact on publicly traded companies. This is because establishing and maintaining necessary internal controls could be very expensive. Section 802 of the SOX Act of 2002 incorporates three rules that have an impact on recordkeeping. The first takes into account the destruction and falsification of records, while the second underpins the retention period for storing records. The third rule highlights particular business records that companies may require for storage, with respect to electronic communications.

The SOX Act of 2002 not only takes into account the financial side of a business, such as audits, accuracy, and controls, but also outlines requirements for IT departments pertaining to electronic records. The act does not mention the business practices in this regard but defines which company records must be kept on file and the specified duration. The rules laid out in the SOX Act of 2002 do not provide details on how a business should store its records, but they mention that it is the company IT department's responsibility to store them.

Some of the important cases which highlighted this act are as follows:

> **Case Study 8: Yates vs. The United States:**

> This well-known case includes a fisherman who did not succeed in saving, as proof, small fish that he had trapped infringing upon federal law [57]. A federal agent saw that the boat's catch obliged small red grouper disregarding bureaucratic protection guidelines while leading an offshore assessment of a commercial fishing vessel in the Gulf of Mexico. The ship's captain, petitioner Yates, was told to keep the undersized fish separate from the remainder of the catch until the ship came back to port. Nonetheless, after the official left, Yates advised a team member to toss the small fish overboard. The captain was charged with violating a provision of the SOX Act, which underpins dealing with the destruction of evidence. Yates was charged disguising, destroying, and concealing undersized fish to disturb a federal investigation. As indicated by SOX Act of 2002, a convict might be fined or detained for as long as 20 years on the off chance that he/she purposely covers, misrepresents, conceals, alters, disfigures, decimates or messes with a document, tangible item or record in order to disrupt, influence or obstruct a federal investigation.

> The captain was found guilty of vandalizing a tangible object, which in this case was considered to be the fish. He was pronounced guilty and condemned to 30 days.

> The Supreme Court found that the charges against the captain were fishy and stated that SOX was intended to address vandalized records and data, not red grouper. With respect to the statute, the destruction of evidence provision specifically refers to corporate fraud and financial audit provisions. Given the context and placement, a tangible object under SOX may be referred to as one that is used to record or preserve information. It may not include all the known objects.

> The term tangible object must refer to something similar to records or documents, else all objects that are tangible. At trial, Yates demanded acquittal and argued that tangible objects may be associated with objects that may be used for storing data, for example, system hard drives, not fish. The District Court denied Yates' motion, and a jury saw him as blameworthy of abusing the law. The Jury pointed that, as objects having physical form, and since fish has a physical form, it falls within the dictionary definition of tangible object.

➤ **Case Study 9: Wiest vs. Lynch:**

> Jeffrey Wiest (defendant) worked in Tyco Electronics Corporation (Tyco) [58]. He was working their accounts payable department. From 2007 to 2009, he refused to process many Tyco expenditures which he believed violated company policy and federal regulations. To scrutinize the spending Wiest sent messages to his manager. He also recommended review to ensure appropriate tax treatment and accounting. Wiest suspected that Tyco's management was becoming disappointed due to his continued questioning of the company's treatment of expenditures. Therefore, in 2009, Tyco opened an investigation concerning certain charges against Wiest. These included inappropriate receipt of baseball tickets, explicitly suggestive remarks, and an ill-advised relationship with a fellow employee. Wiest was terminated by Tyco; however, Wiest brought suit, and asserted that Tyco had abused the SOX Act's forbiddance against retaliatory staff activities. Tyco filed a motion to dismiss for failure to state a claim.

5.3.5 HEALTH INSURANCE PORTABILITY AND ACCOUNTABILITY ACT OF 1996 (HIPAA)

The health insurance portability and accountability act of 1996 (HIPAA) is a government law that depends on the formation of national principles for shielding delicate patient health data from being uncovered without the patient's assent or information. The law gives information protection and security arrangements for defending clinical data. It has two primary purposes:

- Giving persistent medical coverage inclusion to workers who lose or change their employment; and

- Diminishing the authoritative burden and cost of healthcare services by normalizing the electronic transmission of regulatory and money-related exchanges.

Some other responsibilities of HIPAA incorporate countering misuse, extortion, and waste in medical coverage and healthcare services conveyance and improving access to long-term care services and medical coverage. The law has become popular recently due to the growing number of health data breaches. These breaches may be attributed to cyberattacks and ransomware assaults on health insurers and providers. The act was marked into law by President Bill Clinton on 21 August, 1996. It has five sections, or titles, listed as follows:

1. **Title I: HIPAA Health Insurance Reform:** Title I is responsible for securing medical coverage inclusion for people who lose or change jobs. It is additionally answerable for precluding group health plans from denying inclusion to people with explicit illnesses and prior conditions, and from setting lifetime inclusion limits.
2. **Title II: HIPAA Administrative Simplification:** Title II is responsible for directing the U.S. Department of health and human services (HHS) towards establishing national standards for processing electronic healthcare transactions. It commands healthcare associations to actualize secure electronic access to health information. It likewise urges the associations to stay in consistency with security guidelines set by HHS. Clinging to HIPAA Title II is one of the most noteworthy segments of the HIPAA consistency. Title II is otherwise called Administrative Simplification arrangements, and it likewise incorporates HIPAA consistence necessities like National Provider Identifier Standard-Each healthcare element, including people, businesses, wellbeing plans and healthcare services suppliers, must have a unique 10-digit national provider identifier number, or NPI.

    Transactions and Code Sets Standard-Healthcare associations must follow a normalized instrument for electronic data interchange (EDI) for submitting and handling protection claims.
3. **Title III: HIPAA Tax-Related Health Provisions:** Title III takes into account all tax-related provisions and guidelines related to medical care.
4. **Title IV: Application and Enforcement of Group Health Plan Requirements:** Title IV establishes health insurance reforms. This

may incorporate arrangements for people with previous conditions and those looking for recommencing coverage.
5. **Title V: Revenue Offsets:** Title V incorporates arrangements on organization-owned life coverage and the treatment of the individuals who lose their U.S. citizenship for income tax purposes.

Figure 5.2 provides an overall view of the HIPAA components.

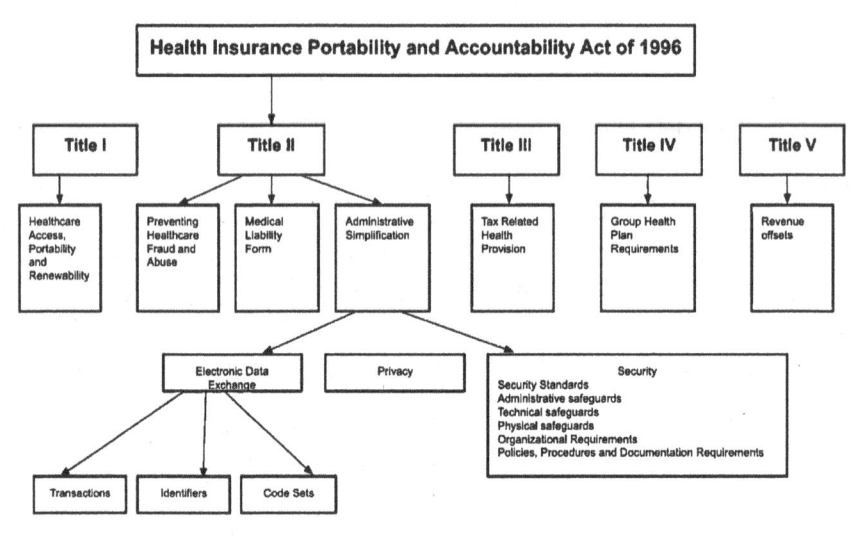

FIGURE 5.2 HIPAA components.

The Standards for Privacy of Individually Identifiable Health Information are referred to as the HIPAA Privacy Rule. This rule defines the first national standards in the United States for protecting patients' personal or protected health information (PHI). PHI consists of a patient's name, address, birth date, and Social Security number. It also includes an individual's physical or mental health condition. There may be information related to any care given to an individual or data concerning the installment for the care given to the person that recognizes the patient. Usually, this information can be held in digital, paper or oral forms. This individually identifiable health information is also referred to as PHI under the Privacy Rule.

To restrict the utilization and exposure of sensitive PHI, the Department of HHS issued a rule for protecting the privacy of patients. The law commands doctors to furnish patients with a record of every entity to which the doctor uncovers PHI for charging and managerial purposes. This might be done while as yet permitting significant health data to move through the best possible channels. The privacy rule further ensures patients the option to get their own PHI, upon demand, from health services suppliers secured by HIPAA.

The HIPAA Privacy Rule applies to organizations that are considered HIPAA-covered entities. It also includes health plans, healthcare clearinghouses and healthcare providers. Also, the HIPAA Privacy Rule requires covered individuals that work with a HIPAA business partner to deliver an agreement that imposes explicit protections on the PHI that the business partner utilizes or unveils. The HIPAA Privacy Rule is responsible for protecting all individually identifiable health information. This information is generally held or transmitted by a business associate of a covered entity.

HIPAA violations are quite common. Some of the famous cases are as follows:

➤ **Case Study 10: The United States vs. Huping Zhou:**

> Huping Zhou was a former cardiothoracic surgeon and Chinese immigrant who worked as a researcher at the UCLA School of Medicine [59]. He was terminated from his position following a HIPAA infringement. It was reported that after his dismissal, he had illegally accessed the UCLA medical records system as many as 300 times. He had seen the health records of his boss, his collaborators, and a few famous people. Due to the HIPAA violation, he was sentenced to four months in jail along with a fine amounting to $2,000. Few names on the list of medical records he accessed include celebrities like Arnold Schwarzenegger, Drew Barrymore, Leonardo DiCaprio, and Tom Hanks.

➤ **Case Study 11: The United States vs. Joshua Hippler:**

> Joshua Hippler was a Texas hospital employee [60]. In 2014, he was convicted of HIPAA violation and was sentenced to an 18-month jail term. Hippler had disclosed private patient medical information and was arrested in Georgia when he was found to be in possession of the medical records. The filing did not mention the number of records he had, in any case, he was accused of improper exposure of private health data for personal gain. Individual charges. Individual charges like this are uncommon since most violations of HIPAA are not intentional. All things considered; this case should fill in as a notice that solitary people are not invulnerable to arraignment.

> **Case Study 12: Nurse Outs STD Patient to Man's Girlfriend, Man Sues:**

> A nurse working at a clinic in New York ended up at the focal point of a revolting HIPAA infringement situation when her sister-in-law's sweetheart was determined to have a STD [61]. When the nurse sent six text messages, to warn the man's girlfriend about the disease, the clinic was sued by the man. The clinic dismissed the nurse from her job. The trial court judge, however, dismissed the claim on several grounds. The nurse's actions were believed to be both unforeseeable and based on personal reasons. The offended party has offered the choice. This is one HIPAA claim model that appears to be unavoidable, with the notice that the facility could have kept the nurse from treating a personal acquaintance.

> **Case Study 13: UCLA HIPAA Violations:**

> Another HIPAA violation case involved six doctors and 13 employees at UCLA Medical Center viewing Britney Spears' medical records after her 2008 psychiatric hospitalization [61]. Most of these representatives were non-clinical support staff, and none of them had a genuine medical need to see the PHI. HIPAA infringement of such kind should be obliterated. This may be achieved by the Principle of Least Privilege, which stresses on permitting access to information just to those representatives who need it to carry out their responsibilities.

> **Case Study 14: Cloud-based HIPAA Trouble:**

> 2016 witnessed a cardiology group of five physicians on staff paying a $100,000 HIPAA settlement involving an online calendar [61]. The surgical and clinical appointments were posted on a public, internet-accessed calendar. This led to HIPAA violation. Although the cloud offers an ever-evolving selection of efficiency-improving tools, along with those new efficiencies come emerging privacy pitfalls.

5.3.6 *THE WASSENAAR ARRANGEMENT (WA)*

The Wassenaar arrangement (WA) is the first global multilateral arrangement on export controls. It was specifically meant for conventional weapons and for sensitive goods and technologies. Although WA was co-founded by 33 countries in July 1996, it began operations only in September 1996. The WA was designed for promoting transparency and exchange of views and information. It was likewise answerable for organizing movement of conventional arms and dual-use merchandise and advancements (technology), along these lines forestalling destabilizing collections. The WA complements and reinforces the current systems for restraint of weapons of mass devastation and their conveyance frameworks, and also ensures that there is no duplication. This is conducted by concentrating on the threats to worldwide and territorial harmony and security which may emerge from the movement of armaments and delicate dual-use products and technologies where the dangers are made a decision about most prominent. This course of action expects to improve participation to forestall the procurement of armaments and sensitive dual-use items for military usage, if the circumstance in an area or the conduct of a state is, or turns into, a reason for genuine worry to the nations in question. The participating countries are responsible for ensuring that movement of arms and double use merchandise and advancements do not add to the turn of events or upgrade of military capacities that subvert global and provincial security and dependability and are not diverted to help such abilities. The measures taken with respect to the arrangement are with respect to the national legislation of the member countries. These measures are executed based on national discretion. The WA urges nations to keep up powerful export controls for the things on the concurred list. Periodical reviews are performed in order to analyze technological developments and lessons learned. Due to transparency and exchange of information and views, it is possible for providers of arms and dual-use items to create common understandings of the threats and risks related to their transfer and assess the scope for planning national control approaches to battle these threats. The arrangement's particular data exchange necessities take into account semi-annual notifications of arms transfers. This spans across seven categories deduced from the UN Register of Conventional Arms. Individuals are answerable for detailing moves or denial of moves of certain controlled dual-use items. Denial reporting assists with bringing to the consideration of individuals the exchanges that may subvert the goals of the arrangement.

➤ **Wassenaar Arrangement (WA) Countries**

> The 42 participating states in the Wassenaar Arrangement are Argentina, Australia, Austria, Belgium, Bulgaria, Canada, Croatia, the Czech Republic, Denmark, Estonia, Finland, France, Germany, Greece, Hungary, India, Ireland, Italy, Japan, Latvia, Lithuania, Luxembourg, Malta, Mexico, Netherlands, New Zealand, Norway, Poland, Portugal, Romania, Russia, Slovakia, Slovenia, South Africa, South Korea, Spain, Sweden, Switzerland, Turkey, Ukraine, the United Kingdom, and the United States.

While WA highlights weapons and arms, it finds it use in the Cybersecurity domain as well. Cybersecurity tools are known to be creative as well as disastrous, in light of how individuals use them. In the correct hands, intrusion tools can reasonably determine weaknesses in the network defenses, while in the wrong hands, they become digital assault weapons and may enable eavesdropping and snooping into other peoples' conversations and steal their data. This led to the WA being updated so that Internet-based surveillance technologies could also be taken into account. The cybersecurity industry however is worried about four major areas:

1. **Effect on Cybersecurity Research:** The standard may make it hard for scientists to test networks and disclose vulnerability information across countries. Subsequently, somebody who needs to introduce their new exploitation technology at an international conference could be in a dubious position. Traveling with exploitation information on devices like phones, tablets, or laptops across a national boundary may be questioned.
2. **Difficulty in Making Cybersecurity Tools Available around the World:** The process of getting an export license could be impossible, and the tools cannot be made available globally.
3. **Difficulty in Developing Perimeter Security Technologies:** If you cannot sell technologies outside a particular country, that makes it harder to legitimize creating them, and could prevent cybersecurity organizations' endeavors to work together.

4. **Difficulty in Cybersecurity Collaboration:** Conveying cybersecurity data to non-US residents would contradict the rules. It may not be conceivable to do a citizenship check for everyone at each and every corporation to whom information is sold. Further, setting up a bug bounty program, or selling the commercial version of a cybersecurity tool [11].

Some of the famous case studies related to the WA are as follows:

> **Case Study 15: The United States and Cybersecurity Controls:**

> The United States once attempted to execute intrusion programming controls which led to Symantec, FireEye, and independent security researchers raising serious concerns about their effect on security software as well as research [62]. Several questions were raised on how security controls would require export licenses for industrially accessible penetration testing items. It was additionally contended that any exploit sent nationally and internationally would require a permit. The Commerce Department disavowed exclusions for economically accessible software items that would have applied to a large number of the recently controlled security items. It was also unable to provide authorization exceptions for security research. All these actions were classified as harmful business and research activities by the security industry too. In any case, the cybersecurity industry likewise had issues with the substance of the Wassenaar language itself. Symantec, FireEye, and many other security software vendors asserted that the definition of intrusion software was too broad, and it also incorporated legitimate items like endpoint security frameworks and different apparatuses that guide into a framework to alter its code. The controls were believed to make it even more complicated for security exploration and vulnerability information sharing. The control on innovation for the advancement of intrusion programming would have secured numerous basic tools for the security research community, for example, exploit proofs-of-concept and automated vulnerability generators. Commerce Secretary Penny Pritzker declared through a letter in March 2016 that the United States would attempt to remove the intrusion software controls at that year's Wassenaar meetings. However, in December 2016, the U.S. negotiating team which incorporated technical experts from the cybersecurity community failed to persuade the other 40 Wassenaar individuals to concur on a narrower language.

➤ **Case Study 16: Israel's Take on Cybersecurity Tools and Weapons:**

Israel's Defense Ministry was dealing with its endorsement procedure on weapons exports. The Ministry manifested that the easing of cyber weapons restrictions was made to encourage compelling support of Israeli businesses while keeping up and securing worldwide standards of export control and management [19]. Israel claimed that any transaction would be allowed following a meticulous evaluation of the nation securing the new digital weapons. Further, the items might be showcased to dependable purchasers, not simply anybody (like terrorists) eager to pay a specific price. With Israeli spyware companies losing market share to their international rivals, it was necessary to take steps. Moreover, it could have possibly taken a year or considerably longer to approve the sale of digital weapons to a firmly verified group of partners, and still, at the end of the day, the organization making the deal expected to have the correct promoting and export licenses. However, with the approval process taking as little as four months, Israeli companies may have significantly become more competitive with respect to the world. The Defense Ministry proposed that export of cyber weapons could be extended to a bigger gathering of possible purchasers. These buyers could be including intelligence agencies, military intelligence groups, or law enforcement authorities, and the list of companies able to obtain the correct licenses for commercialization had to be expanded. Simultaneously, other influential voices within the Israeli government proposed that the deal and export of digital weapons could be an authentic tool in discretion. This may be beneficial specifically when negotiating diplomatically with Middle East nations. Israel required the same number of partners as it might get in the Middle East, and obviously did not need to be worried at all about working with authorities or other distasteful leaders on the off chance that it implied that it could support its military, national security, conciliatory or monetary benefits in the region. Israel's Economy Ministry, turned out on the side of the country's cyber weapons industry. From the point of view of the Economy Ministry, more must be done to help security organizations stay competitive, aware, and serious globally. To make that a reality, the Economy Ministry is right now making another division to deal with the handling of exports of cyber technology, both hostile and protective.

December 2017 Wassenaar meeting witnessed the members agreeing on a set of changes to the intrusion software controls. The revised control list incorporated many additions and alterations. The changes are:

- Supplanting language that controlled programming exceptionally intended to operate or communicate with intrusion software with the terms programming uniquely intended for order and control of intrusion software.
- Including an exemption for programming that carries out updates approved by the proprietor or administrator of the framework.
- Including exclusions for controls, innovation either associated with the improvement of intrusion software or the advancement of software that operates, controls, or conveys intrusion software. These exclusions said the controls do not have any significant bearing for vulnerability disclosure or cyber-IR activities. The list characterizes vulnerability disclosure and cyber-IR as procedures for sharing data about weaknesses and cyber incidents however does not clarify how these exceptions apply to explicit classifications of things.
- Including an explaining note saying that the above-portrayed exceptions do not decrease national authorities' privileges to find consistency with existing controls.

The changes seem to address a portion of the worries related to the Wassenaar language, strikingly the worries from the security research network about vulnerability information sharing. However, it is not yet clear how the new dialect mitigates worries about the wide meaning of intrusion software that may incorporate genuine security devices not utilized for "vulnerability disclosure" or "cyber incident response."

5.3.7 THE FAMILY EDUCATIONAL RIGHTS AND PRIVACY ACT OF 1974 (FERPA)

The Family Educational Rights and Privacy Act of 1974 (FERPA) is a United States federal law that is responsible for governing the access to educational information and records by public entities. These public entities are usually potential employers, publicly funded educational institutions, and foreign governments. The FERPA is also responsible for protecting the confidentiality of student educational records. The act holds true for any private, public, elementary, secondary, or post-secondary schools. It is also applicable to any local or state education office that gets government reserves. Every single government-funded school and every non-public school are secured by FERPA as they get some type of government funding. The act comprises two parts. Firstly, it gives students the option to survey and investigate their own education records. They may demand redresses and prevent the release of PII. In addition, they reserve the option to get a duplicate of their establishment's strategy concerning access to educational records. Secondly, it denies educational foundations the right to divulge PII in education records from without the written assent of the student, or if the student is a minor, the student's guardians. On the off chance that schools neglect to agree to FERPA, they might be at a risk of losing federal funding.

FERPA makes it mandatory for schools to inform parents and eligible students of their rights. The strategy for giving such data is left to the circumspection of the school. For the most part, schools must acquire written assent from guardians and qualified students before revealing any PII from a student's education record other than directory data. Anyway, there are a few special cases to this overall principle. A school may disclose PII from education records without assent under the accompanying conditions:

- Education records might be revealed to class authorities inside the school, for example, instructors, who have a genuine instructive interest for the data. It is the school's obligation to decide when there is an authentic instructive intrigue.
- In a situation where a student is intending to get enrolled in another school, the education records may be disclosed to that particular school, school district, or post-secondary institution.
- Education records might be uncovered to delegates of the Attorney General of the United States, the Secretary of the United States Department of Education, Comptroller General of the United States, or other neighborhood or state experts for motivations behind review or assessment.
- Education records might be uncovered for purposes relating to budgetary guidance for which the student has applied as long as the data is important to make judgments of qualification for aid, amount or conditions of aid, or requirement of terms of aid.
- Education records might be unveiled to state or local authorities or officials inside a juvenile justice framework, as long as the divulgence is made according to a state law.

- Education records might be revealed to associations that are leading investigations for educational offices or organizations regarding the turn of events or organization of predictive tests or student help projects, or studies that are expected to improve educational guidance. Such examinations must not allow identification of guardians or students by anybody other than delegates of the association. Besides, the PII must be obliterated when not required for the investigation.
- Education records might be disclosed to approved associations for reasons for directing accreditation methodology.
- Education records might be unveiled to the guardians of a dependent student as characterized by the internal revenue service (IRS).
- Education records might be disclosed regarding a health or safety crisis.
- Education records might be delivered in consistency with a court request, for example, a summon, however, schools should initially put forth a sensible attempt to give notice to guardians or students. On account of law authorization or government jury summons, the responsible court or organization may, for good motivation, request the school not to uncover the contents of the summon or the records delivered in accordance with the summon. However, if there is no crisis, schools cannot give non-catalog student data to the police without a summon.
- Student directory information may likewise be revealed without the guardian or student's assent. Directory data can incorporate the student's name, address, phone number, date, and location of birth, significant field of study, dates of attendance, interest in school-supported extracurricular exercises, height, and weight of student-athletes, degrees earned, honors, and grants earned, the educational foundation last attended, photos, and email addresses. Schools do not need to deliver directory data, yet in the event that they do, they should give open notification of the categories of information they classify as directory information. The school should then give guardians and qualified students a sensible measure of time to educate the school that they do not need some or all of their directory data unveiled without assent.

A few records kept up by schools are excluded from FERPA. They are as per the following:

- Records that are in the sole ownership of the school authorities;
- Records that are kept up by a law implementation unit of the educational foundation;
- Records of an educational establishment's non-student representatives;
- Records on a student who is at least 18 years old;
- Records of a student who goes to a post-secondary establishment that are kept up by a health professional.

Moreover, FERPA permits, however requires, schools to deliver directory information, similar to students' names and addresses, to the general public. However, this special case was altered in 2002, and secondary schools are currently required to give students' names, addresses, and phone numbers to military selection representatives, except if a student or parent opts out of such disclosure.

Some of the case studies related to FERPA are as follows:

➢ **Case Study 17: The Harvard Case:**

> More than 1.4 million emails that encompassed Harvard students' evaluations, monetary aid information, and at least one person's Social Security number were sent over Harvard computer society (HCS) email records [64]. It was discovered that the data was available to the public. Teaching fellows, tutors, College administrators, and a huge number of students have utilized the email list service for quite a long time. Messages sent over HCS lists contained data identified with the members of certain BGLTQ (bisexual, gay, lesbian, transgender, queer, and questioning) undergraduate societies, financial account numbers for some student associations, advance duplicates final exams, and answer keys to problem sets. Instructors and teachers utilized the list to talk about students' evaluations abusing the family educational rights and privacy act (FERPA). While just Harvard affiliates could get to the directory of the email records, the messages themselves were available to the public. At least two dozen students who oversee HCS records expressed that they never understood their messages were open. All College executives who utilized the lists, including the Dean of the College, were unaware that their messages were open. Any person who had a Harvard email address could, without much of a stretch, set up an email list through HCS. The default setting for HCS list files was open. This means that the users' emails were publicly available unless list owners specified otherwise.

> **Case Study 18: Principle on Naming Students who Failed in a Newsletter:**

> Caterina Lafergola, a principal who left Brooklyn's battling Automotive High School for Long Island, had conveyed newsletters at her new school naming suspended students, infringing upon a federal student privacy law [65]. She issued a weekly staff newsletter that included announcements on upcoming events, meeting notes, and a Kudos Corner. It also had a section titled Suspensions. One newsletter named a student ousted for engaging in inappropriate, threatening behavior and using vulgar language. Another named a student, booted for five days, who engaged in a physical altercation outside the building. There was yet another newsletter that named a female student suspended for disregarding the particulars of her earlier suspension when she attempted to attend a basketball game. The Federal Educational Rights and Privacy Act, or FERPA, strongly disallows the divulgence of student records, including disciplinary choices, but to those straightforwardly associated with the student's education, the principal was convicted. If the information is disseminated to others, it may be termed a violation of FERPA. A parent condemned Lafergola's practice, asserting that everyone in the school can see the newsletter. The case led to many education advocates being stunned. Printing suspension information in a staff bulletin may be deemed both unnecessary as well as invasion of privacy, since the newsletters could easily be passed around.

5.3.8 *GENERAL DATA PROTECTION REGULATION (GDPR)*

The general data protection regulation (GDPR) is a guideline in EU law on information assurance and privacy in the European Union (EU) and the European Economic Area (EEA). Its essential target is to address the exchange of individual information outside the EU and EEA territories. The Purpose of GDPR is to essentially offer control to people over their own information. Further, it plans to streamline the administrative condition for worldwide business by bringing together the guideline inside the EU. The GDPR drills down seven standards for the legitimate handling of individual information. These may include use, collection, communication, organization, restriction, structuring, alteration, storage, consultation, combination, destruction or erasure or personal data. The seven principles are as follows:

- Accuracy;
- Purpose limitation;
- Storage limitation;
- Data minimization;
- Integrity and confidentiality (security);
- Accountability; and
- Lawfulness, fairness, and transparency.

GDPR is based on the above principles as they are the core values of the guideline and compliant processing. Data controllers are liable for following the standards and letter of the guideline. Data controllers are likewise responsible for their preparation and should exhibit their consistency. Personal data must be:

- Processed fairly, lawfully, and in a transparent manner with respect to individuals;
- Gathered for determined, explicit, and genuine purposes and not further handled in a way that is contradictory with those reasons;
- Adequate, relevant, and confined to what is required with respect to the purposes for which they are processed;
- Up to date and accurate;
- Stored in a manner that permits distinguishing of information subjects for no longer than is essential for the reasons for which the individual information are handled;
- Processed in a way that encourages appropriate security of the personal data. This may incorporate security against unapproved or unlawful processing and against unintentional loss, devastation, or harm, utilizing proper technical or hierarchical measures.

GDPR is valid for any association working inside the EU, just as any associations outside of the EU which offer products or administrations to clients or organizations in the EU. In this manner, GDPR compliance methodology is a prerequisite for pretty much every significant partnership in the world. There are two unique sorts of data handlers the enactment applies to: processors and controllers. A controller may be defined as a person, agency, public authority, or other body which, alone or jointly with others, is capable of determining the purposes and means of processing of personal data. A processor is an individual, public authority, organization or other body which processes individual information in the interest of the controller. Data considered personal under the existing legislation may be in the form of name, photos, and address. As per GDPR, IP address can be personal information since it expands the definition of personal data. It also takes into account sensitive personal data such as genetic data, and biometric data that can uniquely identify an individual. GDPR inaugurates one law across the continent and a single set of rules and guidelines which hold true for companies doing business within EU member states. This implies that the reach of the legislation extends beyond the borders of Europe itself, as international associations based outside the region yet with movement in Europe will still need to comply. It is believed slimlining data legislation with GDPR, may possibly be beneficial to businesses. As indicated by The European Commission, having a solitary supervisor authority for the whole EU, will make it cheaper and simpler for businesses to operate within the region. This sort of guideline ensures information insurance safeguards are incorporated with items and services from the earliest phase of advancement, giving information protection by design in new technologies and products. Associations are likewise urged to embrace methods like pseudonymization so as to profit by gathering and analyzing personal information, while the privacy of their clients is ensured simultaneously. Some of the famous cases that involve GDPR are as follows:

> **Case Study 19: Google Fined $57 Million:**

> The French Commission Nationale de l'Informatique et des Libertés (CNIL) forced the fine of 50 million euros (generally $57 million) on Google based on the results of an inquiry into GDPR compliance complaints [66]. The grievances had been asserted by French privacy rights associations None Of Your Business and La Quadrature du Net. As indicated by CNIL, Google neglected to sufficiently educate clients about how their information was gathered. This data was also used in serving advertisements and marketing messages. Google additionally was unsuccessful in appropriately getting client assent to utilize their information to serve them customized ads. With the fine, Google turns into the main major U.S. tech organization to be rebuffed for falling afoul of GDPR guidelines since the EU brought the protection-centered standards into impact in 2018. According to CNIL, Google's violations were primarily caused by uncertainty of data introduced to clients about their information assortment and utilization, in addition to the inability to incorporate data about the information retention period for some data. In addition, the agency found Google's user consent policy for

data use invalid. This is because it is neither specific nor unambiguous. The information was diluted across several documents, which made it hard for users to comprehend what information was collected and how it was used. The French watchdog further highlighted that Google continued to engage in several of these unlawful practices, suggesting that they were a part of a pattern of fundamental GDPR infringement instead of erratic offenses.

➤ Case Study 20: The Twitter GDPR Violation Case:

Under the European Union's GDPR, which became effective in May 2018, European citizens reserve the option to realize what information organizations gather on them, and what they do with that information [67]. London privacy researcher Michael Veale ventured to know if Twitter tracks his web movement when he taps on an abbreviated "t.co" link. So he requested that Twitter give him all the information it has on him. Yet, Twitter did not have that information. Also, Twitter denied the solicitation on the grounds of the unbalanced exertion it would take to collect that data, which the GDPR takes into account. Twitter must not be able to obscure data transparency despite the existence of the provision. Veale considered it as misinterpreting the text of the law and complained to the Irish Data Protection Commission (DPC). DPC responded in a letter saying that it would investigate Twitter.

➤ Case Study 21: The British Airways Data Breach:

The Information Commissioner's Office (ICO) in the UK fined British Airways (BA) with a £183.4 million ($230M) fine over a 2018 data breach [68]. This is one of the ICO's biggest fines since the GDPR came into effect. The breach which is believed to have impacted 500,000 people. The breach was disclosed in September by BA, and witnessed individuals visiting its site being redirected to a counterfeit website. The fraudulent website harvested details of individuals, including name, billing address, email address and payment information. While an initial disclosure asserted that the breach occurred between August and September, affecting 380,000 card payments, the carrier later notified that 185,000 individuals who booked tickets between April and July could have been similarly compromised.

➤ Case Study 22: Google Location Tracking:

In August 2018, an investigation was performed by the Associated Press, which revealed that disabling the 'Location Tracking' feature on an Android smartphone did not stop the device from tracking the user's location [69]. Although Google's support page stated that one could turn off Location History at any time and that, with Location History off, the places one goes to would no longer be stored, it was found out to be false. Google then revised its support page to explicitly attest that some area information may even now be spared with the setting turned off, which raised genuine worries over the transparency and trust with which Google gathers client information, the transparency which is mandated by GDPR. Consumer protection groups in a few countries of Europe have documented grievances against Google with their data regulators.

➤ Case Study 23: The Starwood Hotel Data Breach:

Marriott's data breach sabotaging 500 million Starwood hotel customers has been one of the biggest data breaches [70]. Marriott, which is the parent company of Starwood, discovered the breach of customers' personal data in 2018. This was after GDPR was introduced and came into effect. However, the breach seems to have been open and ongoing since 2014. This predates GDPR as well as Marriott's acquisition of Starwood. Although private class actions have already begun, state and national regulatory bodies may also press charges. Since the breach was found in September 2018, however, was not divulged until late November 2018, GDPR is applicable to the breach and Marriott has certainly violated the regulations. This is outside the 72-hour window for divulgence (disclosure) set by GDPR and may be one of the strongest cases to be treated as GDPR violation.

5.3.9 CALIFORNIA CONSUMER PRIVACY ACT OF 2018

The California Consumer Privacy Act (CCPA) is a state law responsible for enhancing privacy rights and consumer protection for residents of California, United States. It came into effect in January 2020 and is also known as AB 375. CCPA permits any California consumer to request to see all the data an organization has saved on them, as a full list of all the third parties that information is imparted to. Further, the law permits consumers to sue organizations if the privacy rules are disregarded, despite there being no breach. The act intends to provide the residents of California the right to:

- Comprehend what individual information is being gathered about them;
- Understand whether their personal data is disclosed or sold and to whom;
- Disagree to the sale or disclosure of personal data;
- Access their personal data;
- Demand a business to delete any personal information about a consumer collected from that consumer;
- Not be oppressed for practicing their privacy rights.

All organizations that serve California residents and have more than $25 million in yearly income must conform to the law. Furthermore, organizations of any size that have personal information for more than 50,000 individuals or that gather more than 50% of their incomes from the sale of personal information, likewise fall under the law. It is not fundamental for organizations to be situated in California or have a physical presence there to conform to this law. The companies need not be based in the United States. CCPA takes a broader approach to what constitutes sensitive data as compared to the GDPR. For instance, olfactory data is secured, just as perusing history and records of a guest's associations with a site or application. This is what AB 375 thinks about personal data:

- Identifiers, for example, a real name, alias, postal location, unique personal identifier, online identifier IP address, email address, account name, Social Security number, driver's license number, passport number, or other comparative identifiers.
- Qualities of secured categories under California or government law.
- Geolocation data.
- Business data including records of individual property, items or services bought, acquired or considered, or other buying or expending chronicles or propensities.
- Electronic network activity data (incorporates Internet) including, however not restricted to, perusing history, search history and data in regards to a consumer's connection with a website, application or advertisement.
- Electronic, visual, audio, thermal, olfactory or comparative data.

- Professional or employment-related information.
- Biometric information.
- Education data, characterized as data that is not openly accessible PII as characterized in the FERPA.
- Speculations deduced from any of the information identified in this subdivision for creating a profile pertaining to a consumer's behavior, abilities, preferences, attitudes, aptitudes, characteristics, intelligence abilities, and psychological trends.

Since the CCPA is relatively new, the number of cases pertaining to it are limited as of now. The following is the first case to have mentioned violation of CCPA:

> **Case Study 24: Barnes vs. Hanna Anderson LLC:**

A youngsters' dress organization and Salesforce.com Inc. confronted data breach charges in a government court claim that referred to infringement of California's landmark privacy law since it took effect 1 January, 2020 [71]. Hanna Andersson and Salesforce could not protect user data, safeguard platforms or give cybersecurity alerts, according to the grievance documented in the U.S. District Court for the Northern District of California. It was claimed by Barnes that the activities disregarded state laws including the California Consumer Privacy Act. The California law highlights a private right of action that makes it easier for consumers to seek damages under state consumer-protection laws. While Barnes' complaint stops short of seeking fines, she still has the right to amend it later for a potential California class to seek damages and relief under the California law's data-breach provisions. Barnes additionally brought claims under the California Unfair Competition Law. This suit was bought after high-end children's apparel company Hanna Andersson announced 15 January, 2020, that hackers had scratched client names, credit card numbers, and other sensitive data. According to the complaint, the hacked data was found for sale on the dark web, and was hosted by Salesforce on its e-commerce platform. Further, the platform was accepted to have been tainted with malware that drove to the data breach. Barnes asked the court to investigate whether Salesforce and Hanna Andersson violated the CCPA. The act mentions that California residents may be granted up to $750 per consumer, per incident following a data breach due to weak data-security protections. For large data collectors, the damage could be in the form of millions of dollars.

Few limitations of the CCPA are as follows:

- The CCPA does not provide the undefined rights claimed. It does not reinforce consumers rights with respect to data breach. Rather, purchasers can recoup legal harms for a breach, given that, given that certain steps are followed.
- The CCPA is not responsible for creating a duty for maintaining reasonable security procedures. Further, the CCPA does not inflict any obligation on businesses for maintaining reasonable security procedures. Rather, it affirms that in specific situations, customers might be qualified for legal harms in case of a data breach.
- The CCPA permits consumers to bring an activity for legal harms following a data breach inferable from the inability of the business to execute sensible security strategies. Nonetheless, before bringing an activity, it is necessary for the consumer to provide the business a 30 days' written notice identifying the specific violation. If the violation is cured by the business and the consumer is given a written statement regarding the same, statutory damages are not available.
- CCPA's statutory damages for data breaches does not hold true for service providers. The CCPA differentiates between entities acting as businesses and entities acting as service providers. A service provider is recognized by the CCPA as an entity that is responsible for processing individual data for the benefit of a business compliant with a composed agreement that incorporates the necessary language. Organizations that neglect to actualize suitable security methodology might be held subject for the legal harms. Because of this explanation, all entities that process personal data in the interest of a managed business must evaluate whether it meets the meaning of service provider and, provided that this is true, update their agreements in like manner. The agreement must explain that there are no third-party recipients to the understanding, no uncommon connections and, explicitly, no obligation to consumers. From the organizations' viewpoint, the agreement must require service providers to arrange in case of an episode, which may help stay away from the numerous issues.
- The responsibility of maintaining appropriate security procedures only applies to organizations that own, permit, or maintain personal data. California law just expects entities to keep up sufficient security techniques to the degree that they own, permit, or maintain

personal data. In the event that an entity does not participate in the prior exercises, it is ostensibly under no commitment to keep up sensible security methods. Without such commitment, the entity cannot be at risk for the CCPA legal damages (Figure 5.3).

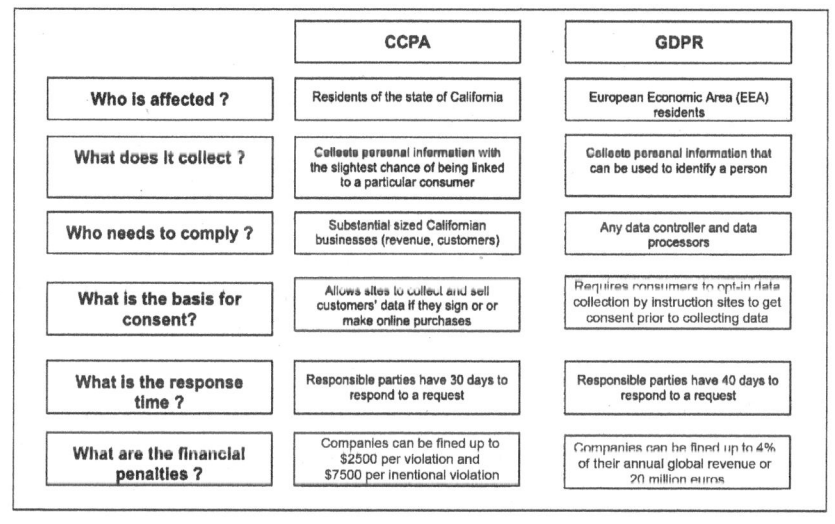

	CCPA	GDPR
Who is affected ?	Residents of the state of California	European Economic Area (EEA) residents
What does it collect ?	Collects personal information with the slightest chance of being linked to a particular consumer	Collects personal information that can be used to identify a person
Who needs to comply ?	Substantial sized Californian businesses (revenue, customers)	Any data controller and data processors
What is the basis for consent?	Allows sites to collect and sell customers' data if they sign or or make online purchases	Requires consumers to opt-in data collection by instruction sites to get consent prior to collecting data
What is the response time ?	Responsible parties have 30 days to respond to a request	Responsible parties have 40 days to respond to a request
What are the financial penalties ?	Companies can be fined up to $2500 per violation and $7500 per inentional violation	Companies can be fined up to 4% of their annual global revenue or 20 million euros

FIGURE 5.3 GDPR vs. CCPA.

5.3.10 CHILDREN'S ONLINE PRIVACY PROTECTION ACT (COPPA) 1998

The children's online privacy protection act (COPPA) is a law that was passed by the U.S. Congress in 1998 and took effect in April 2000. It is responsible for protecting the privacy of children under the age of 13. COPPA, which is governed by the FTC specifies:

- That sites must require parental assent for the assortment or utilization of any personal data of young users.
- What all needs to be included in a privacy policy. It also includes the requirement that the policy itself must be posted anywhere data is collected.
- The method of seeking verifiable assent from a parent or guardian.
- What duties the administrator of a Website lawfully holds as for children's privacy and safety on the web. This likewise highlights

limitations for the sorts and techniques for advertising focusing on those under 13.

COPPA was liable for tending to the fast development of web-based marketing methods during the 1990s that were targeting children. It was found that various websites collected personal data from children without parental knowledge or consent. In addition, children did not comprehend the likely antagonistic results of uncovering personal data on the web. While COPPA does not obviously characterize how parental assent ought to be picked up, the (FTC) has built up rules that may help website administrators guarantee consistency with the act. These proposals include:

- Downloadable consent forms that are to be mailed or faxed to an operator must be clearly displayed.
- Parents must use credit cards to authenticate age and identity.
- Parents must call a toll-free phone number.
- Accepting an email from a parent must necessitate digital signatures.

COPPA makes it essential for site administrators to permit guardians to survey any data gathered from their children. This infers any significant site must give full access to all client records, profiles, and sign-in data when a parent demands it. The FTC has specified that guardians may erase certain data; however, they may not, in any case, modify it. Any Website that gathers data from children younger than 13 must submit to COPPA. The act influences numerous well-known websites like MySpace.com, Facebook.com, Friendster.com, Xanga.com, and other networking platforms.

➢ **Case Study 25: Google and YouTube COPPA Violation:**

> Google and YouTube paid $170 million to settle claims by the FTC and the New York Attorney General that the YouTube video-sharing feature wrongfully gathered sensitive data from children without their guardian's assent [72]. FTC and New York Attorney General affirm that YouTube abused the COPPA Rule by gathering personal data. This information was primarily in the form of persistent identifiers that may be used to follow clients over the Internet from viewers of child-directed channels, without first telling guardians and getting their assent. YouTube earned millions by utilizing cookies. Cookies were used to deliver targeted ads to viewers of these channels. While YouTube is believed to be a general audience site, some of YouTube's individual channels, like the ones operated by toy companies are child-directed and therefore must comply with COPPA. The respondents

knew that the YouTube platform had various child-directed channels. Many channel proprietors notified YouTube and Google that their channels' substance was targeted to children. Moreover, in several instances, YouTube's own content rating system identified content as directed to children. As per the complaint, YouTube manually reviewed children's content from its YouTube platform to feature in its YouTube Kids app. Despite this knowledge of channels directed to children on the YouTube platform, YouTube served targeted advertisements on these channels. The complaint mentioned that YouTube even revealed to one publicizing organization that it did not have clients under the age of 13 on its platform, and accordingly channels on its platform did not have to consent to COPPA.

> **Case Study 26: Google's G-Suite COPPA Violation:**

Google was accused of collecting data of students using its G Suite for Education program [73]. This feature offers its productivity services to students for free and also offers free tools for K-12 students. Some of the services include free access to its Calendar, Drive, Gmail, Docs, and other applications. In this Education program, 25 million students were also using Chromebook, Google's laptop that was targeted for classrooms. The lawsuit alleged that Google had utilized the feature to gather information of children utilizing the service. The gathered information consolidated their physical location, sites they visit, terms utilized in Google's internet searcher and videos viewed on YouTube. In addition, Google was likewise blamed for gathering personal contact records, voice recordings, stored passwords and other behavioral information. The claim showed that until April 2014, Google additionally mined students' email accounts and separated that information for advertising purposes. In all of these cases, Google had not properly disclosed to users that their data is being collected. The claim asserted that when students signed into their Chromebook, the Chrome Sync function was utilized by Google to synchronize applications, auto-fill data, and more was turned on by default. The feature then automatically started uploading Chrome usage data to Google servers. This includes online browsing habits, web searches and passwords. This level of data collection is a blatant violation of the children's online privacy protection act (COPPA), since it requires parental assent for the assortment and utilization of that personal information for users under the age of 13. It also disregards the FERPA, a federal law that oversees the access to educational data and records by public elements.

5.3.11 DIGITAL MILLENNIUM COPYRIGHT ACT (DMCA)

The digital millennium copyright act (DMCA) is a 1998 United States copyright law. It actualizes two 1996 treaties of the World Intellectual Property Organization (WIPO) and condemns the creation and dispersal of innovation, gadgets, or services expected to bypass measures that control access to copyrighted works. Further, it likewise condemns the act of bypassing access control, regardless of whether there is genuine encroachment of copyright itself, and uplifts the punishments for copyright encroachment on the Internet. The DMCA is partitioned into five titles:

1. **Title I:** The WIPO Copyright and Performances and Phonograms Treaties Implementation Act of 1998, which actualizes the WIPO settlements.
2. **Title II:** The Online Copyright Infringement Liability Limitation Act, which is answerable for making restrictions on the obligation of online service providers (OSP) for copyright encroachment while taking part in specific sorts of exercises.
3. **Title III:** The Computer Maintenance Competition Assurance Act, which is answerable for making an exclusion for making a duplicate of a computer program by actuating a computer for reasons for fixing and maintenance.
4. **Title IV:** It takes into account six arrangements, which are identified with distance education, the elements of the Copyright Office, webcasting of sound recordings on the Internet, the exemptions in the Copyright Act for libraries and for making ephemeral recordings, and the implementation of collective bargaining agreement commitments on account of transfer of rights in movies.
5. **Title V:** The Vessel Hull Design Protection Act, is answerable for making another type of assurance for the structure of vessel hulls.

The DMCA focuses on two primary areas. These two segments have been a source of specific contention since they became effective in 2000. The anti-circumvention provisions bar circumvention of access controls and specialized assurance measures, while the safe harbor provisions protect service providers who meet certain conditions from fiscal harms for the encroaching activities of their clients and third parties on the net. DMCA safe harbor is basically alluded to as the arrangement of DMCA.

It gives place of refuge to OSP and other internet intermediaries by excluding them from direct copyright encroachment. There are four safe harbors endorsed by Congress, and in these cases, there is constrained to no copyright encroachment obligation for OSP. Coming up next are the allowed safe harbors under DMCA:

- System caching;
- Information location tools;
- Temporary digital network communication;
- Storing information at the user's direction on system or network.

The DMCA safe harbors are answerable for extending the web and for improving the quality and assortment of services that must be given on the web. It may not be conceivable to accomplish it by restricting the risk for ISPs. ISPs would have ended up submitting copyright encroachments while making duplicates of copyrighted content with the end goal of improved speed, facilitating websites, or basically guiding clients to destinations that perhaps contain infringing or encroaching content. Therefore, to stay away from these issues and for the upgraded productivity and extension of the web, there is a constrained obligation for ISPs and sites falling under one of the safe harbor classifications.

Since DMCA is a part of the United States copyright law, it is pertinent just to the websites facilitated in the US. All sites facilitated in the US will undoubtedly comply with the law. Consequently, regardless of whether the copyright proprietor is outside of the US, they can even now issue DMCA notice if the facilitating website is situated in the US. Some prominent cases are as follows:

> **Case Study 27: Viacom vs. YouTube:**

> Viacom sued YouTube and Google in 2007, and asserted that they ought to be considered answerable for the copyright encroachments performed by YouTube clients [74]. The claim looked for more than $1 billion in harms and went ahead the impact points of Viacom's conveyance of in excess of 100 000 takedown notices focusing on recordings supposedly possessed by Viacom (which YouTube expeditiously conformed to). Not long after the Viacom claim, various class actions were likewise recorded in the interest of sports groups, music distributors, and other copyright proprietors against YouTube, all based on a similar hypothesis.

➤ **Case Study 28: Apple DMCA Violations:**

Apple sued an iOS (iPhone operating system) virtualization seller called Corellium for trafficking under the DMCA [75]. The organization was at first sued by Apple for copyright encroachment, charging that Corellium's virtualization of iOS was abusing Apple's ownership of code. The later filing expands the case, charging that Corellium's offer of the virtualization programming counts as trafficking with copyright-ensured products. The product is equipped for running an exact copy of iOS in a controlled system environment. In the event that there is no conventional connectivity, the program cannot be utilized as a telephone. In any case, it permits enthusiasts to research how explicit software performs on iOS in minute detail. It is explicitly valuable for malware-based examination, and was most recently used to reveal observation-related conventions in the United Arab Emirates' ToTok application.

➤ **Case Study 29: Photographer Baffoli vs. Facebook:**

Boffoli, a professional photographer, created a series of photographs which incorporated tiny figures photographed against real food backdrops [76]. He named the series Big Appetites.' The complaint states that photos are copyrightable and have been enrolled with the United States Copyright Office. Not just the photos have been highlighted in the publications including the New York Times, the Washington Post, NPR (National Public Radio), and CBS (Columbia broadcasting system) This Morning, yet in addition his work has been included in exhibitions in the United States and globally. Boffoli has been employed and appointed by organizations for his work, and his pictures have been authorized. Boffoli's business and salary depend on the work he licenses and sells. Boffoli claimed that Facebook's clients posted photos from Big Appetites without permit or consent from Boffoli Facebook webpages. The protest likewise expressed that Facebook did not keep clients from posting the copyrighted material on its platform. Additionally, it neglected to keep the content from being open over the Internet in spite of notice from Boffoli. This infringement content was not expelled by Facebook. Nor were the clients kept from posting this kind of content, in spite of supposedly having the ability to do as such. Boffoli likewise referenced that he did not give consent for his work to be posted on Facebook. A DMCA was submitted regarding the infringing posts. In spite of the fact that Facebook affirmed that the content being referred to was removed, anyway over 100 days in the wake of accepting the notification, Facebook had still not expelled or restricted access

> to infringing duplicates of Boffoli's work and just removed or disabled access to the content after being reached out by his lawyer following few weeks. Facebook admitted to have neglected to previously remove the material because of a technical error. Facebook did not promptly expel the infringing content regardless of information on the infringement from Boffoli's notice. This prompted Boffoli contending that Facebook penetrated DMCA. Boffoli likewise blamed Facebook for copyright infringement itself, and contributory copyright infringement. Boffoli expressed that without his authorization or assent, photos from Big Appetites were recreated, subsidiary works were produced using, duplicates were conveyed, and the photos were distributed on Facebook's platform. This led to violation of his exclusive rights in the photographs in Big Appetites.

5.3.12 UNLAWFUL INTERNET GAMBLING ENFORCEMENT ACT OF 2006 (UIGEA)

The unlawful internet gambling enforcement act (UIGEA) is liable for removing the progression of income to unlawful Internet betting organizations. It forbids receipt of checks, credit card charges, electronic fund movement, and so forth by such organizations. It depends on the help of banks, credit card issuers and other payment system participants and members to help stem the progression of assets to unlawful Internet betting organizations. The last guideline which was received by organizations to actualize the arrangements of the UIGEA, addresses the achievability of distinguishing and prohibiting the progression of unlawful Internet betting proceeds in five payment systems: cash transmission systems, card systems, check collection systems, wire transfer systems, and the automated clearing house (ACH) systems. It recommends that, excluding financial institutions that are concerned with illegal Internet gambling operators, monitoring the flow of revenue within the wire transfer, ACH systems, and check collection is not practical, following which it exempts them from the regulations' requirements. The act charges those with whom illegal Internet gambling operators deal directly by means of the three systems. It also holds true for members in the card and cash transmission systems, to embrace approaches and methods to empower them to recognize the nature of their client's business, to utilize customer understandings accepting corrupted exchanges, and to set up and keep up therapeutic strides to manage tainted exchanges when they are identified. The last guideline gives non-exclusive instances of reasonably designed policies and techniques to forestall limited exchanges. The Agencies announced that adaptable, risk-based due determination methods led by members in the payment systems, in order to establish and maintain commercial customer relationships, is the most effective method for preventing or prohibiting the restricted transactions.

> **Case Study 30: The PokerStars Case:**

> A report from Forbes mentioned that PokerStars Co-founder Isai Scheinberg was arrested in New York City after intentionally making a trip to the US to deal with indictments [77]. Scheinberg had been blamed on betting, bank fraud and money laundering charges in 2011. He had invested energy in Canada and the Isle of Man and was arriving in the US from Switzerland. The US Government and Scheinberg had been negotiating a deal and had made significant progress in those negotiations. PokerStars was founded by Isai and his son Mark Scheinberg in 2001. In 2019, the US Government had made a deportation request to Switzerland, where Scheinberg had been on a trip. Isai surrendered his passport and was released with a bail figure of $1 million. However, the main issues date back to 2006, when the UIGEA was implemented, resulting in a federal ban on online gambling. PokerStars was one of the few operators in the market after the law came into effect. This led to 'Black Friday' in 2011. On 15 April 2011, the US Department of Justice disclosed allegations against chiefs of PokerStars, Full Tilt Poker and Absolute Poker which also included Isai but not Mark Scheinberg. The highlight of the accusation was Scheinberg being charged with the violation of UIGEA and the Illegal Gambling Business Act. He was accused of conspiracy to commit bank fraud and tax evasion. None of those blamed due to Black Friday were accused of abusing the Federal Wire Act of 1961. Isai ventured away from any administration job inside his organization. PokerStars was sold to Amaya Gaming by Mark Scheinberg in 2014 for $4.9 billion.

In this chapter, we acquainted ourselves with different types of cyber laws, the cybercrimes pertaining to them, as well as their consequences. We also got a glimpse of the cyber laws that make up the current legislative framework of the United States. Some interesting case studies related to popular laws have also been mentioned. The case studies not only provided information on how the laws may be applied to different scenarios, but also highlighted some of the loopholes that exist in the legal system. Should similar crimes ever occur in the future, it may be easier to anticipate if an entity is convicted or acquitted, keeping in mind the loopholes.

5.4 SUMMARY AND REVIEW

- Cybercrime laws can be classified into two categories: Substantive cyber laws and procedural cyber laws.
- Substantive cybercrime laws are the laws that are concerned with prohibiting online identity theft, hacking, intrusion into computer systems, child pornography, intellectual property (IP), online gambling, etc.
- Procedural cybercrime laws might be characterized as laws concerned about the authority to protect and acquire electronic information from outsiders, including internet service providers, authority to intercept electronic communications or, the authority to search and seize electronic evidence.
- The current legislative framework of the United States includes: "The Counterfeit Access Device and Computer Fraud and Abuse Act of 1984, The Electronic Communications Privacy Act of 1986 (ECPA), The Computer Security Act of 1987, The Paperwork Reduction Act of 1995, The Clinger-Cohen Act of 1996, The Homeland Security Act of 2002, The Cyber Security Research and Development Act of 2002, The E-Government Act of 2002, and, The Federal Information Security Management Act of 2002."
- The computer fraud and abuse act (CFAA) prohibits accessing a computer without authorization, or in excess of authorization.
- According to the Gramm-Leach-Bliley Act (GLBA), it is mandatory that financial institutions communicate to their customers regarding how the customers' sensitive data is being shared, and inform customers of their right to opt-out if they prefer that their personal data not be shared with third parties.
- The Electronic Communications and Privacy Act 1986 was enacted by the United States for extending government restrictions on wiretaps from telephone calls for including transmissions of electronic data by computer.
- The Sarbanes-Oxley (SOX) Act was passed on 30 July, 2002, with the aim to protect investors from fraudulent financial reporting by corporations.

- The health insurance portability and accountability act of 1996 (HIPAA) is a government law that depends on the formation of a national standard for shielding confidential patient health data from being uncovered without the patient's assent or knowledge.
- The Wassenaar arrangement (WA) is the first global multilateral arrangement on export controls. The 42 participating states in the WA are Argentina, Australia, Austria, Belgium, Bulgaria, Canada, Croatia, the Czech Republic, Denmark, Estonia, Finland, France, Germany, Greece, Hungary, India, Ireland, Italy, Japan, Latvia, Lithuania, Luxembourg, Malta, Mexico, Netherlands, New Zealand, Norway, Poland, Portugal, Romania, Russia, Slovakia, Slovenia, South Africa, South Korea, Spain, Sweden, Switzerland, Turkey, Ukraine, the United Kingdom, and the United States.
- The family educational rights and privacy act of 1974 (FERPA) is a United States federal law which is responsible for governing the access to educational information and records by public entities.
- The general data protection regulation (GDPR) is a regulation in EU law on data protection and privacy in the European Union (EU) and the European Economic Area (EEA). Its primary objective is to address the transfer of personal data outside the EU and EEA areas.
- The children's online privacy protection act (COPPA) is answerable for securing the privacy of children under the age of 13.
- The digital millennium copyright act (DMCA) is a 1998 United States copyright law. It actualizes two 1996 arrangements of the World Intellectual Property Organization (WIPO) and condemns the creation and spread of innovation, gadgets, or administrations proposed to dodge measures that control access to copyrighted works.
- The unlawful internet gambling enforcement act (UIGEA) is responsible for cutting off the flow of revenue to unlawful Internet gambling businesses. It prohibits receipt of checks, credit card charges, electronic funds transfers, and the like by such businesses.

QUESTIONS TO PONDER

1. Robert Morris created the Morris worm to highlight security flaws and was convicted of the Computer Fraud and Abuse Act 1986. Many big companies pay hackers to identify security flaws in their systems. How has the law evolved over the last few decades?

2. Will a post-secondary establishment reveal budgetary records of an eligible student with the student's guardians?
3. What are the most widely recognized reasons for HIPAA infringement?
4. Under the current ECPA, is there any action one can make to shield Google from giving over the contents of one's entire Google account (Gmail, informing, schedule, Android telephone use, and so on.) upon demand? Given Google's total control of numerous individuals' electronic lives, and the organization's oft-questioned history of consistency, are there steps Google clients can take to guarantee their data privacy?
5. In the off chance that the email on my system is shielded from government search (without a warrant), isn't that equivalent bit of email additionally at that point protected if it lives in the cloud?
6. What was Lori Drew indicted for? Do you concur with the adjudicators' decision on the Lori Drew case? Why/Why not? Are there any changes required to the act being referred to?
7. Normally, organizations take about 200 days to identify a breach. How does the GDPR handle this?
8. Does GDPR compliance vary depending on the quantity of workers an organization has?
9. The Wassenaar arrangement (WA) is the principal worldwide multilateral arrangement to export controls. For what reason is it considered as a cybersecurity law?
10. Would I be able to play online poker legitimately?

KEYWORDS

- **child pornography**
- **cyberethics**
- **cybercrimes**
- **financial establishments**
- **telecommunications**
- **Wassenaar arrangement**

PART II
Cybersecurity: Risks and Policies

CHAPTER 6

Risks in Cybersecurity

In the previous section, we acquainted ourselves with the idea of ethics and law in cyberspace. While ethics and law promote good cybersecurity, security is never guaranteed completely. This is because security risks can never be completely mitigated. In this chapter, we familiarize ourselves with the concept of cyber risks. Analyzing risks, assessing risks, and managing risks are significant for ensuring security in organizations and cyberspace. We also delve into risk remediation, and some use cases involving cyber risks.

6.1 INTRODUCTION TO CYBER RISKS

The cyber risk may be defined as any risk that may result due to financial losses, any kind of disruption, or damage to the reputation of an organization. This often leads to the failure of an organization's information technology (IT) systems. Cyber risk could materialize in a variety of ways. Some of these ways are listed as follows:

- Deliberately breaching security to gain access to information systems;
- Unauthorized access to information;
- Unintentional breaches of security;
- Accidental breaches of security;
- Poor system integrity promoting operational IT risks;
- Unsatisfactory management of cyber risks that may support an assortment of cybercrimes. The consequences may be in the form of data disruption and economic destitution.

It is a well-known fact that businesses commonly find themselves in the middle of unpleasant situations when they struggle to recover lost assets in order to prevent theft in the future. Regardless of whether a business is big or small, cybercrime could be lurking just around the corner. Recently, more and more small businesses are being targeted by cybercriminals, therefore, without the right preventative measures in place, it could aggravate even further. Hence it is necessary to get more familiar with the cyber risks that organizations may face. It is often seen that the more sophisticated and extensive a business' digital operations are, the greater the chances of cyber risks being involved. The following are some elements that contribute to elevation of cyber risks, many of these have the potential to affect an organization:

- Employees or customers accessing the organizational systems from remote locations.
- Staff making utilization of organization-owned devices at their homes or while traveling.
- Employee access to administrative privileges on the organization's network or computers.
- A policy emphasizing on the use of bring your own device (BYOD) in the workplace.
- Accessing public buildings without authorization.
- Employees using computers for accessing bank accounts or initiating monetary transactions.
- Employees neglecting certain policies regarding regular updation of passwords.
- Critical information that may be lost as a result of a network disaster.
- Neglecting to review the company's cybersecurity policies from time to time.

All businesses come across the risks of cyber breaches at some point during their life cycle. However, understanding the level of risks, the factors that led to the risks, and the origin of threats, i.e., source of the threats may be significantly effective in preparing a productive response.

> **Definition 6.1: Cybersecurity Risk**

> Cybersecurity risk may likewise be characterized as the likelihood of exposure or loss due to cyber-attack or information breach in an organization. It is regularly viewed as the possible loss or damage identified with technical infrastructure, utilization of technology, or reputation of an organization.

With organizations becoming more and more technology and business driven, they become more prone to attacks and threats. This is a direct result of the expanding dependence on networks, systems, programs, social media, and information globally. This in turn makes data breaches an increasingly common affair in cyberspace [15, 16, 23]. Cyber-attacks may have a negative business sway, and these frequently emerge from deficiently secured information. Global connectivity and increased use of cloud services with poor default security parameters often lead to an increased risk of cyber-attacks outside the organization [13, 78]. IT risk management may effectively address how access control needs to be complemented by sophisticated cybersecurity professionals, software, and cybersecurity risk management. With the attacks and attackers becoming more and more sophisticated, it may not be sufficient to bank upon traditional IT professionals and security controls for information security. There is an urgent need for sophisticated threat intelligence tools and security programs that may assist in reducing an organization's cyber risk and potential attack surfaces. Decision-makers in the organizations must take into account assessments when dealing with third-party vendors. There is likewise a need to present risk mitigation techniques and cyber incident response (IR) plans in associations for when a breach may take place. Figure 6.1 is a manifestation of how cyber risks may be initiated.

Threat Agent might be characterized as an individual or group that can exhibit a threat. It is necessary to determine who could exploit the assets of an organization and how they might use them against the organization. Therefore, a threat agent may also be thought of as a technique utilized in penetrating the security of an office, operation, or framework by abusing a weakness [14]. Cyber threats, or simply threats, are cybersecurity circumstances or events that are capable of causing harm by way of their outcome. A very common threat is a social-engineering or phishing attack in which an attacker installs a trojan and steals sensitive information from a user's system, applications. Similarly, DDoS is yet another threat (distributed denial of service (DDOS) [14]). While DDoS-ing a site, an administrator can unintentionally leave information unprotected on a production system. This may lead to a data breach, or a storm flooding the ISP (internet service provider)'s data center. Cybersecurity threats are actualized by threat actors. These threat actors may be individuals or entities who have the potential to initiate a threat. Although natural disasters, and other environmental and political events,

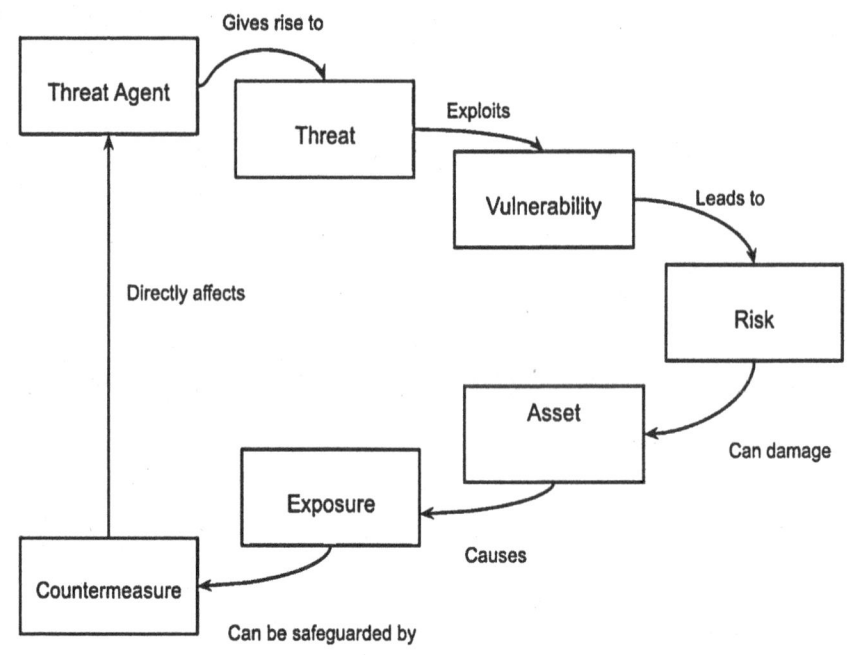

FIGURE 6.1 Understanding cyber risks.

may often constitute threats, they are not considered threat actors. Some basic threat actors are monetarily propelled lawbreakers (cybercriminals), politically spurred activists, nation-states and competitors. In an office environment, careless employees and disgruntled employees are threat actors. Cyber threats can also become increasingly perilous if they can leverage vulnerabilities for gaining access to systems, which are usually operating systems. Vulnerabilities may be thought of as weaknesses in a system that have the potential to make threat results conceivable and are possibly significantly riskier. A single vulnerability is enough to exploit a system. Consider a simple structured query language (SQL) Injection attack that could give an assailant full command over delicate information. An attacker can possibly chain several exploits together, to take advantage of more than one vulnerability. This will allow the attacker to gain more control. Instances of basic weaknesses are SQL injection, Cross-site Scripting, server misconfigurations, delicate information communicated in plaintext, and that is just the beginning. Ordinarily, risks are mistaken for threats despite the fact that there is a slight contrast between them.

A cybersecurity risk may be a combination of a threat probability and loss/impact. This loss impact is usually in the financial terms, however, evaluating a breach is not easy. Essentially, this translates to the following:

$$\text{Risk} = \text{Threat Probability} \times \text{Potential Loss}$$

The situation of risk must be avoided due to the losses that result from that scenario. Let us consider a hypothetical example showing that the risk construction is a given scenario.

A very common vulnerability is SQL Injection, which enables data theft, which is a significant threat. Attackers who are motivated financially could serve as threat actors here. Also, the loss of sensitive data will not only impact the organization in monetary terms, but will also affect its reputation. SQL injection also happens to be an easy access, widely exploited vulnerability, due to which the probability of such an attack is high. Hence, such a vulnerability in this situation may be regarded as a high-risk vulnerability. Thus it is easy to comprehend how threats influence risks. This may be effective in the prevention and mitigation of security breaches. Effective risk assessment and risk management rely on good understanding. It takes into account designing efficient security solutions based on threat intelligence, and may also be responsible for building a viable security strategy and a cybersecurity system. Threat agents may be responsible for generating threats which often leads to vulnerabilities being exploited. Once vulnerabilities are exploited, it leads to cyber risk.

Cyber risks have the capability to damage assets, and this may lead to exposure. While Vulnerability is a shortcoming inside the framework, for example, software package defects [9], unlocked doors, or an unprotected framework and system port leaves things open to an assault or harm, exposure is a solitary occasion when a system is vulnerable to harm. This calls for safeguarding the system, which may be performed using countermeasures.

6.2 CYBER ATTACKS AND THEIR BUSINESS SIGNIFICANCE

Regardless of being helpful, numerous IT arrangements are lacking for giving cyber-attack assurance from complex attacks and poor configuration. With the uncontrolled multiplication of innovation and growth of technology, there is more unapproved access to the association's information than ever before. There is a tremendous increase in information provided to third parties by means of supply chains, customers, and other providers. With organizations increasingly storing, associations are progressively storing enormous volumes of personally identifiable information (PII). For this, they are depending on external cloud providers that should be designed accurately so as to adequately secure information. This leads to added risk; additionally, there is an increase in the number of devices that are always connected in data exchange. With organizations globalizing and the web of employees, customers, and third-party vendors increasing, there is also an increase in expectations of instant access to information. Consumers expect instant real-time access to data from anywhere. The consequence is an exponential increment in the attack surface for malware, weaknesses, and every other exploit. Unexpected cyber threats may emerge from threatening unfamiliar powers, contenders, organized hackers, insiders, poor configuration, and third-party vendors. Cybersecurity guidelines are getting progressively perplexing as orders and administrative principles around the revelation of cybersecurity occurrences and data breaches continue to grow. This makes organizations rely on software that may not only assist their third-party vendors but can also continuously monitor for data breaches. The need of recognizing, tending to, and conveying a potential breach is more noteworthy than the preventive estimation of customary, repetitive IT security controls. Data breaches have a massive, negative business impact. This is because of the insufficiently protected data. A good risk management strategy is external monitoring of risks through third and 4th-party vendor risk assessments. But, there is also a need for a comprehensive IT security management, without which an organization is prone to financial, legal, and reputational risk.

Cybercrimes are additionally known to target businesses [79]. Emerging new technologies based on IoT give rise to some of the biggest cyber threats. As the network keeps on expanding and more devices are added to the network for developing greater connectivity, cyber risks also increase. Thus security measures need to evolve, too. Some common reasons why businesses fall victim to cyber-attacks are as follows:

- Staff shortcomings may make an organization vulnerable. Cybercriminals could attack from anywhere in the world. More company employees are carrying out cyberattacks. This is because they have access to sensitive information, and they can lead to significant damage. However, even good-intentioned representatives can be

a weak connection in your business. Basic attacks like phishing tricks and malware assaults can spread immediately when email attachments are opened and shared randomly.
- Cloud computing provokes security. Workforce, as well as devices, are mobile, and when operations move off-site, traditional security measures become insufficient. With more businesses being connected to the cloud, data may become difficult to defend with firewalls. Not to overlook, cybercriminals are getting progressively refined and are additionally attracted to potentially lucrative targets consolidating delicate information. Hence, Wi Fi and cloud based security services must be protected.
- Ransomware can infiltrate networks. Irrespective of whether businesses are associated with the cloud, ransomware is a critical threat that can rapidly wreck important operations. The WannaCry ransomware infestation around the globe is proof of the extent of damage that can be done by a computer worm. Opening an attachment in the phishing email allows the worm to spread through the local network. This worm can also attach itself to remote hosts and encrypt data and files until the ransom is paid.

6.3 KEY CYBER RISKS AND THREATS

Cybersecurity is applicable to all systems that contribute to an organization's business operations and objectives. It may also be applied to an organization's compliance with regulations and laws. Associations essentially plan and actualize cybersecurity controls for ensuring the integrity, confidentiality, and availability of information assets. While no single reason can be attributed to cyberattacks, they may be committed as financial frauds, data theft, activist causes, to refuse assistance, upset basic infrastructure and indispensable administrations of government or an association. The common sources of cyber threats are as follows:

- Nation-states;
- Cybercriminals;
- Hacktivists;
- Insiders;
- Service providers;
- Developers of substandard products and services.

To understand the cyber risk profile of an organization, it is necessary to identify the kind of information that would be valuable to outsiders. Now and then hackers target data that can cause critical disturbance. It is inexorably critical to figure out what data has the capability of causing monetary or reputational harm to the association if it somehow happened to be obtained or made open. One such information is PII that includes names, social security numbers, and biometric records. The following may be considered as potential targets to cybercriminals:

- Customer data;
- Employee data;
- Intellectual property (IP);
- Third and 4th-party vendors;
- Product quality and safety;
- Contract terms and pricing;
- Strategic planning;
- Financial data.

6.4 WHO SHOULD OWN CYBERSECURITY RISK?

Cybersecurity risk management is usually set by leadership which may incorporate an organization's board of directors in the planning processes. Most organizations will also include a chief information security officer (CISO). The CISO establishes and maintains the enterprise vision. He/she also plans the strategies and programs to ensure that the information assets and customer data may be adequately protected. Some common cyber defense operations that are undertaken by a CISO are as follows:

- Administering security procedures;
- Training and testing;
- Maintaining secure device configurations;
- Up-to-date software;
- Vulnerability patches;
- Deploying intrusion detection systems and establishing penetration testing;
- Configuring secure networks to manage and protect business networks;
- Deploying programs that may be useful for data protection and loss prevention;
- Deploying programs that will perform system monitoring for risks;

- Restricting access to least required privilege for employees;
- Encrypting data when required;
- Configuring cloud services;
- Implementing vulnerability management programs;
- Recruiting and retaining cybersecurity professionals.

In a situation where an organization cannot support a CISO or other cybersecurity professionals, it may be possible for board members with experience in cybersecurity risk to step up. Therefore, it is significant for all degrees of an association to comprehend its role in managing cyber risk. Since vulnerabilities can come from any employee, it is fundamental to an organization's IT security to continually educate employees. The employees must be educated on how to avoid common security pitfalls that may cause data breaches or other cyber incidents. The National Institute of Standards and Technology's (NIST) cybersecurity framework (CSF) provides some of the best practices to manage cybersecurity risk. We will understand its cybersecurity policy framework in Chapter 8.

6.5 CYBERSECURITY RISK IDENTIFICATION AND MANAGEMENT

Risk identification may be thought of as the process of identifying and assessing threats to an organization, its operations, and its workforce. It takes into account assessing IT security threats. These threats may be in the form of malware and ransomware, accidents, natural disasters, and other potentially harmful events. The aim of such threats is to disrupt business operations. Hence it is important for companies to develop risk management plans. Companies that develop robust risk management plans conveniently minimize the impact of such threats, whenever they occur.

The process of risk identification and management includes five core steps which are risk identification, risk analysis, risk evaluation, risk treatment, and risk monitoring. We will be discussing about these in detail:

1. **Risk Identification:** The primary objective of risk identification is to unveil what, where, when, why, and how something could affect an organization's ability to operate. For example, 'the possibility of wildfire' may have an impact over a business located in central California. The consequences could be disruption in business operations.
2. **Risk Analysis:** The procedure of risk analysis is utilized for researching the likelihood that a risk event may happen. Risk

analysis additionally investigates the possible results of the related events. Considering the California wildfire example, it may be asserted that safety managers could be responsible for assessing how much rainfall has occurred over the past few months. This may give an idea with respect to the degree of harm the organization may confront if a similar situation takes place in the future.

3. **Risk Evaluation:** It is responsible for comparing the magnitude or intensity of each risk. It also ranks them according to prominence and consequence. Considering the same example, the effects of the wildfire could be weighed against the effects of a possible earthquake or landslide. Undergoing risk evaluation, the event that is identified to have a higher likelihood of occurring and causing harm, would rank higher.
4. **Risk Treatment:** Or risk response planning takes into account risk mitigation strategies, preventative care, and emergency courses of action which are made based on the surveyed estimation of each risk. Considering the wildfire example, risk managers may decide to house extra network servers offsite, so business activities could in any case continue if an onsite server is harmed. In addition, the risk manager may likewise create evacuation plans for employees.
5. **Risk Monitoring:** Risk monitoring is a constant procedure that adjusts and changes after some time. Repeating and continually monitoring the procedures can help guarantee maximum coverage of known and obscure risks.

There are numerous ways to identify risks. Project managers (PM) often combine techniques to identify risks. Some of the risk identification techniques are as follows:

1. **Interviews:** Choosing key partners and arranging the meetings. This technique also incorporates defining specific questions and documenting the results of the interview.
2. **Brainstorming:** These sessions must be planned in advance. They should include project objectives (schedule, budget, quality, or scope) as well as project tasks (Significant risks, requirements, coding, testing, training, implementation).
3. **Checklists:** It is necessary to see if the company has a list of the most common risks. If not, such a list may be created. After each project, a post review must be conducted for capturing the most consequential risks. Also, this list may be used for subsequent

projects in the future. However, it is a known fact that no checklist can incorporate all the risks.
4. **Assumption Analysis:** According to the project management body of knowledge (PMBOK), assumption may be defined as factors that are considered to be true, real, or certain. They may not be supported by any proof or demonstrations. Assumptions are sources of risk. PM should ask stakeholders, regarding the assumptions for the project. Furthermore, these assumptions and associated risks must be documented.
5. **Cause and Effect Diagrams:** These are very useful. This is because PM can use these to help identify causes of risks. Addressing the causes may assist in reducing or eliminating the risks.
6. **Nominal Group Technique (NGT):** This technique is a brainstorming technique such that inputs are collected and ranked in order of priority. Similarly, the output of NGT is another prioritized list of risks.
7. **Affinity Diagram.** This technique is fun, creative, and beneficial exercise. Participants are asked to brainstorm risks which are then sorted into groups or categories. Lastly, each group is given a title (Figure 6.2).

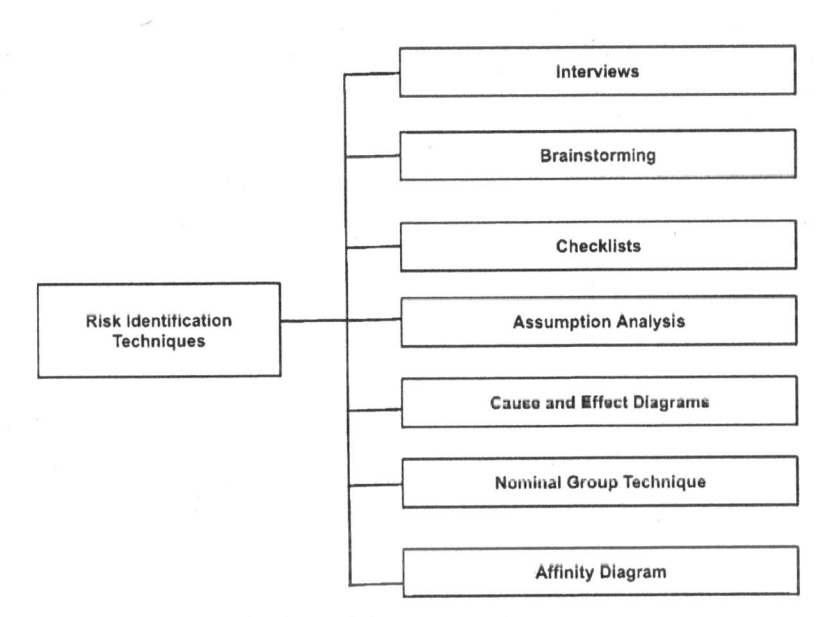

FIGURE 6.2 Risk identification techniques.

In order to ensure voluminosity of risk identification, an organization must identify the factors contributing to risks. The factors can be both internal as well as external. Risk identification can be performed in many different ways, some of which we will be discussing in this chapter:

1. **Strategic Risk Identification:** This may be used to identify risks that arise from strategic choices made by an organization. The choices are made keeping in mind whether they weaken or strengthen the organization's ability to execute its constitutional mandate:
   - Strategic risk identification must precede the finalization of strategic choices. This is to guarantee that potential risk issues are factored into the decision-making process for choosing the correct alternatives.
   - Risks inherent to the chosen key decisions must be archived, surveyed, and overseen through the typical working of the arrangement of risk management;
   - Strategic risks must be reviewed from time-to-time following changes in strategy, considering new and emerging risks.

2. **Operational Risk Identification:** This is targeted to recognize risks inside the organizations' functioning and operations:
   - Operational risk identification is responsible for establishing vulnerabilities which are generally introduced by employees. Many times they are introduced by internal processes and systems, contractors, regulatory authorities and external events.
   - Operational risk identification must be an embedded continuous process. This is because it is responsible for identifying new and emerging risks and considering shifts in the already existing risks through mechanisms like management and committee meetings, environmental scanning, process reviews.
   - Operational risk identification must be repeated time and again, quarterly or yearly, especially when changes occur. This is helpful in recognizing new and rising risks.

3. **Project Risk Identification:** It is responsible for identifying risks that are inherent to particular projects: project risks ought to be distinguished for every significant task, covering the entire lifecycle:
   - Project risks must be identified for all major projects. They should cover the whole lifecycle of the project.

- In the case of long-term projects, reviewing must be performed at least once a year for identifying new and emerging risks.

Before starting with the risk identification process, it is necessary to have knowledge of the business. For identifying risks that an organization may be prone to, there may be a need to learn from past experience as well as experiences of others. Only then can one come up with the best strategy for responding to the risks. The first thing to consider is the objectives of the institution, both implicit and explicit. There is a need to distinguish unpleasant events, undesirable results and rising threats. Identifying emerging opportunities may also play an important role in risk identification. This is because risk also has an opportunity component. Although an organization will never be 100% risk free, it is necessary to establish controls. While working on the risk identification process, it is equally important to pay attention to identifying potential opportunities. Exploiting such opportunities may also improve an organization's performance. For identifying risks, consideration must also be given to risks associated with not pursuing an opportunity. For example, failure to implement an IT system for collecting municipal rates. Risk identification exercises must not be limited to conceptual or theoretical detail. It should not be restricted to a fixed list of risk categories. Anyway, lists might be compelling in recognizing risks. There are several steps to effectively identify risks in an organization. Some of these have been mentioned as follows:

1. **Understanding What to Consider When Identifying Risks:** In order to develop an extensive list of risks, there is a need to come up with a systematic process that incorporates defining objectives as well as key success factors for their achievement. This is helpful explicitly for giving certainty that the procedure of risk identification is finished and significant issues have not been missed.

2. **Gathering Information from Different Sources to Identify Risks:** In order to identify risks, there is a need for good quality information. To start with risk identification, one may consider historical information regarding the corresponding organization. This may be followed by a discussion with a number of stakeholders regarding the historical, current, and evolving issues, performance indicators, economic information, scenario planning, or any other information that can produce important information related to risk. There are many processes that are relied on during strategic

planning like strength weakness opportunity and threat (SWOT) analysis, political economic social technological environment and legal (Pest (EL)) Analysis and benchmarking. These techniques may be used to identify important risks and opportunities that must not be neglected, hence, they must be listed in a risk register. Certain disciplines like IT, Strategic Management, Health, and Safety, etc., rely on risk identification methodologies which are considered mandatory as per their professional norms. The outputs of these techniques are analyzed in order to identify risks. The sooner risks are identified, the better it is for the organization.

3. **Applying Risk Identification Tools and Techniques:** Organizations have different objectives. Hence they must apply a set of risk identification tools and techniques that complement their objectives, capabilities, and risks. The appropriate and latest information regarding risks can be useful in identifying them. This information may also incorporate background details. People with sufficient experience and knowledge of risks must be involved in identifying them. There are several approaches that may be used to identify risks. Few of these are utilizing agendas and checklists, decisions dependent on experience and records, use of flow charts, brainstorming sessions, analyzing systems and scenarios and many other system engineering techniques. The approach used may rely on the nature of the activities under review, types of risks, the organizational context, and the purpose of the risk management exercise. A popular approach is team-based brainstorming since it supports commitment, takes into account alternate points of view, and embraces contrasting experiences. There are other structured techniques like flowcharting, system design review, frameworks investigation, hazard and operability (HAZOP) contemplates, and operational modeling. Nonetheless, these must be utilized where the potential results are disastrous and the utilization of such concentrated procedures are cost-effective. Risk workshops are answerable for separating and screening of potential risks. Therefore, the workshops could be supplemented by more sophisticated or structured techniques. For situations that are not defined, it may be necessary to perform a what-if and scenario analysis. Here, resources available for risk identification and analysis may be constrained. Moreover, the structure and approach may have to be adapted for achieving efficient outcomes within budget limitations.

If there is limited time for identifying risks, a smaller number of key elements may be taken into account at a higher level. This may also be accompanied by a checklist.

4. **Documenting the Risks Identified:** The risks identified during the risk identification are regularly recorded in a risk register that incorporates risk description and causes and results of the risks. It may also incorporate existing internal controls that are responsible for reducing the likelihood or consequences of the risks. In order to identify risk, the following techniques may be used:

   i. **Description of an Event:** It may be necessary to describe the occurrence of an event or a specific set of circumstances that can provide information about risk.

   ii. **Causes:** Identifying factors that contribute to a risk is equally important. Sometimes, the causes can identify if the occurrence and increase in the risk.

   iii. **The Likelihood of a Risk Occurring Consequences:** This defines the outcome(s) or consequences of an event. identifying outcomes may also lead to identification of risks.

   A combination of the elements mentioned above can contribute to the identification of risks. These elements provide information on factors that make up risk. An in-depth analysis of such factors and corresponding details will enable an organization to better understand its risks.

5. **Documenting the Risk Identification Process:** Along with documenting identified risks, it is likewise important to report the risk identification procedure. This may help with controlling future risk identification activities and guaranteeing good practices are kept up. Utilizing past activities and experiences, lessons can be learnt for better risk identification. Hence documentation is necessary. Documentation usually incorporates information like:

   i. The methodology adopted for identifying risks;
   ii. How extensive is the process of identification;
   iii. The participants in the risk identification;
   iv. Information sources taken into account.

Frequently the board ignores well-controlled risks while archiving the risk profile of the association. Consequently, a well-controlled risk should be recorded in the risk profile of the association. This is because in order to identify risks, there is a need to identify factors that can be ignored and mitigated. However, the factors will be considered when the risk is being assessed.

After the process of risk identification is completed and existing controls have been assessed, there is a need to assess if controls are adequate. If it is established that controls are inadequate, there needs to be an assessment of whether the risk is acceptable or not. There may also be an option of whether the risk needs to be treated or not.

6.6 RISK IDENTIFICATION USING SWOT ANALYSIS

Risk Identification has been characterized as a procedure of recognizing risks. Risks have the potential to prevent an institution, organization, or investment from achieving its goals. It incorporates documenting and communicating the concern. If a risk is poorly identified, it may not be possible for the project manager to properly and efficiently communicate the corresponding risk to team members as well as to other high-level stakeholders. Poor identification of risks makes it extremely difficult for the parties involved as well as stakeholders and partners to appreciate the measurement or force of the risk. As a consequence, instead of the risk materializing, the project manager and the team will be manifested ill-prepared. However, the issue in the first place is with the risk identification, since it was not understood or communicated properly. SWOT analysis is a risk identification technique which is performed using the SWOT matrix. It is popularly used by PM. Often organizations implement SWOT analysis in their risk management strategies. SWOT stands for Strengths, Weaknesses, Opportunities, and Threats, and it aims to identify any opportunities for the project that may emerge from organizational qualities, and any threats emerging from organizations' shortcomings. The SWOT framework is used by organizations and teams for assessing the internal and external influences. These influences may have an impact on a project, product, or an institution in a positive or negative way. The analysis is useful in identifying internally-generated risks that arise from within the organization. SWOT examination is more about the association than the task. It contributes immensely in strategic decision making and is commonly used to manifest important project information to management. This technique has proven to encourage management for altering some environmental factors from the strengths and weaknesses sections that may have an impact on the project.

A SWOT examination can likewise be used for deciding the internal and external factors that may be beneficial or detrimental to a business venture. SWOT analysis for the team may be used to address the team's challenges for sustainment and growth. A SWOT analysis is usually performed with a team leader or an entire team. During SWOT analysis, it is important to capture all ideas so that everyone on the team must have a good understanding regarding the risks to the team. Once the analysis is performed, the team presents action plans for addressing the issues that were identified during the analysis. This can significantly improve the likelihood of success and sustainment for the team. The Strengths and Weakness analysis focuses on issues within the organization (internal), while Opportunities and Threats are usually focused on external factors. A SWOT analysis consists of items captured in a four-box layout. Strengths and Weaknesses are listed in the top 2 boxes, while opportunities and threats are listed in the bottom 2 boxes (Figure 6.3).

STRENGTH	WEAKNESS
OPPORTUNITY	THREAT

FIGURE 6.3 SWOT matrix.

6.7 RISK ANALYSIS

Risk analysis refers to the survey of the risks related to a specific occasion or activity. Risk analysis is important for ventures, innovation, technology, security issues, and any activity where risks need to be analyzed quantitatively and qualitatively. Risk analysis may be thought of as a component of risk management as risks are part of technology-driven projects and business endeavors. Therefore, it is imperative that risk analysis occurs on a recurring basis. There is a need to update risk analysis for accommodating new potential threats. Strategic risk analysis can adequately diminish harms and future risk probabilities. Risks can be associated with individuals who use computers incorrectly or inappropriately, as this may lead to security risks. There is risk associated with ventures that are not finished in a timely manner, since this leads to significant costs. In order to identify risks, quantitative and/or qualitative risk analysis is applied. A contingency plan may also be utilized during risk analysis. In the event that a risk is introduced, emergency courses of action help limit harm.

Risk analysis may be of two types: qualitative and quantitative. Quantitative risk analysis (QRA) measures the expected risk likelihood to gauge assessed money-related misfortunes from possible risks. Qualitative risk analysis does not utilize numbers however surveys threats and decides and sets up risk mitigation techniques and arrangements. QRA relies on pre-existing relevant and verifiable data. This data is used to produce a numerical worth which is then used to foresee the likelihood of a risk event outcome. Qualitative risk analysis relies on subjective assessment of risk occurrence likelihood (probability). This probability is measured against the potential severity of the risk outcomes (impact) to determine the overall severity of a risk.

6.7.1 QUALITATIVE RISK ANALYSIS

A qualitative risk analysis highlights the identified project risks through a rating scale such that risks are given scores. These scores depend on their likelihood or probability of happening. The scores additionally consider the effect on project objectives in the event that they happen. Probability/likelihood is featured on a zero to one scale. The impact scale is organizationally defined based on budget, schedule, or quality. Qualitative risk analysis may likewise consolidate the suitable classification of the risk, either source-based or effect-based.

The process of Risk Management is not only continuous but is collaborative too. Hence it takes into account the implementation of both Quantitative and Qualitative Risk Analysis techniques. Numerous ventures weigh on including numerous obligatory QRA studies in their scope, but managing the day-to-day risks inherent in every project is often overlooked with respect to formal QRA requirements. In order to manage these types of risks, an ongoing on-going collaboration between team members is required. Sometimes, this is followed by regular risk review workshops. The various techniques used in Qualitative Risk Analysis can differ significantly, owing to the type of project being run. This is likewise controlled by the risk management assets and resources accessible to the project. Some of the most useful QRA techniques applied in project management are discussed in subsections.

6.7.1.1 DELPHI TECHNIQUE

It is a common form of risk brainstorming. However, there is a slight difference between Delphi Technique and brainstorming. The Delphi Technique relies on the opinion of experts for identifying, analyzing, and evaluating risks on an individual and anonymous basis. Every expert at that point audits each other expert's risks, and a risk register is created through consistent survey and agreement between the experts [80].

The Delphi technique is a multistep strategy and it is prominently used to gauge future demand for an item or service. In Delphi strategy, a unique gathering of specialists and experts trade their perspectives on risk, cost, and schedule of a given project. Each expert submits individual estimates and assumptions to an analyst who then reviews all the data received and issues a summary report. This summary report is considered and reviewed individually. The review is conducted by the group members who analyze the reports and submit reworked forecasts to the analyst. The analyst considers this second review issues a secondary report. The process is carried on until all participants find a convincing solution. The Delphi technique is useful when other methods are either inadequate or inappropriate for data collection. The Risk Management process ordinarily utilizes Delphi method. According to Lidstone and Turoff [81], Delphi Technique may be applied if the problem cannot be applied for exact investigative methods however can profit by subjective decisions on a collective basis. In order to carry out the Delphi method, the individuals must have experience with expertise and must be able to interact effectively. It may not be possible to arrange for frequent group meetings due to time and cost involved, hence Delphi technique may be very useful. SA supplemental gathering correspondence forms are known to expand the proficiency of up close and personal gatherings. Scheele showed a procedure that recorded a typical sequence of occasions in the Delphi procedure in six stages [82]:

1. **Recognizing the Group Members Whose Agreement Suppositions are Looked for:** If the review goes past an intact group the end goal that delegates must be chosen, it must be guaranteed that all the different publics or positions are proportionately examined.
2. **Questionnaire One:** Having every member produce a rundown of objectives, concerns, or issues toward which agreement assessments are wanted. Altering the outcomes to a sensible outline of things introduced in random requests. Setting up the second survey in a suitable format for rating or ranking.
3. **Questionnaire Two:** Having each member rate or rank the resulting items.
4. **Questionnaire Three:** Introducing the consequences of Questionnaire Two as Questionnaire Three, demonstrating the preliminary degree of gathering agreement to every item. If the opinion of an individual differs from the group, and the individual chooses to remain so on Questionnaire Three, the respondent needs to provide an explanation for the same.
5. **Questionnaire Four:** Questionnaire four is basically the results of Questionnaire Three. These questions may manifest a new level of group agreement for each item. This may be followed by repeating the member's latest ranking, along with a listing by item of the major reasons' members had for dissent from the prevailing group position. Finally, each member is given the opportunity to rate each, owing to the emerging pattern of group agreement and the reasons for the objection.
6. **The Results of Questionnaire Four:** These are considered the final statement of group agreement and are presented in a tabular form.

There are many risk factors involved in using the Delphi technique for risk management. They have been listed as follows:

- The judgments formed are passed by specific those group of people and hence, may not be representative;
- There is a tendency to terminate extreme positions as well as force a middle-of-the-road agreement;
- Delphi technique is time-consuming;

- Delphi technique cannot be viewed as a complete solution;
- It requires written communication skills;
- Performing the Delphi technique means adequate time and participant commitment; it may take months to conduct an analysis.

6.7.1.2 SWIFT ANALYSIS

SWIFT analysis or structured what-if technique, is a simplified version of a HAZOP and applies an orderly, group-based methodology in a workshop environment. In this, the group explores how transforms from an affirmed structure, or plan, may influence a project through a progression of what-if scenarios. This method is especially compelling in assessing the suitability of opportunity risks [83]. The Structured What-If Checklist Technique (SWIFT) uses a combination of checklists along with brainstorming what-if approach. In order to provide an effective hazard identification technique where HAZOP fails, the SWIFT procedure was proposed as a productive alternative. SWIFT may likewise be utilized related to or corresponding to a HAZOP. SWIFT is an intensive, methodical, multi-disciplinary group-orientated explanatory strategy. While HAZOP looks at the facility item-by-item, technique-by-technique, and so on by applying guidewords, SWIFT is a system-oriented procedure which analyzes complete systems or subsystems. SWIFT guarantees extensive identification of hazards and considers organized conceptualizing effort by a group of experienced experts. This team is supported by a list of supplementary questions that are developed specifically for identifying risks. The team incorporates specialists, or subject matter experts who can assist in evaluating the consequences of hazards. These hazards are the result of potential failures or errors that have been identified. For distinguishing risks, the inquiries raised are essentially with respect to the plan or activity of a system, deviation of system from normal operations, assessing the likelihood of an incident, the consequences, the safeguards involved to prevent or mitigate the risks, etc. The viability of SWIFT in recognizing hazards depends on the inquiries raised concerning important areas. SWIFT is capable of conveniently converging different types of failures or errors that may have been the result of a hazard within the system being examined. The what-if questions can be posed by any team member. These questions are organized by different question categories. The SWIFT investigation is additionally reinforced by the utilization of category-specific checklists and agendas. These category specific checklists are also used at the conclusion of each question category. This ensures an additional level of meticulousness. The results incorporating conclusion and information of the SWIFT meeting are recorded on log sheets in columns as 'What If,' 'Consequences,' 'Existing Safeguards,' and 'Recommendations.'

SWIFT may be suitable for different types of risk assessment applications, specifically the ones that are dominated by relatively simple failure scenarios. It may also be used separately, but mostly is used in conjunction with other, more structured techniques.

> **Advantages:**
- The technique is efficient since there is no need to perform lengthy discussions in domains where risks are surely known or where earlier analysis has manifested that no hazards exist.
- It is adaptable and applicable to different kinds of installation, operations, or processes. It can be performed at any stage of the life cycle.
- It does not consider deviations redundantly, hence it is quick.
- While performing SWIFT analysis, the experience of operating personnel is considered as part of the team.
- SWIFT sessions can take place even if the subject matter experts are not present. This is because the questions can be collected in advance and appended to the checklist.
- Since the inquiries take into account historical events that have occurred previously, the checklists used are very robust.

> **Disadvantages:**
- For achieving completeness, there is a need to be prepared beforehand. This includes preparation of a checklist.
- SWIFT analysis can be conducted only if the leader is experienced and the team is knowledgeable.
- SWIFT takes into account the knowledge of the participants to identify potential issues. If important questions are not raised by the team, the analysis may not be efficient.
- It is difficult to review a what-if analysis.
- What-if reviews mostly produce qualitative results, hence there are no quantitative estimates with respect to risk-related characteristics.

- Although it is a simplistic approach and offers great value for a minimal investment, it can answer more confounded risk related inquiries just if some level of measurement is included (for instance utilizing Risk Matrices).

6.7.1.3 BOW-TIE ANALYSIS

Bow-tie analysis is one of the most practical methods accessible in distinguishing risk alleviations. It is performed by observing a risk event and then projecting this in two different columns. One section records all the expected reasons for the event that are recorded. The other column includes all the potential consequences of the event. Hence, identifying, and applying mitigations (or barriers) to each of the causes and consequences separately, becomes easy. This method can be useful for effective mitigation of both probability of risk occurrence and the subsequent impacts, respectively.

6.7.1.4 PROBABILITY/CONSEQUENCE MATRIX

This is one of the standard methods in building up risk severity in Qualitative Risk Analysis. Risk Matrices may change in size; however, they all basically do something very similar. They provide a practical means for positioning the general seriousness of risk. This is performed by obtaining the product of the likelihood of risk occurrence with the impact of the risk, should it still occur. By ranking risk probability against risk consequence, it is possible to identify the overall severity of the risk. Moreover, the main driver of the risk severity can also be identified whether it is, probability or consequence. This information is essential for identifying suitable mitigations for managing the risk, based on its prominent drivers. In qualitative risk analysis, a probability impact (PI) Matrix is normally used to depict the severity of the hazard, utilizing the presumption that risk severity or magnitude is the product of probability (likelihood) and result (consequence). In semi-quantitative terms:

$$\text{Risk Exposure} = \text{Probability} \times \text{Impact}$$

Risks are assessed for probability along the vertical axis, and impact is assessed along the horizontal axis. However, the impact of units and thresholds are different for different category consequences. This is the thing that empowers risks of varying outcome categories to be consolidated in the one PI matrix and ranked in the one qualitative risk analysis register. A risk assessment matrix (RAM) might be characterized as a tool that may be useful in determining which risks one needs to develop a risk response for. In order to develop a RAM, the first step would involve defining the rating scales with respect to likelihood and impact. For qualitative analysis, likelihood or probability is measured using a relative scale. An example of a likelihood scale is given in Figure 6.4.

Rating	Likelihood	Description
1	Very Low	Highly Unlikely to occur. May occur in exceptional situations
2	Low	Most likely will not occur. Infrequent occurrence in past projects
3	Moderate	Possible to Occur
4	High	Likely to occur. Has occurred in Past projects
5	Very High	Highly likely to occur. Has occurred in past projects and conditions exist for it to occur on this project

FIGURE 6.4 Likelihood scale.

A low likelihood of occurrence for one project could imply that a risk event is probably not going to happen inside the next 10 arrangements. Another type of project 'low' may refer to a risk event that is unlikely to occur within the next year. The impact scale for a project may also incorporate different contemplations, for example, scope, political, and worker impacts (Figure 6.5). With the rating scales prepared, it is possible to create a RAM that may assist in categorizing the risk level for each risk event.

A risk event with a Moderate Likelihood of occurring and a High impact, is considered a Moderate Risk using the RAM shown in Figure 6.6.

6.7.2 QUANTITATIVE RISK ANALYSIS (QRA)

A QRA is used to assess the highest priority risks. In QRA, a numerical or quantitative rating is allocated so as to build up a probabilistic investigation

Risks in Cybersecurity 265

of the project. The underlying idea is to quantify the possible outcomes for the project. This is followed by assessing the probability of achieving specific project objectives. It likewise gives a quantitative way to deal with settling on choices when there is unpredictability and creates realistic and achievable cost, schedule, or scope targets. For conducting a QRA, high-quality data is needed along with a well-developed project model, and an organized list of project risks. This information is acquired generally from performing a qualitative risk analysis.

Rating	Impact	Cost	Schedule
1	Very Low	No increase in budget	No change to schedule
2	Low	<5% increase in budget	<1 week delay to schedule
3	Moderate	5-10% increase in budget	1-2 weeks delay in schedule
4	High	10-20% increase in budget	2-4 weeks delay in schedule
5	Very High	>20% increase in budget	>4 weeks delay in schedule

FIGURE 6.5 Impact scale.

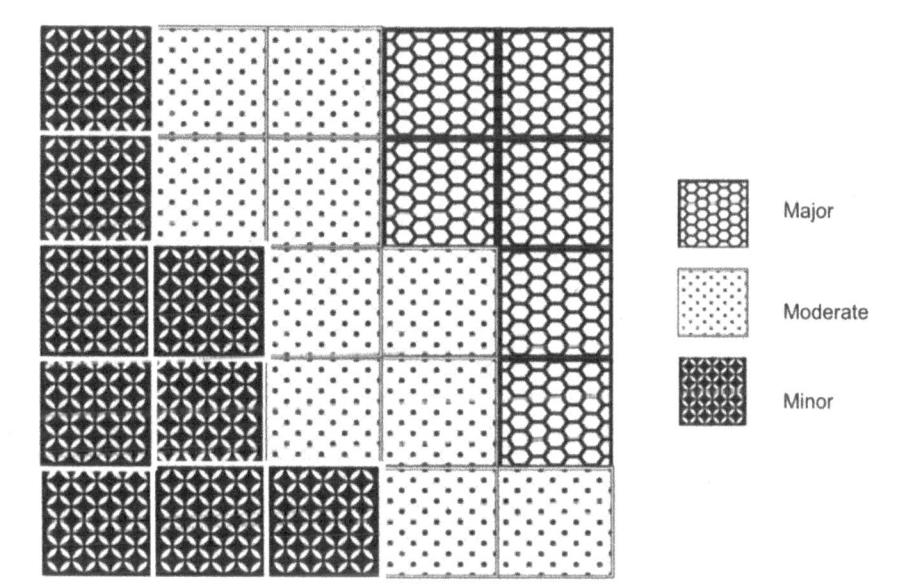

FIGURE 6.6 Risk assessment matrix.

Henry Gantt developed the first known quantitative analysis in 1917 in the form of the Gantt Chart [84] which, at that point, was utilized only for planning risk investigation. Gantt charts form the basis of most scheduled applications used nowadays. However, taking into account QRA, we have an expansive scope of choices that are explicit to various risk types and their effects. This is the result of the evolution of the available methods of the last few years. In QRA there are several methods like: Bayesian analysis, Markov analysis, Monte-Carlo analysis, failure mode and effect analysis (FMEA), etc. It is mandatory to practice carrying out a technical safety QRA. This will be useful for evaluating the risk to personnel who work on the facility. The main objectives of this analysis are establishing the individual risk per annum (IRPA) and potential for loss of life (PLL). The analysis may further be used for recommending measures that can vouch for keeping the risks as low as reasonably practicable (ALARP). IRPA is the product of the location specific individual risk (LSIR) and the extent of time an individual spends in that area. PLL is the product of IRPA and the quantity of workforce working inside the area. LIST is given by the overall summation of the products of frequency of each anticipated major accident event (MAE) and probability of fatality due to an MAE at that location. These calculations may be defined mathematically as follows:

$$LIST = \sum (F \times P)$$
$$IRPA = LSIR \times T$$
$$PLL = IRPA \times N$$

Once the results for LSIR, IRPA, and PLL are found, they are compared with a set of Risk Tolerability Criteria. If the results fall outside the acceptable range, mitigation measures have to be considered in order to ensure that the results fall within the acceptability criteria.

QRA may be thought of as a numeric estimate describing the overall effect of risk on the project objectives. The numeric estimate may be in the form of cost and schedule objectives. The results may also comment on the likelihood of project success and may also be used to develop contingency reserves. In the case of critical decisions, QRA provides objective information and data than with respect to qualitative analysis. Although quantitative analysis is more objective, it is still an estimated value. Therefore, it might be useful to consider other factors in the decision-making process. Quantitative risk assessment tools and techniques QRA apparatuses and procedures incorporate however, are not constrained to the following subsections.

6.7.2.1 THREE-POINT ESTIMATE

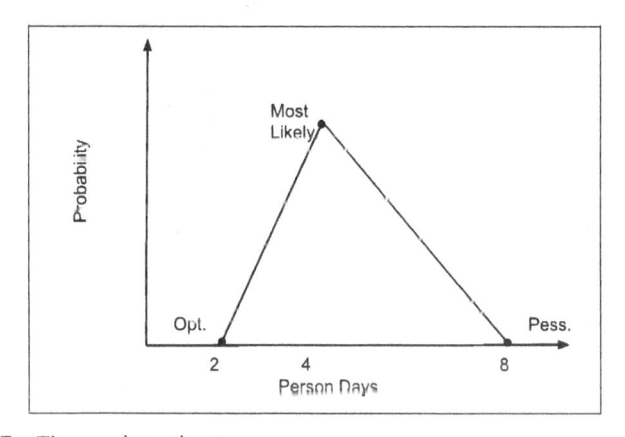

Three-point estimate technique uses the optimistic and pessimistic values to determine the best estimate. This method is used for constructing an approximate probability distribution which can represent the outcome of future events, despite having limited information. Hence, it is a widely accepted technique in management and information systems applications. The distribution used for the approximation is usually normal distribution, but that is not always the case. A much simpler triangular distribution can also be used, based on the application. In a three-point estimation, three figures are created at first for each distribution that is necessary, in view of related knowledge or best-guesses. These figures are then converged to get a full likelihood distribution. The accuracy ascribed to the outcomes inferred can be no better than the accuracy characteristic in the initial points, and there are clear risks in utilizing an assumed form for an underlying distribution that itself has little basis. Three-point estimation looks at three values, the most optimistic estimate (O), a most likely estimate (M), and a pessimistic estimate (least likely estimate (L)). Three-point Estimate (E) depends on the straightforward average and follows the three-sided or triangular distribution (Figure 6.7):

$$E = (O + M + L)/3.$$

FIGURE 6.7 Three-point estimate.

In triangular distribution, Mean = (O + M + L)/3
Standard deviation = $\sqrt{[((O - E)^2 + (M - E)^2 + (L - E)^2)/2]}$.

Three-point estimation steps:

- **Step 1:** Arriving at the work breakdown structure.
- **Step 2:** For each task, finding the most optimistic estimate (O), and a pessimistic estimate (L) and a most likely estimate (M).
- **Step 3:** Calculating the mean for the three values. Mean = (O + M + L)/3.
- **Step 4:** Calculating the three-point estimate of the task. Three-point estimate is the mean. Hence, E = Mean = (O + M + L)/3.
- **Step 5:** Calculating the standard deviation (SD) of the task.
  Standard Deviation (SD) = $\sqrt{[((O - E)^2 + (M - E)^2 + (L - E)^2)/2]}$.
- **Step 6:** Repeating Steps 2, 3, 4 for all the tasks in the work breakdown structure.
- **Step 7:** Calculating the three-point estimate of the project. E (Project) = $\sum$ E (Task).
- **Step 8:** Calculating the SD of the project.

$$SD\ (Project) = \sqrt{(\sum SD\ (Task)^2)}$$

Converting the Project Estimates into Confidence Levels. The Three-point Estimate (E) and the SD that have been determined might be utilized for transforming the project estimates to Confidence Levels. The transformation is based on the end goal that the Confidence Level in E +/SD is roughly 68%. Confidence Level in E value +/− 1.645 × SD is approximately 90%. Confidence Level in E value +/− 2 × SD is approximately 95%. Confidence Level in E value +/− 3 × SD is approximately 99.7%. Ordinarily, the 95% Confidence Level, i.e., E Value + 2 × SD, is utilized for all venture and task estimates.

6.7.2.2 EXPECTED MONETARY VALUE (EMV)

This method is useful for establishing the contingency reserves with respect to a project budget and schedule. The formula for EMV of risk is as follows expected monetary value (EMV) = Probability of the Risk (P) × Impact of the Risk (I) or simply:

$$EMV = P \times I$$

EMV alludes to the average outcome when the future incorporates questionable situations positive (opportunities) or negative (threats). Positive qualities signify Opportunities while negative qualities express Threats. The two qualities will be considered by including them together.

6.7.2.3 DECISION TREE ANALYSIS (DTA)

Decision tree analysis (DTA) can be performed using a diagram. This chart can show the ramifications of picking one or different other options. DTA can help identify the best course of action in situations of uncertainty. There might be unpredictability in the result of potential occasions or proposed plans. DTA investigation is performed by beginning with the initial proposed decision which is mapped to different pathways and outcomes. These pathways and outcomes are the different results of events occurring from the initial decision. After the establishment of all pathways and outcomes, their probabilities are evaluated. This is followed by a course of action that may be selected based on multiple features. These features can be associated events, probability of success, and a combination of the most desirable outcomes. DTA relies on EMV analysis internally. Since decisions are to be made, there are multiple options. here, future uncertain events are taken into account and the event names are put into rectangles. Option lines are drawn from here, which serve as decision points or decision nodes and multiple chance points or chance nodes when a decision tree is drawn. Each point may be represented by different symbols. These symbols may be in the form of: a filled up small square node which represents a decision node; a small, filled-up circle which is indicative of a chance node; and a reverse triangle is where the branch terminates in the decision tree. These are noted in Figure 6.8.

This structure results in a diagram that looks like a tree branching from left to right. Hence it is called a decision tree. For analyzing a decision tree, we start from the decision node and then move from left to right. At the decision node, branching begins and each branch can prompt a chance node. Similarly, at every chance node, there can be further branching. Finally, a branch terminates with the end-of-branch symbol. The node of the branch typically includes the probability value. The cost value (impact) is usually the end. Next, we perform some calculations on the branches of the tree. For calculation, we move from right to left on the tree. The cost value can be either at the end of the branch or on the node. We generally follow the branch to do the calculation. The best decision is the option that gives the highest positive value or lowest negative value, depending on the scenario.

Notation	Shape	Meaning
■	Filled up square	Decision Node
●	Filled up circle	Chance (Condition) Node
◀	Reverse Triangle	End of Branch

FIGURE 6.8 Decision points.

6.7.2.4 MONTE CARLO ANALYSIS

Monte Carlo analysis is used for determining the total project cost and project completion dates [85]. This is done by using optimistic, and pessimistic estimates. The Monte Carlo simulation is basically a QRA methodology that aims at identifying the risk level of achieving objectives. To conduct a Monte Carlo Analysis, a manager needs to create three estimates for the duration of the project. The first estimate is the one that has the most likely duration or the best-case scenario, the second one is the worst-case scenario. A probability of occurrence is associated with each estimate. The project involves three tasks such that the first task is likely to be completed in three days (70% probability). However, it may also take two days or even four days for completion. The probability of it getting complete in two days is 10% and its likelihood of taking four days to complete is 20%. Essentially, the subsequent assignment has a 60% likelihood of taking six days to complete, a 20% likelihood each of being finished in 5 days or 8 days. The last undertaking has an 80% likelihood of being finished in four days, a 5% likelihood of being finished in three days, and a 15% likelihood of being finished in five days. In Monte Carlo analysis, a series of simulations are performed on the project probabilities such that the simulation is to run for a 1000-odd time. For every simulation performed, an end date is observed. Once the Monte Carlo Analysis is finished, there will be multiple project completion dates in the form of a likelihood curve portraying the reasonable dates of completion and the likelihood of accomplishing each. This probability curve may be used to derive the expected date of completion. The manager would pick the date with a 90% possibility of accomplishing it. Thus, by utilizing the Monte Carlo Analysis, the venture has a 90% possibility of being finished in a given number of days. Moreover, the project manager can consider the estimated budget for a project using probabilities to simulate different end results. The findings in a probability curve can be used to analyze the completion of projects in a specific number of days. In this method, risk analysis involved in a project is manifested using probability distribution or a model of possible values. Commonly used probability distributions or curves for Monte Carlo Analysis are discussed as follows:

1. **The Normal or Bell Curve:** A probability curve, in which the values in the middle are most likely to occur.
2. **The Lognormal Curve:** A curve in which values are skewed. Monte Carlo Analysis uses such distribution for project management in the real estate industry or oil industry.
3. **The Uniform Curve:** In this curve, all instances have an equal chance of occurring. This kind of graph is common in analyzing manufacturing costs and future sales revenues for a new product.
4. **The Triangular Curve:** To obtain a triangular curve, the manager enters the minimum, maximum or most likely values. A triangular curve displays values around the most likely option.

Using the information obtained by Monte Carlo Analysis, PM may give senior administration the measurable proof for the time required to finish a task. A suitable budget can also be proposed using this analysis.

6.7.2.5 SENSITIVITY ANALYSIS

This procedure is used to figure out which risks have the greatest effect on an undertaking. Sensitivity analysis is a sort of quantitative risk assessment. It determines how changes in a specific model variable can have an impact on the output of the model. It is based on Monte Carlo simulations of project schedules. Also known as Tornado chart, this method is capable of depicting which task variables like Cost, Start, and Finish Times, Duration, etc., have the greatest impact on project parameters. Sensitivity Analysis is performed to analyze the various risks to the project. Several aspects of the project and their corresponding potential impact is also studied. It is necessary to know the level of the impact various elements have on a project. Such information may be useful in assisting management in setting priorities to achieve results quickly [86]. A tornado diagram is extremely effective in conveying the results of sensitivity analysis for a specific project. Since the values are quantitative, it is easier to analyze risks by studying the differences. This is indicative of the comparisons between the various elements so as to comprehend which risks are worth taking and which can be ignored. Project management relies on sensitivity analysis for creating priorities to manage essential risks to the project. By realizing which influences the target the most, more endeavors can be concentrated to diminish that risk. Lowering risk potential can lead to projects being completed in a smoother fashion with fewer unexpected delays. Firstly the base case yield is characterized. For instance, the Net Present Value at a specific base case input esteem (V1) for which the sensitivity is to be estimated is characterized. Here, every single other contribution of the model is steady. Then the value of the output at a new value of the input (V2) while keeping different data sources steady is determined. The following stage includes finding the percentage change in the yield and the rate change in the input. To calculate the sensitivity, the percentage change in output is divided by the percentage change in input. The process of testing sensitivity for another input while keeping the rest of the input's constant is repeated several times until the sensitivity figure for every one of the inputs is acquired. Higher the estimation of sensitivity, the more sensitive the yield (output) is to any adjustment in that input and the other way around.

6.7.2.6 FAULT TREE ANALYSIS (FMEA)

Fault tree analysis (FTA) or Failure Mode and Effects Analysis explores a structured diagram that identifies elements that can cause a system failure. It is a top-down, deductive analysis that visually depicts a failure path or failure chain. FTA is loosely based on the concept of Boolean logic.

Boolean logic witness's creation of a series of statements based on True/False. When these statements are linked logically, they form a logic diagram of failure. Events can also be arranged in sequences of series relationships (OR) or parallel relationships (AND). Logic symbols are used to depict dependencies among events, and the results of these events are presented in a tree-like diagram. Thus, FTA may be explained as a logical breakdown from the Top-level undesired event, cascaded to the Base-level event (root cause). Each path is associated with a probability. The paths that compare to the highest severity/highest probability blends are distinguished and may require alleviation. Beginning at the Base-level event (at the base of the FTA) and stirring the way up to the undesirable Top-level event is known as a Cut Set. There can be many cut sets within the FTA such that each cut set has an individual probability assigned to it. The Base-level event is usually color-coded for distinguishing the risk level depicted. The five fundamental strides to perform FTA are as per the following:

- **Step 1:** Identifying the hazard.
- **Step 2:** Obtaining the understanding of the system being analyzed.
- **Step 3:** Creating the fault tree.
- **Step 4:** Identifying the cut sets.
- **Step 5:** Mitigating the risk.

We identified some qualitative and QRA methods. Based on the methods, the difference between them may be summarized as follows (Table 6.1).

TABLE 6.1 Difference between Qualitative and Quantitative Risk Analysis Methods

Qualitative Risk Analysis	Quantitative Risk Analysis
Risk-level	Project-level
Subjective evaluation of probability and impact	Probabilistic estimates of time and cost
Quick and easy to perform	Time-consuming
No special software or tools required	May require specialized tools

6.8 RISK ASSESSMENT

Every organization requires a risk management strategy. This makes cyber risk assessment a crucial part of any company or organization. Every organization these days depends on innovation, technology, and information systems to lead business. Also, there are risks incorporated in that. Cyber risk assessment refers to performing an assessment of the cyber risks faced by companies or organizations. According to NIST, Risk assessments can identify, estimate, and prioritize risks for organizational operations. These operations may be missions, functions, image, and reputation. Risk assessment also estimates and prioritizes risks for individuals, other organizations, organizational assets, and the Nation. Cyber risk assessment may be used to assist decision-makers and to support proper risk responses [89]. Since chiefs and executives cannot dive into security risks on an everyday basis, cyber risk evaluation may act as a kind of executive summary to enable those parties to make informed decisions about security. Some possible ways to do that is by identifying the significant threats to your organization:

- Internal and external vulnerabilities;
- Impact if those vulnerabilities are exploited;
- Likelihood of exploitation.

There are various purposes behind performing a cyber risk assessment, some of them are as follows:

- **They Can Reduce Long-Term Costs:** Recognizing likely threats and attempting to eradicate them can possibly forestall security incidents, which sets aside the association money in the long run.
- **It Gives a Layout to Future Evaluations:** Cyber risk appraisals are not a one-time thing; they need persistent updating. It is possible to make a repeatable procedure that can be handed over to another person in case of staff turnover.
- **They Provide an Organization with Greater Self-Awareness:** Knowing where your organization's weaknesses lie may provide a better idea of what areas the organization needs to grow and invest in.
- **It Helps Avoid Breaches and Other Security Incidents:** A well-done cyber risk analysis can improve security implementations. It may also contribute in mitigating attacks and personal data breaches.
- **It Can Improve Communication:** Cyber risk assessment relics on input from many different departments and stakeholders. This may ensure organizational visibility and strengthen communication.

Ideally, an association would have a workforce in-house that could deal with this sort of evaluation. There may be a need for IT staff with an understanding of how the digital and network infrastructures are set up. Further, there must also be information that high-level executives understand and potentially proprietary organizational information that may be useful during the assessment. Cyber Risk assessment does not ignore organizational visibility. However, small or medium-sized businesses may need to outsource the assessment because they might not have the right people to do the job in-house. In that case, a third party might have to do it. Such services can be provided by companies as well as individual consultants.

Over the last few years, many IT risk-assessment frameworks have been introduced with the objective of guiding security and risk executives. We will discuss some of these key frameworks and identify their strengths and weaknesses. We will emphasize on input from those who have used them in real-world settings.

6.9 RISK ASSESSMENT FRAMEWORKS

6.9.1 OCTAVE

Operationally critical threat, asset, and vulnerability evaluation (OCTAVE) is a risk assessment methodology that aims to recognize, oversee, and assess data security risks. This approach is specific to organizations and is concerned with [87]:

- Creating qualitative risk assessment standards that portray the association's operational risk;
- Identifying resources that are essential to the strategies of the association;
- Identifying vulnerabilities and threats to those assets;
- Determining and evaluating the possible consequences to the organization in case the threats are realized;
- Initiating continuous improvement actions to mitigate risks.

OCTAVE system is centered around people who are answerable for overseeing an organization's operational risks. It incorporates IT department, risk managers, personnel in an organization's business units, persons involved in information security within an organization, and all staff participating in the activities of risk assessment with the OCTAVE method.

The latest version Octave Allegro has decreased a ton of necessities and difficulties that were remembered for the past adaptations of OCTAVE.

Allegro boasts of an 8-step process which is divided in four categories. The primary objective is to ensure that an organization is capable of identifying, analyzing, assessing, and mitigating potential risks. The connection between the activity areas and the actual steps of the approach can be seen from the chart below:

1. **Establish Drivers:** This area incorporates the initial step through which the association creates risk measurement criteria that are steady with organizational drivers. These drivers are instrumental in evaluating the risk effects of an organization's mission and its objectives.
2. **Profile Assets:** This area contains steps 2 and 3, through which the information asset profiles are created. After the process of profile identification and creation, the assets' containers are identified. This is followed by the profile for each asset being captured on a single worksheet. A profile indicates an information asset. It describes its unique features, value, qualities, and characteristics.
3. **Identify Threats:** This area considers steps 4 and 5. This section is responsible for the identification of threats to the information assets. This is followed by documentation through a structured process. In this segment, zones of concern emphasize on real world situations that can happen in associations, and threat situations that contain extra threats and risks are distinguished.
4. **Identify and Mitigate Risks:** In this last stage of risk assessment, identification of risks is ensured. These risks are examined dependent on threat information, and mitigation techniques created to address those risks. Here, threats recognized in the previous category are investigated and alleviated. This process is unique because the outputs from previous steps in the process are captured on worksheets. These outputs are then used as inputs to the next step in the process. The information security community has accepted OCTAVE methodologies as one of the effective standards for conducting risk assessments (Figure 6.9).

For managing operational risk effectively, OCTAVE methodologies have proven to be a clever choice for many organizations that insist on implementing a successful risk assessment strategy. For implementing the OCTAVE methodologies effectively, organizations need to ensure two things; preparing for OCTAVE and performing an assessment:

Risks in Cybersecurity 277

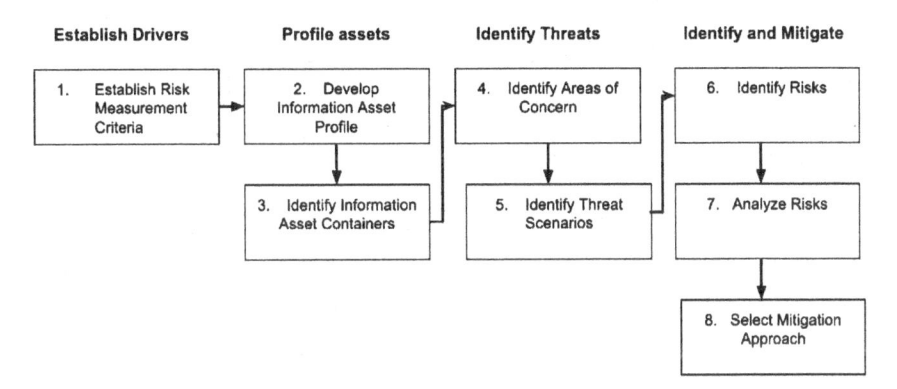

FIGURE 6.9 OCTAVE methodology and activity areas.

1. **Preparing for Octave:** Getting sponsorship from the top management is one of the most critical factors for implementing OCTAVE and performing OCTAVE methodologies. Senior management needs to be convinced that OCTAVE is necessary for the organization needs. They may likewise require activities from the implementer to show nonstop improvement, for example:
   i. Supporting OCTAVE activities;
   ii. Encouraging staff participation;
   iii. Delegating roles and responsibilities to the analysis team;
   iv. Committing to allocate resources;
   v. Presenting ideas on how to continually improve.

   Once the senior management approves for sponsorship, organizational resources need to be allocated for implementing OCTAVE. A team of professionals known as the analysis team must invigilate the implementation. Furthermore, best industry practices have shown that organizations whose assessment teams have received training are capable of successfully implementing OCTAVE methodologies.

2. **Performing an Assessment:** The OCTAVE Allegro methodology was developed by CERT (community emergency response team). It contains guidance, worksheets, and questionnaires that are essential for performing an OCTAVE Allegro assessment. Organizations that need to perform an assessment with OCTAVE must understand all the materials that will assist them in successfully implementing the assessment. It is necessary that before the judgment of the assessment team, organizations must identify and

select information assets that will form the basis of the implementation. Organizations must also set the risk measurement criteria that reflect the management's risk tolerance. The assessment must be repeated every time there is a significant change in the information asset.

OCTAVE methodologies bring a unique perspective that takes into account collaboration between risk identification, assessment, and mitigation. By amalgamating the importance and sensitivity of data to the IT teams, by ensuring proper communication to the top management, OCTAVE guarantees the organizational connection within companies that highlights communication. Because of such collaborations, it is possible to identify any gaps that minimize the ability to eliminate risks. Once these gaps are exposed, organizations may witness a diversity of understanding, opinions, and experiences. This, in turn, will strengthen the quality of the risk assessment in the future. Different advantages of utilizing OCTAVE approaches for risk assessment are discussed below:

- OCTAVE approaches have profoundly subjective and qualitative contemplations and depictions against risk assessment strategies, with respect to quantitative ones.
- Analyzes the risks formally and systematically, organizations can adapt to this easily.
- OCTAVE empowers associations to execute controls just where they are required.
- OCTAVE methodologies are more cost-effective since only risks that are out of the risk measurement criteria (unacceptable risks) get addressed, and fewer expenses in incidents will occur.
- OCTAVE risk management permits senior administration to perform due diligence and comprehend the real condition of the association while being educated about the entire evaluation technique also.

6.9.2 FAIR

Factor analysis of information risk (FAIR) is a risk assessment framework promoted by the open group. The aim of FAIR is to ensure that organizations can analyze, measure, and understand risk [88]. The FAIR model works by assessing factors that add to IT risk and how they impact. This assessment is performed alongside breaking down risk by recognizing and characterizing the risk model. FAIR is utilized for setting up probabilities for the recurrence and extent of information loss. The system supplements existing risk analysis procedures and data security programs. FAIR can strengthen the overall analysis of risk and functions efficiently when an organization understands risk. Once the organization is well acquainted with risk, FAIR discusses operational risk concepts for analyzing the risk better. It also explores the probability of the risk related to cybersecurity.

The FAIR framework is based on four primary components-threats, assets, the organization itself, and the external environment. Any risk scenario can be defined with these components. Each of these categories have attributes, or factors, that contribute positively or negatively to risk.

1. **Threats:** These can be anything from objects and substances. Threats act against an asset in a manner that can result in harm. Natural disasters and hackers may be thought of as threats. The basic idea is that when threats apply force against an asset, it can lead to the occurrence of a loss event. Anyone and anything can, under specific circumstances, be a threat agent. For example, the well-intentioned, however unskilled system operator who wastes a daily batch job by composing an inappropriate command, etc.
2. **Assets:** These may be defined as any data, device, or other components of the environment which reinforce data-related exercises. Assets can be affected in a manner that results in a loss. Some of the characteristics of assets are related to value, liability, and controls strength that represents risk factors.
3. **The Organization:** As we know, a risk exists within the context of an organization or entity. When assets are harmed, it affects one or more of the organization's value propositions. The organization loses assets or the capacity to work. Characteristics of the organization may also assist in attracting the attention of certain threat communities. This may lead to an increase in the frequency of events.
4. **The External Environment:** The environment in which an association works assumes a critical role in risk. There are many external characteristics, like regulatory landscape, competition within the industry, etc., that may lead to the probability of loss.

Basic FAIR Analysis is comprised of 10 steps in four stages:

➢ **Stage 1:** Identifying scenario components:

- Identifying the asset at risk;
- Identifying the threat community under consideration;

➤ **Stage 2:** Evaluating loss event frequency (LEF):
- Estimating the probable threat event frequency (TEF);
- Estimating the threat capability (TCAP);
- Estimating control strength (CS);
- Deriving vulnerability (Vuln);
- Deriving LEF.

➤ **Stage 3:** Evaluating probable loss magnitude (PLM):
- Estimating worst-case loss;
- Estimating probable loss.

➤ **Stage 4:** Deriving and articulating risk:
- Deriving and articulating risk.

Some of the advantages of the FAIR Risk Assessment framework are as follows:

- FAIR plugs in and upgrades existing risk management frameworks (RMF);
- FAIR has a robust taxonomy and technology standards;
- FAIR influences analytics to decide risk and risk rating;
- FAIR takes into consideration associations to make decisions when hard information is missing dependent on the formula, risk = (Threat × Vulnerability)/Controls.

Some of the disadvantages of FAIR Risk Assessment framework are as follows:

- FAIR is not a technique for performing organizational or individual risk assessments;
- FAIR does a lot of assessing, which is on a par with speculating;
- FAIR requires a firmly characterized taxonomy to work;
- FAIR technical standards depict risk and connections yet need genuine estimations and appraisal techniques.

6.10 RISK MANAGEMENT

Cybersecurity risk management is basically real-world risk management applied to the cyber world. It includes recognizing risks and vulnerabilities in an association and applying authoritative actions and comprehensive solutions to ensure that the association is satisfactorily secured. Risk management basically involves identifying potential risks. However, it also refers to assessing the impact of the risks, and planning how to respond if the risks become reality. Therefore, every organization or industry, despite its size, must develop a cybersecurity risk management plan.

However, it is also necessary to understand that not all risks, even if identified in advance, can be eliminated.

It is possible to reduce the potential impact of risks. Risk management deals with four important spheres:

1. **Accepting Risks:** Acknowledging that risks exist, but applying no safeguards;
2. **Transferring Risk:** Shifting responsibility for the risk to a third party;
3. **Avoiding Risks:** Eliminating the asset's exposure to risk or eliminating the asset; and
4. **Mitigating Risk:** Modifying the asset's risk exposure and applying shields (Figure 6.10).

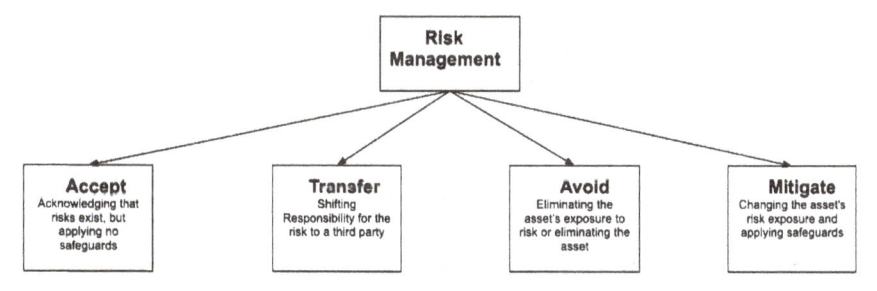

FIGURE 6.10 Risk management spheres.

> **Definition 6.2:**

Risk management may be defined as the progressing procedure of distinguishing, evaluating, and reacting to risk. It must take an expansive perspective on risks over an association to inform resource allocation, better manage risks, and empower responsibility.

For managing risks, organizations need to perform an assessment of the likelihood and potential impact of an event. Once it is done, the next thing to do is to find out the best approach to deal with the risks: avoid, transfer, accept, or mitigate. For mitigating risks, an association should ultimately figure out what sorts of security controls must be applied. Security risks cannot be eradicated totally, and no association has an unlimited budget or enough staff to battle all risks. Therefore, risk management must be thought of in terms of dealing with the impacts of vulnerability on organizational goals so that it makes the best and effective utilization of constrained assets. Clear communication and situational awareness can contribute immensely to a good risk management program. This permits risk decisions to be well informed, well-considered, and made with regards to organizational targets, for example, changes to help the association's mission or for acquiring business rewards. Risk management is responsible for identifying risks early and implementing appropriate mitigation strategies for preventing incidents or attenuating their impact. In order to plan risk management for an organization some points must be considered:

1. **Building a Company Culture:** When planning the organization's cybersecurity risk management program, the company's culture plays a vital role. With the average cost of a cyberattack exceeding $1.1 million, 37% of companies observed a diminution of their reputation following the attack, it is important to set up a cybersecurity-centered culture all through the association, from the part-time staff up to the executive suite.
2. **Distributing Responsibility:** Maintaining cybersecurity is not easy and must not be confined to the IT or security departments. It is necessary to ensure that every employee in the organization is aware of potential risks and is responsible for preventing security breaches. Company-specific security plans must take into account not just the hardware and software, but also human factors. Most of the data breaches are caused by phishing or social engineering. There is a need to present the correct tools, techniques, and training to employees for safeguarding against these human-related intrusions. Employees must be adept in recognizing malware, phishing emails, and other social engineering attacks. This is necessary for developing an organizational culture of security.
3. **Training Employees:** For implementation of the cybersecurity plan, it is necessary to train staff at all levels. Staff must be adept

in identifying risks and must be aware of the strategies and frameworks intended to eliminate those risks. Employee training can lead to spreading of and encouraging a security-aware culture. It can likewise guarantee that all employees know about cybersecurity systems and tools that are to be implemented.

4. **Sharing Information:** Confining cybersecurity will result in failure. Along these lines, data with respect to cybersecurity risks must be shared over all offices and at different levels. Activities and operations concerning cybersecurity must be imparted to all the appropriate partners, particularly those included in the company's decision making. All appropriate parties must be informed about the potential business impact of relevant cyber risks. They must also be aware of the ongoing activities.

5. **Implementing a Cybersecurity Framework (CSF):** It is essential to actualize the fitting CSF for an organization. This is usually asserted by standards adopted in the industry. Some widely adopted CSFs are NIST Framework for Improving Critical Infrastructure Security, PCI/DSS, ISO 27001/27002, CIS Critical Security Controls, etc.

6. **Prioritizing Cybersecurity Risks:** The organization does not have an infinite number of staff or an unlimited budget. Thus, not everything can be protected against all possible cyber risks. Consequently, there is a need to organize risks with respect to both likelihood and the degree of impact, and afterward organize security arrangements in a like manner.

7. **Encouraging Diverse Views:** Too frequently cybersecurity staff and the board see risk from a solitary perspective, regularly dependent on personal experience or organization history. However, cybercriminals seldom share this same viewpoint and are more likely to think better and identify loopholes that may have not been seen before or even considered. Therefore, it is useful to urge colleagues to consider and contend various perspectives. Such a diversity will help identify more risks and more possible solutions.

8. **Emphasizing Speed:** As soon as a security breach occurs, there is a need for an immediate response. The longer it takes to address the threat, the more damage may be done. Most of the IT managers take more than an hour to get information about an ongoing

cyberattack. An hour is capable of bringing a lot of damage to an organization. The rapid response must be consistent with the security-forward culture. That means there is a need to develop a system that may contribute to early recognition of the potential risks. This is synonymous with an immediate identification of the attacks and breaches, followed by a rapid response to security incidents. When it comes to risk containment, speed is critical.

9. **Developing a Risk Assessment Process:** Risk assessment is a significant and indispensable part of any cybersecurity risk management plan. It is necessary to:
    i. Identify the company's digital assets. This may also include all stored data as well as IP.
    ii. Identify all potential cyber threats. Threats can be external like hacking, attacks, ransomware, etc., as well as internal like accidental file deletion, data theft, malicious current or former employees, etc.
    iii. Identify the impact if any of the assets were to be stolen or damaged.
    iv. Rank the probability of each potential risk occurring.

10. **Incident Response (IR) Plan:** Finally, there is a need to develop an IR plan, that focuses on the priority of risks that were previously identified. There is a need to think about what can anyone do if a threat is identified and who needs to do it. It is necessary to codify this plan so that even if an incident occurs, other team members will have a roadmap for how to respond (Figure 6.11).

Although specific methodologies vary, a risk management program typically follows steps along these lines:

- Identifying the risks that might compromise cybersecurity. This generally includes distinguishing cybersecurity vulnerabilities in a framework and the threats that may misuse them.
- Analyzing the severity of each risk. This might be performed by evaluating if it is prone to happen, and how noteworthy the effect maybe if it does.
- Evaluating how each risk fits within the risk appetite predetermined level of acceptable risk).
- Prioritizing the risks.

- Concluding how to react to each risk. There are commonly four alternatives:
  - **Treat:** Changing the probability and the impact of the risk, normally by executing security controls.
  - **Tolerate:** Making an active decision to retain the risk.
  - **Terminate:** Avoiding the risk altogether by closure or totally changing the action causing the risk.
  - **Transfer:** Sharing the risk with another party, which is usually done by outsourcing or taking out insurance.
- Since cyber risk management is a continual procedure, monitoring the risks to make sure they are still acceptable, reviewing controls to ensure they are as yet fit for purpose, and making modifications as needed. Risks change continuously as the cyber threat landscape evolves, and as systems and activities change (Figure 6.12).

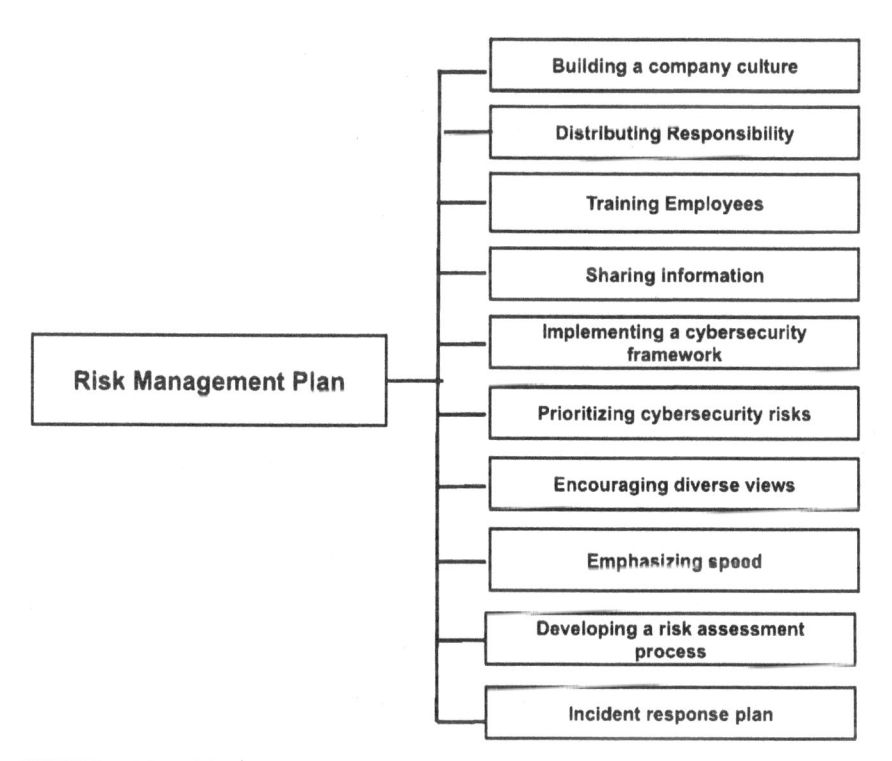

Risk Management Plan
- Building a company culture
- Distributing Responsibility
- Training Employees
- Sharing information
- Implementing a cybersecurity framework
- Prioritizing cybersecurity risks
- Encouraging diverse views
- Emphasizing speed
- Developing a risk assessment process
- Incident response plan

FIGURE 6.11 Risk management plan.

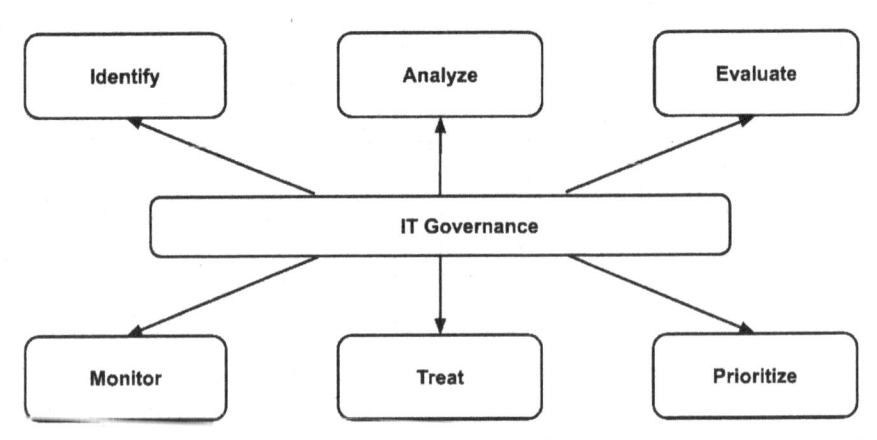

FIGURE 6.12 Risk management program.

Risk management Services may assist in predicting, prioritizing, preventing, and minimizing the financial impact of potential risks to the most valuable assets. It takes into account the following considerations:

- The need to identify and assess risks with respect to their consequences to the business as well as likelihood of their occurrence;
- The need to establish communication lines with stakeholders. This is done to inform them about the probability and results of recognized risks and risk statuses.
- The need to build up priorities for risk treatment just as acknowledgment.
- The need to build up priorities for decreasing the chance of risks happening.
- The need to establish risk monitoring and risk review processes.
- The need to inform stakeholders, partners, and staff with respect to risks to the association just as the moves being made to relieve those risks.
- The need to establish internal and external risk context, scope, and boundaries, and choice of the RMF.

6.11 RISK MANAGEMENT FRAMEWORK (RMF)

Risk management framework (RMF) is responsible for integrating cybersecurity activities into existing processes. These include requirements, program security planning, trusted frameworks and systems examination, developmental, and operational test and assessment, money-related administration and cost evaluating, and sustainment and removal.

> **Definition 6.3: Risk Management Framework (RMF)**

> A risk management framework (RMF) is a structure that brings a risk based, full-lifecycle way to deal with the usage of cybersecurity. RMF takes into account the integration of cybersecurity in the systems design process. This results in a more reliable framework that can constantly work even with a proficient cyber adversary.

PM must ensure that cybersecurity risks are managed consistently with framework execution prerequisites and are adequate to the service-assigned authorizing officials (AO). These authorities are liable for giving the framework authorization to operate (ATO). Close coordination between the PM and AO is basic to the administration of cybersecurity risks all through the whole securing process. RMF weight on the accompanying key focuses [90]:

- Cybersecurity is risk-based, strategic, and must be tended to ahead of schedule and consistently.
- Cybersecurity requirements must be treated like other system requirements.
- System security architecture and data flows need to be developed early and must be continuously updated throughout the system lifecycle as the system and environmental change. This is done to keep up the ideal security posture dependent on risk evaluations and alleviations.
- Cybersecurity must be implemented for increasing a system's capacity to secure, distinguish, respond, and reestablish, even when under attack from an adversary.
- A modular, open systems approach may be used for implementing system and security architectures. This architecture must be capable of supporting the quick advancement of countermeasures to rising vulnerabilities and threats.
- Cybersecurity risk assessments need to be conducted early and often and must be integrated with other risk management activities.

Once the system matures, the security controls are chosen, executed, surveyed, and checked. The PM then collaborates with the authorizing official (AO). the AO must ensure that the cybersecurity risk posture of the framework is overseen and kept up during activities. This prompts continued alignment of cybersecurity in the technical baselines, framework security architecture, information flows, and plan. Reciprocity or interdependency must be ensured through sharing and reuse of test and evaluation products, i.e., tested once, used by all approaches.

The accompanying three security goals ought to be viewed as when attempting to balance explicit cybersecurity necessities with different prerequisites that apply to the framework (Figure 6.13):

1. **Confidentiality:** The property that information is not unveiled to system entities (users, processes, devices) except if they have been approved to get to the data.
2. **Integrity:** The property whereby an entity has not been modified in an unauthorized manner.
3. **Availability:** The property of being accessible and usable upon demand by an authorized entity.

Some RMFs are discussed in subsections.

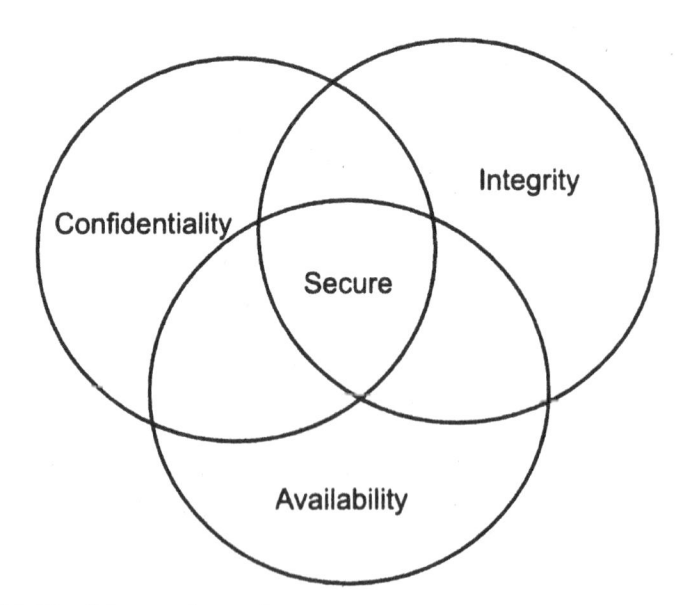

FIGURE 6.13 Cybersecurity requirements.

6.11.1 NIST RISK MANAGEMENT FRAMEWORK (RMF)

The NIST RMF consolidates a lot of data security strategies and standards for the federal government. It was created by The NIST. The organizational risk may be thought of as an efficient, organized method of distinguishing, evaluating, and rating the risks an association faces. In many cases, it is within the context of systems operations. The RMF is liable for providing an effective framework to encourage decision-making for choosing proper security controls. The RMF applies a risk-based methodology that underpins policies, efficiency, regulations, effectiveness, and restrictions due to directives, executive orders, and other rules. The RMF has distinguished the few exercises. These exercises can be applied to both new and legacy frameworks, that are implementable with a powerful information security management system (ISMS). The RMF approach in six stages:

1. **Categorize:** Classifying and labeling the data handled, stored, and shared, and the frameworks that are utilized. The same is performed on the basis of an impact analysis.
2. **Select:** Reviewing the categorization and selecting baseline security controls. There is a need for revising and adding to the security control baseline as necessary. This must be in light of association appraisal of risk and local conditions.
3. **Implement:** Introducing the security controls and integrating with legacy systems. There is a need to record how the controls are displayed inside the framework and their consequences for the environment.
4. **Assess:** Evaluating the security controls to decide if they are actualized accurately, and their quality and adequacy.
5. **Authorize:** Top management is responsible for testing and approving the secured framework dependent on the acknowledged risk appetite to activities and resources. This refers to the amount of risk the organization is willing to tolerate. Management also takes into account the system's operational impact on individuals and other organizations. Therefore, it must recognize how much risk is as yet present, and either approve it or choose on the required changes.
6. **Monitor:** Setting up a continuous checking and evaluation plan for security controls to gauge adequacy. It is necessary to document system or operation adjustments and incorporate impact

analyses of changes made and Report discoveries to data security authorities.

Since organizations must be continuously evaluated for cyber-attacks and risks, RMF is meant to be a continual cycle. It should be possible to start again from stage one completely through to stage six in case there are modifications in the environment or to the system itself (Figure 6.14).

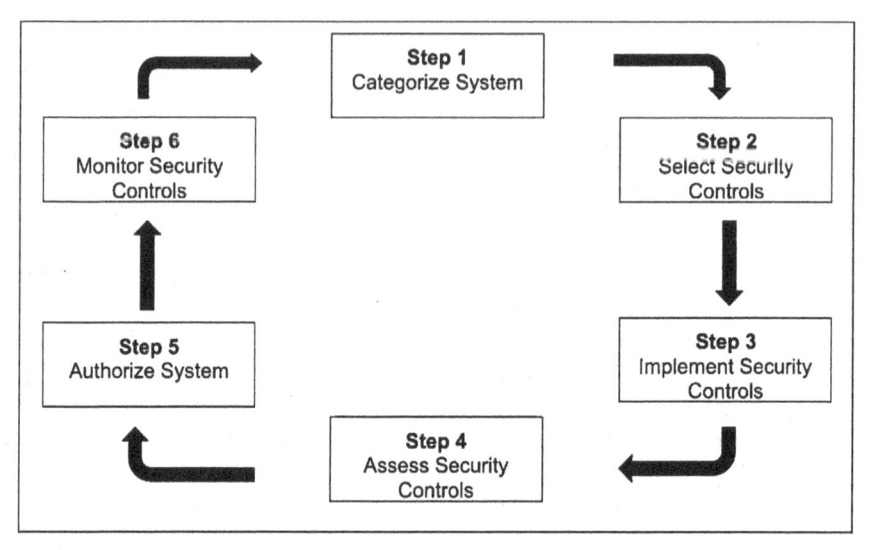

FIGURE 6.14 NIST risk management framework (RMF).

6.11.2 COSO RISK MANAGEMENT FRAMEWORK (RMF)

The Committee of Sponsoring Organizations (COSO), initially focused on sponsoring research for the causes of fraudulent financial reporting. COSO is responsible for providing thought leadership through the development of comprehensive frameworks [20]. It is also accountable for providing direction on fraud deterrence, internal control and enterprise risk management (ERM) intended to improve organizational execution and administration. Its objective is to decrease the degree of misrepresentation and fraud in associations. COSO's direction is non-required, in any case, it has been profoundly persuasive. This is on the grounds that it gives frameworks against which risk management and internal control frameworks can be surveyed and improved. Corporate outrages, that are common in companies demand risk management and internal control, which are deficient in many organizations. COSO attempts to regulate corporate behavior due to these scandals. COSO's ERM model is a broadly acknowledged system for organizations to use, despite attracting criticisms. This is because the framework has been set up as a model that can be utilized in various situations around the world. COSO relies on a cubical structure for illustrating the links between objectives. The objectives are shown on the top and the eight components are shown on the front. The components manifest what is expected to accomplish the goals. The third dimension communicates with the organization's units, which is indicative of the model's ability to focus on parts of the organization as well as the whole. The interior control of the COSO model is investigated by top managerial staff, the board, and other workforce intended to give sensible confirmation of the accomplishment of objectives in the accompanying categories:

- Operational effectiveness and efficiency;
- Financial reporting reliability;
- Appropriate laws and regulatory compliance.

In a successful internal control framework, the accompanying five segments work to support the accomplishment of an entity's mission, plans, and related business targets [91]:

1. **Control Environment:**
   - Exercise integrity and ethical values;
   - Make a commitment to capability;
   - Use the board of directors and the audit committee;
   - Facilitate the management's philosophy and operating style;
   - Create an organizational structure;
   - Issue task of power and duty;
   - Utilize human resources (HR) approaches (policies) and methods.
2. **Risk Assessment:**
   - Make company-wide goals;
   - Consolidate process-level goals;
   - Perform risk identification and analysis;
   - Manage change.
3. **Control Activities:**
   - Follow policies and procedures;
   - Improvise security (network and application);

- Conduct application change management;
- Plan business congruity/reinforcements;
- Perform redistributing.

4. **Information and Communication:**
   - Measure the nature (quality) of the Information;
   - Measure the adequacy of correspondence.
5. **Monitoring:**
   - Perform continuous checking;
   - Lead separate assessments;
   - Report insufficiencies.

These components are responsible for working together in order to establish the foundation for efficient internal control within the company. Accountability is achieved through directed leadership, shared values, and a culture. The organization is prone to several risks, which are recognized and surveyed routinely at various levels inside all capacities in the association. There is a need to design Control activities and other mechanisms for addressing and mitigating the significant risks. Established channels across the company are relied on for communicating information critical to identifying risks. There is continuous monitoring of the internal control and problems are addressed timely (Figure 6.15) [91].

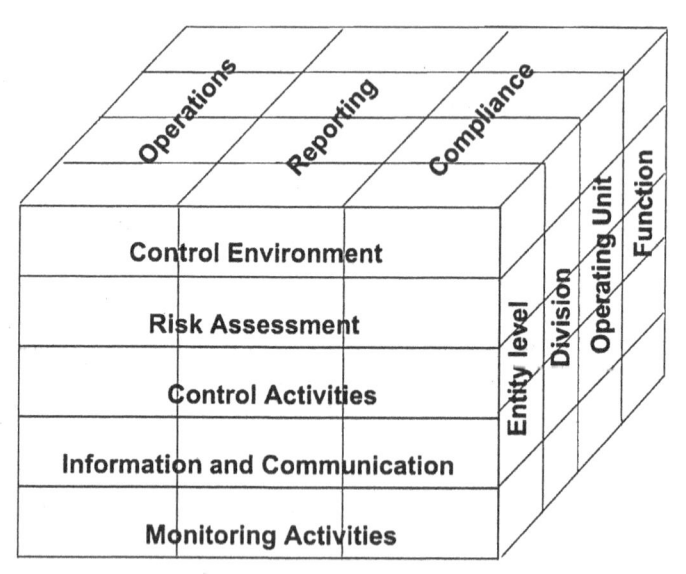

FIGURE 6.15 COSO risk management framework (RMF).

6.11.3 ISO 31000 STANDARDS

ISO 31000 is an international standard that is responsible for providing principles and guidelines for effective risk management [92]. It adopts a conventional way to deal with risk management, which can be applied to various kinds of risks related to domains like financial, safety, project risks, etc. ISO 31000 can be used by any type of organization. The standard gives a uniform vocabulary and ideas for talking about risk management by highlighting guidelines and principles that may assist in undertaking a critical review of the organization's risk management process. The standard, however, does not detail instructions or requirements on how to manage specific risks. It does not provide any advice related to a specific application domain and is applicable at a generic level. Contrasted with past norms on risk management, the 31,000 principles are extemporized in a few territories. It gives another meaning of risk as to the impact of uncertainty on the possibility of accomplishing the association's objectives. It features the significance of characterizing targets before endeavoring to control risks and emphasizes the role of uncertainty. Moreover, it introduces the idea of risk appetite or the degree of risk which the association acknowledges to take on in return for the expected value. It likewise defines a RMF with various organizational strategies, jobs, and duties in the administration of risks. ISO 31000 outlines a management philosophy where risk management is seen as an integral part of strategic decision-making and the management of change (Figure 6.16).

The risk management procedure defined in the ISO 31000 standard incorporates the accompanying exercises [92]:

1. **Risk Identification:** Identifying what could prevent an organization from achieving its objectives.
2. **Risk Analysis:** Understanding the sources and causes of the identified risks and inspecting probabilities and consequences of the existing controls. This stage also involves identifying the level of residual risk.
3. **Risk Evaluation:** The aftereffects of risk analysis and investigation are compared against risk criteria for deciding if the residual risk is acceptable or not.
4. **Risk Treatment:** In order to achieve a net increase in benefit, the extent and probability of outcomes, both positive and negative, must be changed.

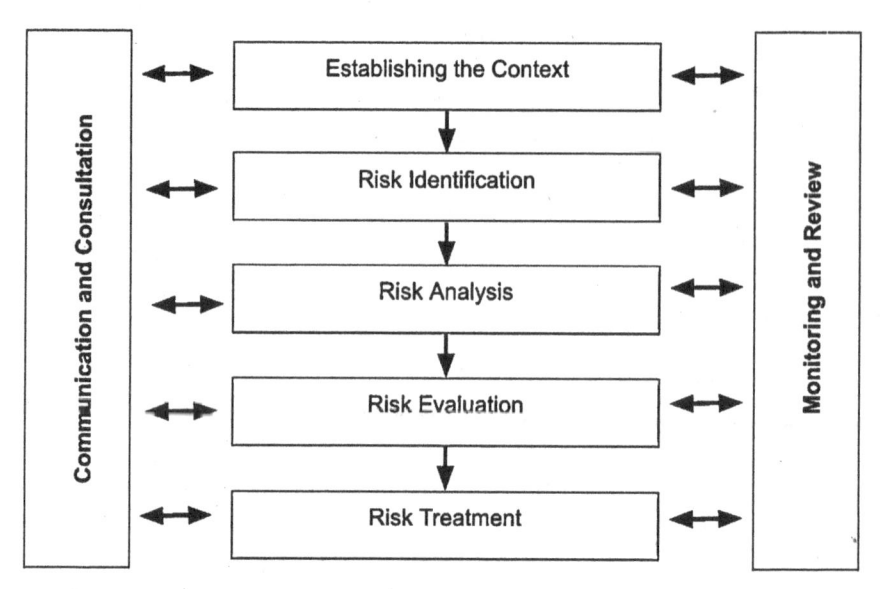

FIGURE 6.16 Risk management process for ISO 31000.

5. **Establishing the Context:** This activity was not included in earlier versions of risk management processes. It defines the scope for the risk management process, as well as the organization's objectives, and is responsible for establishing the risk evaluation criteria. The context comprises both external elements like regulatory environment, market conditions, stakeholder expectations, etc., and internal elements like the organization's governance, culture, standards, and rules, capabilities, existing contracts, worker expectations, information systems, etc.
6. **Monitoring and Review:** In this task risk management performance is measured with respect to indicators, which are periodically reviewed for appropriateness. There is a need to check for deviations from the risk management plan. There is also a need to check on the appropriateness of policies, the RMF, and plan considering organizations' external and internal context. This is followed by giving an account of risk, progress with the risk management plan and how well the risk management strategy is being adjusted to, and evaluating the viability of the RMF.
7. **Communication and Consultation:** This operation is performed for understanding stakeholders' inclinations and concerns in order

to make sure that the risk management process is considering the right elements. It further helps explain the basis for decisions and for particular risk treatment options. The standard incorporates several principles that risk management must verify and creates and protects value that is based on the best information of organizational processes. Decision-making considers human and social factors unequivocally. It also addresses uncertainty and promotes transparency. It is not only systematic and structured but is also timely, dynamic, iterative, and responsive to the changing environment. It facilitates continuous improvement of the organization.

6.12 RISK REMEDIATION

Threat remediation is popularly used for fighting cybersecurity compromises. The process of remediation involves the treatment of a security breach. Using remediation techniques, it is possible to dispense with dubious exercises and malevolent assaults as malware, ransomware, phishing, and such.

> **Definition 6.4:**

> Remediation in cybersecurity refers to tending to a breach and restricting the extent of damage that breaches can possibly cause to your business.

If timely actions are not taken against breaches, they can grow massive such that it may be impossible to contain them. Thus the valuable data that belongs to a particular business can be compromised. Moreover, some security breaches can render systems inoperable, thus costing billions of dollars. Remediation is responsible for detecting and containing such breaches before they figure out how to spread and hurt your frameworks. Many remediation procedures these days act as elementary techniques to the gargantuan security incidents. For example, many remediation solutions trigger an automated kill process. While killing the problematic process may seem like the ultimate solution, it must be viewed as what sort of threat as well as breach is answerable for the problematic process. Regardless of whether the dubious activity is killed, the attackers can remain in the system and network. In order to end the problem, the proposed remediation processes must consider the detection of the cause.

Numerous remediation procedures cannot totally confirm if the threat is completely eradicated. Therefore, the remediation processes employed by the security teams must include gathering exact and sufficient data concerning the incident. Remediation is extremely essential for cybersecurity. Most of the security compromises take more than a month to detect. The inability to determine a security compromise implies that the sensitive and valuable information that belongs to an organization can be out there anytime. Failing to detect compromises significantly lower the chances of eradicating them. That is the reason threat remediation must be an essential part of cybersecurity operations.

Organizations encounter cybersecurity risks from various threat vectors. These threat vectors may range from common vulnerabilities to environment-specific threats. With the volume and complexity of cyber threats increasing exponentially, several organizations are required to address these difficulties in an all the more ad-hoc way, trying to stay up with the threats [93]. There are a few of the operational difficulties that associations face in building up and maintaining a reasonable cybersecurity risk remediation program. These difficulties include:

- Characterizing risk baselines;
- Setting up a technique for risk prioritization;
- Optimizing risk remediation processes;
- Developing reliable and valuable metrics.

Organizations must take into account both long-term opportunities that can lead to improvement in every one of these test zones, along with immediate steps that can rapidly improve risk remediation practices over the complete cyber threat landscape.

6.12.1 DEFINING RISK BASELINES

Large enterprises like banks and financial institutions, energy, and utilities, communications, etc., require risk mitigation strategies. Understanding the baseline against which a risk must be evaluated is a common challenge faced by these enterprises. Implementing risk assessment methodologies for ensuring that security policies and standards are tested and validated may not be a simple task for many organizations. However, having the correct policies, standards, guidelines, and procedures in the first place can be equally challenging (Figure 6.17).

Risks in Cybersecurity 297

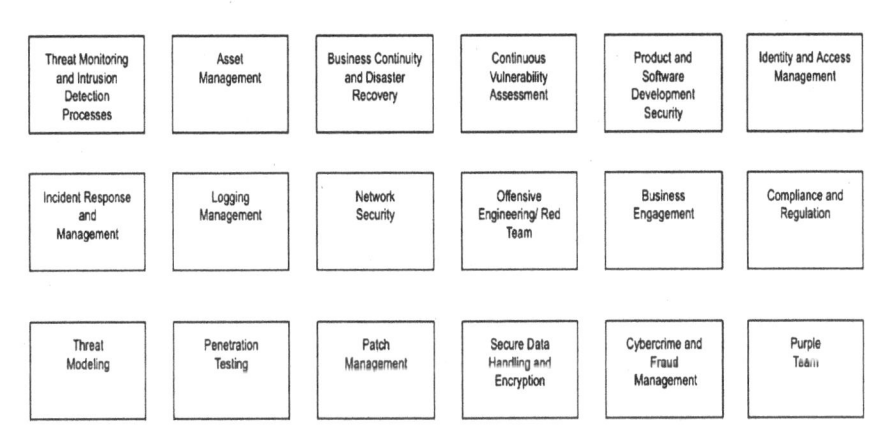

Threat Monitoring and Intrusion Detection Processes	Asset Management	Business Continuity and Disaster Recovery	Continuous Vulnerability Assessment	Product and Software Development Security	Identity and Access Management
Incident Response and Management	Logging Management	Network Security	Offensive Engineering/ Red Team	Business Engagement	Compliance and Regulation
Threat Modeling	Penetration Testing	Patch Management	Secure Data Handling and Encryption	Cybercrime and Fraud Management	Purple Team

FIGURE 6.17 Risk baselines.

An organization may consider hundreds of security procedures within the scope of a defined policy. All of these documents may be benchmarks against which an organization may evaluate risk and must be drafted with respect to the organization's risk tolerance and needs:

1. **Defining a Set Hierarchy of Rules:** Clear limits between strategies, principles, and security controls may help with smoothing out the risk assessment procedure and help with executive reporting or accountability into risk assessment findings. Although an association may claim a significant level of Risk Remediation Policy, for appraisals of a wide range of risks, granular must be considered.
2. **Configuring Security Assessment Tools Appropriately:** In the case of automated technologies like vulnerability scanners and configuration checkers, guaranteeing appropriate guidelines and settings is necessary for information security teams. Stock configuration may be useful in evaluating systems or assets against all possible checks, without taking into account the organization's risk tolerance and defined security standards. Creating a manageable backlog of risk findings may lead to successful remediation.
3. **Re-Assess and Re-Evaluate:** Risk tolerance has the capability to change over time. Strategies may be easily altered by new attack vectors or external threats. Therefore, information security teams must perform re-evaluation of their security standards and controls. There is also a need to validate how those will affect risk

assessments/findings. A versatile procedure to risk management is a long-term methodology.

6.12.2 RISK PRIORITIZATION

Remediation of even minor cybersecurity risks is not an immaterial cost for the information security team. Thus organizations of all degrees of security development have a chance to drive time and cost-efficiency in their remediation process. The maturing enterprise risk remediation program encourages various methods for identifying cybersecurity risks.

Consider a situation such that a vulnerability scanning tool identifies thousands of known Common Vulnerability Exposures across an enterprises' IP space with time and money needed to remediate all of those individual items. It is necessary to understand that vulnerabilities are just one class of risks that financial and IT organizations face. Several other cybersecurity risks like regulatory reviews, security assessments, or internal audits also exist. There is a need to fix these as they are a part of the cybersecurity remediation management program (Table 6.2).

TABLE 6.2 Sample Risk Identification (Specific to Organization)

Operations	Tools	Types
Vulnerability scans	Qualys, Nessus	Authenticated/remote
Configurability checks	Symantec control compliance suite	Windows/Linux
Application security tests	Metasploit, Burpsuite	Pen tests, fuzzing, white box
Code analysis scans	Veracode, Checkmarx	Static, dynamic
Regulatory audits	Self-assessment	PCI/DSS, SOX, GDPR
Internal audits	Internal audit team	Security policy reviews
Architecture reviews	Internal team	Pre-production reviews/services
Threat models	Internal team	Pre-production reviews/services

- It is necessary to simplify the risk scoring process. The holy grail of risk prioritization, money-based risk evaluation, can be a long-term aspiration, yet the intricacy and guesswork expected to allot a dollar value to each risk can restrict the underlying viability of simple risk frameworks. Beginning with an industry-standard like CVSS v2 is a key building block of a risk remediation process.

- Layer in company-specific factors using other scores may be another way to identify risks. CVSS Base scores are a great starting point, as the security team matures, however, adding in environmental scoring (CVSS or self-developed) aids the teams in highlighting the most pressing risks to a specific organization. This may add a layer of complexity, but IT teams must know their assets best and must have a genuine picture of alleviating factors or other defenses set up that a base scoring framework will not consider.
- Developing an effective framework may assist in risk prioritization. If everything for the risk management team is a priority, then nothing ends up actually being a high priority. Any risk scoring framework must be developed in a way that teams have the opportunity to be successful. Prioritizing only Critical or Urgent as per the common vulnerability scoring system (CVSS) terminology may allow for full completion during a set time period since there ought to be less of these items over a full conveyance of distinguished vulnerabilities.

6.12.3 RISK REMEDIATION PROCESS

Design bugs in applications, products, or systems may also qualify as risks. The risk remediation process likewise considers bugs in applications since that can be an obstacle to an effective cybersecurity risk management program. Risk management is specific to organizations, with respect to governance, but the process may be generalized in the form of Findings, followed by remediation followed by completion. Although bugs in the intricacies within each of those steps can be inherently flawed, there is still the opportunity to improve. Frequently, the absence of straightforwardness of risk management results or inputs can be a challenge for risk remediation. Data security teams may not be passing along all the fundamental data required for risk remediation or the reciprocal may likewise be valid:

1. **Putting the Risk Prioritization Framework to Use:** Once an association concurs upon a risk scoring framework, straightforward or complex, it is important to create awareness and utilize this prioritization framework. Groups may get a blanket list of risks or vulnerabilities and not have the context of the framework that a different data security group has chosen.

2. **Practicing DevSecOps:** Issues related to the separation of team and process that historically vandalized development and operations teams may also be found in security processes. DevSecOps is capable of bringing security practices into normal businesses. It can ensure that product and service teams lead to a more streamlined form of the process flow above.
3. **Embracing Automation:** Automation of risk remediation workflows may be costly, although it may provide a positive Return on Investment for mature cybersecurity organizations. It is possible to hold back Risk remediation throughput time by dead periods between remediation workflows. In addition to tools, it may be necessary to develop a culture of automation with resources that are dedicated to this function. For example, automating the identification of high-volume low complexity transactions, estimating benefits, implementing, etc.

6.12.4 RELIABLE AND USEFUL METRICS

With the teams streamlining the risk remediation process, one challenge for several organizations is to demonstrate to the executives that their activities have effectively decreased the risk profile of their association. Many teams are unaware of where to turn in terms of useful metrics that justify the facts. On the other hand, painting a bad picture is easy for security reporting rather than one of transformative and continuous improvement. Any reporting will aim to promote transparency and clarity since even terrible outcomes can be a chance to comprehend where obstructions or clashing needs might be hindering advancement:

1. **Determining Key Metrics:** Although the essential step for risk remediation reporting for most associations is a status of burndown to all risks, it may not justify the facts. Taking into account specific metrics like average time to remediate vulnerabilities or zero-day specific burndown can be more useful in determining areas for process improvement. It may also determine an actual reduction to information security risk.
2. **Separate Reporting by Role:** The reporting metrics that are useful for executive leadership may not be as useful practically. Therefore, there is a need to focus on metrics that allow for much more

organized capabilities. This may assist groups with concentrating on their own particular issues and how those may be combated for successful risk mitigation.

3. **Mixing in Proactive Reporting:** Lagging indicators like remediated risks may be good for showing progress, however, the next development of detailing may consolidate metrics for activities performed for risk root causes. Featuring the advancement made on fixing the underlying cause of vulnerabilities can paint an extraordinary picture for initiative that a data security team is being ground breaking in their security exercises.

Information security teams across all industries face similar challenges and come up with various strategies and solutions. The best answer to develop a practical and effective risk remediation program is a long-term initiative and requires several teams across the organization to agree and contribute to a full risk management lifecycle. However, there are key advances that can be taken by data security groups in the short term likewise that can show quantifiable outcomes as far as an association reducing its risk profile:

- Ensuring any manual risk identification methods are baselined against the company security policy:
  o There is a need to ensure that company security policies and standards are up to date and practical;
  o Associations should be certain they are trying against these approaches;
  o Policies should be designed keeping in mind the risk tolerance of the organization.
- Ensuring risk evaluation techniques and prioritization strategies are distributed and comprehended:
  o Organizations may miss risk findings due to a lack of awareness by the proper remediation teams;
  o Organizations must understand all methods of risk identification for efficient risk management and remediation.
- Ensuring that the security team is using an industry-standard risk scoring system:
  o It may not be enough to quantify a risk as simply "High, Medium, Low;"

- Industry-standard scores like the common vulnerability scoring system (CVSS v2) provide a granular 0–10 scale;
- CVSS 2 can likewise give the granularity required within High-level vulnerabilities.
• Simplifying reports to include only role-based metrics:
  - Senior leadership may insist upon binary forms of reporting or want to view simplified trend lines over time;
  - A more useful form of reporting may incorporate detailed data custom-made explicitly to their risk profile and attack vectors;
  - Automating these reports, where possible, may assist information security teams operating better.

In this chapter, we have explored the risks that exist in cyberspace and how they influence a few associations that are identified with different domains. We studied risk identification and management by means of risk identification, risk analysis, risk evaluation, risk management, and risk remediation. Specific methods for many of these have been discussed in detail. Many industry-specific RMFs have also been discussed so as to highlight the practical significance of the topic. In the next chapter, we will be exploring yet another interesting concept that underpins the business aspect of cybersecurity in terms of insurance and liability.

6.13 SUMMARY AND REVIEW

- The cyber risk might be characterized as any risk of money-related loss, or harm to the notoriety of an association as a result of the failure of its technological frameworks.
- With the attacks and attackers becoming more and more sophisticated, it is no longer enough to depend on customary information technology (IT) experts and security controls for data security. There is a reasonable requirement for threat intelligence devices and security programs to lessen an organization's cyber risk and highlight potential attack surfaces.
- To understand the cyber risk profile of an organization, it is necessary to figure out what data would be important to outsiders or cause noteworthy interruption if inaccessible or corrupt. One such data is personally identifiable information (PII) that incorporates names, social security numbers, and biometric records.

- Cybersecurity risk management is generally set by initiative which may fuse an association's directorate in the planning process.
- Risk identification might be thought of as the way toward recognizing and evaluating threats to an association, its activities, and its workforce. It considers evaluating IT security threats, for example, malware, and ransomware, mishaps, natural disaster events, and other possibly hurtful occasions that could disturb business activities.
- Risk Identification has been characterized as the way toward deciding risks that might forestall the program, enterprise, or investment from accomplishing its targets. It incorporates documenting and communicating the concern.
- SWOT (Strengths Weaknesses, Opportunities, and Threats) analysis is performed using the SWOT matrix. It is one of the procedures project supervisors and associations have executed in their risk management practices.
- Risk analysis alludes to the audit of the risks related to a specific occasion or action. Risk analysis may be of two types: Qualitative and Quantitative.
- A qualitative risk analysis highlights the identified project risks through a rating scale with the end goal that risks are scored depending on their likelihood or probability of happening and the effect on project targets if they occur. Common Techniques include Delphi technique, SWIFT analysis, Bow-Tie Analysis and Probability/Consequence Matrix.
- In quantitative risk analysis (QRA), a numerical or quantitative rating is assigned in order to develop a probabilistic analysis of the project. It measures the potential results for the project and surveys the likelihood of accomplishing explicit project targets. Common Techniques include Three-point estimate, expected monetary value (EMV), decision tree analysis, Monte Carlo analysis, sensitivity analysis, and fault tree analysis (FTA).
- Several formal IT risk-assessment frameworks have emerged over the years to help guide security and risk executives through the process.
- Operationally critical threat, asset, and vulnerability evaluation (OCTAVE) is a risk assessment methodology that aims to identify, manage, and evaluate information security risks.
- Factor analysis of information risk (FAIR) is a risk assessment framework advanced by the open group that empowers associations to break down, measure, and comprehend risk.

- Risk management may be defined as the methodology of recognizing likely risks, surveying the effect of those risks, and planning how to react if the dangers become reality. To manage risk, organizations need to survey the probability and possible effect of an occasion and afterward decide the best way to deal with managing the risks: avoid, transfer, acknowledge, or alleviate.
- A risk management framework (RMF) is a framework that brings a risk-based, full-lifecycle way to deal with the execution of cybersecurity. RMF takes into account the integration of cybersecurity in the systems design process.
- The NIST risk management framework (RMF) incorporates a set of information security policies and standards for the federal government. It was developed by The National Institute of Standards and Technology (NIST).
- COSO (Committee of Sponsoring Organizations) Framework is liable for giving idea-based administration through the advancement of thorough frameworks.
- ISO 31000 is an international standard that is responsible for providing principles and guidelines for effective risk management. It adopts a generic approach to risk management, which can be applied to different types of risks related to domains like financial, safety, project risks, etc.
- The process of remediation involves the treatment of a security breach. Using remediation techniques, it is possible to eliminate suspicious activities and malicious attacks in the form of malware, ransomware, phishing, and such.

QUESTIONS TO PONDER:

1. You are supposed to make a prototype for your project; however, you are uncertain about whether to continue with this model prototype. In the event that you do the model, it will cost you $100,000; and, obviously, on the off chance that you do not pursue it, there will be no expense. If the model is created, there is a 30% chance that the model will fail and for that the cost effect will be $50,000. However, if the model succeeds, the task will make $500,000. In the event that you do not create any model, you are taking a risk, the possibility of which is 80% with a failure impact

of $250,000. In any case, without a model, should you succeed, the task will make the same money as referenced previously. What would it be advisable for you to do?
2. There is a negative risk with a 10% likelihood of disallowing the execution of a work package. On the off chance that the risk occurs, the effect of not executing the work is assessed at $40,000. For the same work package, there is a positive risk with a 15% likelihood and impact assessed at a positive $25,000. Would it be advisable for you to execute the work package?
3. Perform a SWOT (Strengths Weaknesses Opportunities Threats) Analysis for the following situations:
   i. **SWOT Example:** Ice Cream for Dinner;
   ii. **SWOT Example:** Encouraging employees to work during pandemic;
   iii. **SWOT Example:** Jogging 10 miles every day;
   iv. **SWOT Example:** Consuming Caffeine every day.
4. Perform a detailed SWOT analysis for Amazon, Starbucks, and Coca-Cola.
5. Figure 6.1 describes the steps to understand cyber risks. Healthcare industry and Finance industries are often struck with cyber-attacks. Based on the illustration, identify the components of cyber risks in the Healthcare and Finance industry.
6. Why must threat remediation be an essential part of cybersecurity operations?

KEYWORDS

- bring your own device
- cyber risk
- cybercrimes
- cybersecurity
- information systems
- internet service provider

CHAPTER 7

Cyber Risks and Cyber Insurance

In the previous chapter, we discussed cybersecurity risks, their identification, analysis, management, and remediation. In this chapter, we study a domain that is closely associated with cyber risks. Cyber risks like identity theft and data breaches lead to huge financial losses within organizations. The effect of cyber incident damage might be limited by presenting cyber insurance, which we will investigate in this section.

7.1 CYBER INSURANCE AND THE NEED FOR IT

Cyber risk alludes to the danger of budgetary damage to an association that might be credited to the disruption or failure of its computer frameworks. Then again, it might be characterized as the danger that organizations face from their dealing with information and depending on technology in their everyday activities. As per insurers, 'computer systems' might be characterized extensively to incorporate numerous information technology (IT) systems.

As we know, the specialty lines insurance market takes into account the insurance industry where the more difficult or irregular risks are composed. Cyber insurance is one such protection item. The main objective of cyber insurance is the protection of businesses and individuals who provide services for such businesses. The risks involved may be Internet-based risks, or general risks related to IT infrastructure. Some risks may also be related to privacy, information governance liability, and corresponding activities. Risks of this nature may be excluded from traditional commercial general liability policies. The risks may not explicitly be examined in conventional insurance products. Cyber-insurance policies provide coverage that may consider first-party coverage against a few losses. These losses may be due to a variety of cyber-attacks that are disguised in the form of information demolition, blackmail, robbery, hacking, denial of service attacks [15,

16, 23, 38]. Liability coverage may indemnify organizations for losses to others caused, for instance, by blunders and exclusions, inability to defend information, or defamation. Different advantages of cyber insurance incorporate intermittent security-audit, post-occurrence publicizing and insightful costs, and criminal prize funds.

Cyber insurance is basically protection for organizations and people. It is used to protect them against cyber risks. Data breaches happen to be the most common risks that are insured. Cyber insurance encourages compensations from lawsuits with respect to data breaches, in the form of omissions and errors. It is also responsible for covering losses that are a result of network security breaches, loss of privacy, and loss of intellectual property (IP). Previously, some prominent information breaches have guaranteed that insurance agencies offer cyber insurance policies for protecting customers from the effects of network threats and data breaches. Cyber insurance strategies join first-party inclusion against misfortunes identified with cyberattacks. Blackmail, malware, hacking, and identity theft, are a few assaults revealed by customers may result in indemnification against lawsuits. Indemnification may also extend to the failure to protect a system. Standards and Guidelines likewise frequently incorporate public relations (PR) reactions to an attack. The drawback of cyber insurance is that insurers constantly need to limit risks, in this manner, potential clients are exposed to broad assessments of their security techniques before an insurer covers them.

The foundation, the clients, and the services offered on computer networks today are on the whole subject to a wide assortment of risks presented by threats that incorporate, intrusions, and interruptions of different sorts, distributed denial of service (DDOS) attacks, hacking, phishing, eavesdropping, worm attacks, spams, viruses, and so on. So as to counter the risk presented by these threats, network users have generally depended on antivirus and anti-spam programming, firewalls, intrusion-detection systems (IDSs), and other additional items to decrease the probability of being influenced by threats [11]. Practically, industry and research are presently based on creating and conveying devices and methods to recognize threats and inconsistencies so as to shield the cyberinfrastructure and its users from the subsequent negative effect of the irregularities. In spite of enhancements in risk protection procedures over the last few years, because of hardware, programming, and cryptographic strategies, it is difficult to accomplish perfect or near perfect cybersecurity protection. The inconceivability emerges because of various reasons:

- Insufficient technical solutions;
- Difficulty in designing tailored solutions to thwart network attacks;
- Skewed impetuses between network clients, security item merchants, and administrative specialists with respect to protecting the organization;
- Network users who take advantage of the positive security effects that are generated by other users' investments. Moreover, these users never invest in security which results in the free-riding problem;
- Client lock-in and first-mover impacts of weak security products;
- Challenges in measuring risks which may further make it difficult to design risk mitigation solutions;
- Liability shell games played by item merchants;
- Naiveness of users in optimally exploiting feature benefits of technical solutions.

Henceforth, given the obstructions to approach 100% risk relief, the need emerges for elective strategies for risk mitigation in the internet or cyberspace. In order to justify the significance of cybersecurity and for improving the current state of cyber-security, a cyber-security executive order was issued in February of 2013 to emphasize the need to reduce cyber-threats. Numerous security specialists in the recent past have recognized cyber-insurance as a potential method for successful risk management. Therefore, cyber-insurance may be believed to be a risk management technique that makes it possible for network user risks to be transferred to an insurance company. This is carried out in return for a fee which is referred to as the insurance premium. Potential cyber-insurers could be cloud providers, ISPs, or any insurance organizations. Advertisers of cyber insurance may contend that cyber insurance is liable for formulating insurance contracts which may helpfully move the proper amount of self-defense liability to the customers. The general effect is making the internet more robust. The term self-defense' alludes to the endeavors of network clients in protecting their system utilizing technical solutions. Some of these technical solutions may be in the form of secure operating systems, anti-virus software, firewalls, etc. Cyber-insurance can possibly be a market arrangement that can line up with the economic incentives of cyber-insurers, users, policymakers, and security software vendors. The cyber-insurers may earn profit from appropriately pricing premiums. Then again, network users would support the limitation of likely losses by mutually purchasing insurances and putting resources into self-defense procedures. Policymakers would guarantee the expansion in overall network security. For Security software vendors, there may be a huge spike in their product sales as a result of forming alliances with cyber-insurers [94]. Another challenging aspect of cyber insurance and risk management is determining the acceptable risk for each organization. A 'duty of care' approach may be essential for protecting all interested parties like the judges, regulators, executives, and the public who can be influenced by those risks. The duty of care risk analysis standard (DoCRA) lists principles and practices for balancing security, business objectives and compliance, while developing security controls.

7.2 CYBER RISK INSURANCE AS PART OF RISK MITIGATION

A solitary cyber attack can make an ill-equipped organization bankrupt [95]. Therefore, it is necessary for all organizations to enforce a cyber risk management program. The main objectives of a cyber risk management program must encompass:

- Observing the risk in an association;
- Assisting an organization in avoiding breaches;
- Assisting an organization to recover following a breach.

Cyber risk insurance deals with helping an association to recuperate following a breach. If an organization has suffered a breach, cyber risk insurance may assist the association in recuperating losses, paying charges and damages, and proceeding with the usual business. Every one of these exercises can be performed while reconstructing and fixing frameworks and networks of the organization.

7.3 ADVANTAGES OF CYBER INSURANCE

The popularity of the cyber-insurance market in many countries is less with respect to other insurance products [8]. Therefore, its general effect on developing cyber threats is hard to evaluate. Moreover, the impact on individuals and organizations from cyber threats is likewise moderately expansive as for the extent of assurance gave by insurance items. Thus, insurance agencies keep on building up their services. With insurers paying on cyber-losses, and as cyber threats developing and evolving continuously, insurance products are being purchased progressively along with already existing security services. The guaranteeing measures for insurers for offering cyber insurance items are at a beginning stage. Therefore, the underwriters are diligently partnering with security firms for developing their products. In spite of the fact that cyber insurance may not improve security, it might be essentially valuable during large-scale security breaches. Cyber insurance can give a smooth financing part to associations to recover from major losses. They can also help businesses to function normally and can also reduce the need for government assistance. Due to cyber insurance, cyber-security risks get dispersed reasonably. The cost of premiums is directly proportional to the size of the expected loss. This may ensure that there are no potentially dangerous concentrations of risk while also preventing free-rides.

7.4 DISADVANTAGES OF CYBER INSURANCE

All modern businesses rely on IT. The need for cyber insurance arises because these businesses are driven by technology and are associated with cyber risks. Existing insurance practices are known to follow either the Flood or Fire model [97]; however, cyber incidents do not give off an impression of being demonstrated by both of these incident types. This has prompted a circumstance where the extent of cyber iInsurance is additionally confined to diminish the risk to the underwriters. The scarcity of data further aggravates the situation such that although there may be actual damage associated with a given event, there may be a lack of proper standards that can lead to efficient classification of events. Often the absence of proof related to the efficacy of Industry best practices makes it difficult to categorize cybersecurity incidents. Insurance takes into account sound actuarial data against a largely static background of risk. Given that these do not exist at present, it may not be possible for either of the buyers of these products to accomplish the value outcomes that they want. This perspective is reflected in the current market state where standard prohibitions bring about a circumstance where an insurer could contend they apply to practically any data breach.

7.5 TYPES OF CYBER INSURANCES

The following are the types of cyber insurances [96]:

1. **Network Security Insurance:** These are insurances targeting cyber-attacks as well as hacking attacks.
2. **Theft and Fraud Insurance:** This protection considers pulverization or loss of the policyholder's information. Loss of data may be attributed to criminal or fraudulent cyber events, which may also include theft and transfer of funds.
3. **Forensic Investigation Insurance:** This insurance takes into account aspects related to legal, technical, or forensic services. It is necessary to assess if a cybersecurity event has occurred, for assessing the impact of the attack, and for stopping the attack.
4. **Business Interruption Insurance:** This insurance supports lost pay and related expenses. In such situations, policyholders may be unable to conduct business because of cybersecurity events or data losses.
5. **Extortion Insurance:** This insurance provides coverage for the costs incurred due to the investigation of threats for committing cyber-attacks. It likewise considers installments to extortionists who intimidate to acquire and divulge sensitive information.
6. **Reputation Insurance:** This insurance provides coverage for reputation attacks as well as cyber defamation.
7. **Data Loss and Restoration Insurance:** This insurance is responsible for covering costs incurred due to computer-related assets, physical damage and loss of data. It also covers the cost involved in retrieving and restoring data, hardware, software, or other information that may have been destroyed or damaged due to cyber-attacks.
8. **Information Privacy Insurance:** This insurance covers organizational liabilities that may be the result of real or affirmed non-compliance with any overall cyber, data privacy, or identity-related guideline, statute, or the law (Figure 7.1).

7.6 CYBER INSURANCE COVERAGE AND ITS ASPECTS

Cyber insurance may be thought of as a financial risk transfer product that assists in shielding associations from digital and IT risks by moving those dangers from the insured. Insurance inclusion nowadays exists for an assortment of possible misfortunes and liabilities brought about by cyber risks. Insurance for specific organizations may incorporate a few or the entirety of the following:

Types of Cyber Insurances:
- Network Security
- Theft and Fraud
- Forensic Investigation
- Business Interruption
- Extortion
- Reputation Insurance
- Data Loss and Restoration
- Information Privacy

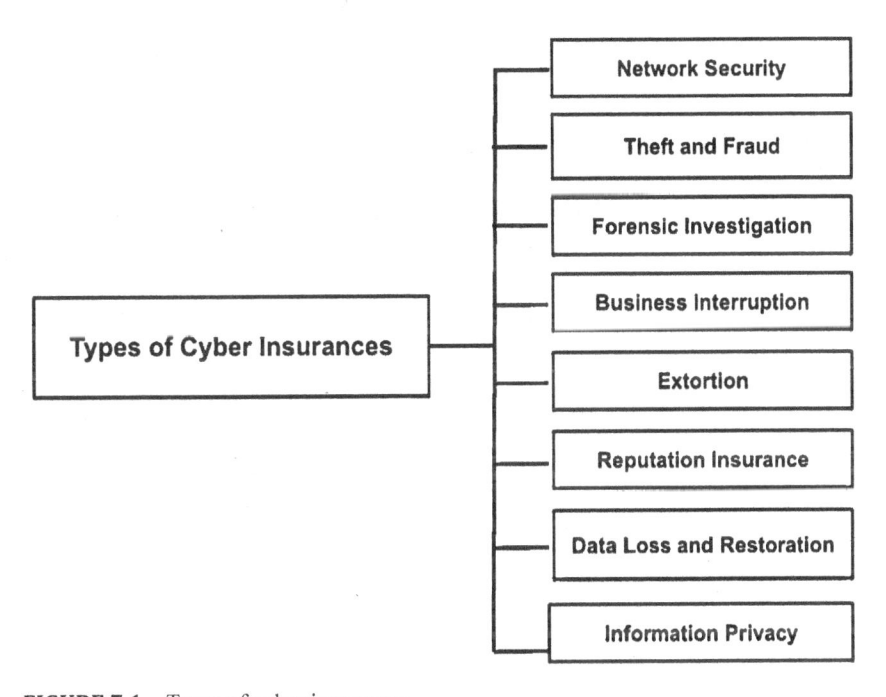

FIGURE 7.1 Types of cyber insurances.

1. **Breach and Incident Response (IR) Coverage:** This inclusion takes care of the accompanying expenses coming about because of a security breach:
   i. Forensics, measurable, and analytical administrations;
   ii. Breach warning assistance (call center, legal fees, mailing of materials);
   iii. Expenses related to fraud monitoring and identity;
   iv. PR and event management.
2. **Regulatory Coverage:** Big organizations like the Federal Trade Commission (FTC), the department of homeland security (DHS), Securities and Exchange Commission, and many other local, state, federal, and non-U.S. controllers presently take a gander at all parts

of cyber risk This coverage is responsible for reimbursing the costs of defending an activity by controllers because of a security breach, yet there is no impediment with regards to what caused the security breach. coverage may likewise apply to insufficient security that results in privacy breach, someone losing a laptop, or someone emailing a document to the wrong person. Coverage applies to non-governmental regulations, specifically, the payment card industry (PCI) and its principles. It additionally covers any fines and punishments forced by a court or controller or forced by certain nongovernmental associations, if insurable by applicable law.

3. **Liability Coverage:** Liability insurance is responsible for protecting the policyholders and insured individuals from the risks of liabilities forced by claims and comparative cases. Liability coverage is responsible for protecting the insured in a situation where the person in question is sued for claims that are secured by the insurance policy. Risk protection and liability insurance may likewise take care of certain expenses acquired by the policyholder in reacting to the case or claim yet not cost the policyholder causes to fix the issue. Basic sorts of cyber insurance liability coverages include:
   i. **Privacy Liability:** This inclusion covers safeguard and liability for inability to forestall unapproved use/admittance to confidential data. Coverage stretches out to personally identifiable information (PII) and delicate information of an outsider or third party.
   ii. **Security Liability:** This coverage covers protection and obligation for the disappointment of framework security for anticipation or moderation of a cyber-attack including yet not restricted to the spread of a virus or a denial of service. Disruption of framework security incorporates the failure of composed policies and systems tending to technology use.
   iii. **Multimedia Liability:** This coverage covers protection and obligation for media misdeed from online distribution, including plagiarism, copyright infringement, negligence in content, defamation, misappropriation of name or similarity disparagement, etc. Coverage stretches out to the establishment's own sites, email, and media exercises, for example, writing for a blog and tweeting.

4. **Internal Costs and Court Participation Cost to Protect Claims:** Under a cyber insurance policy, the insurer may have the privilege

and obligation to safeguard any case brought against an insured or may reimburse the insured for sensible expenses acquired by the insured to guard a case. An insured will be needed to help out the insurer in the defense of the claim and give to the insurer all data and help that the insurer sensibly demands, including going to hearings, statements, and trials, and help with influencing settlements, securing, and giving proof, getting the participation of witnesses, and leading the protection of any case secured by the policy.

5. **Cyber Extortion:** Or coercion typically appears as a ransomware assault, whereby a cybercriminal will encode a casualty's documents or compromises the release of delicate information unless a payment is made. Cyber extortion coverage takes care of the expenses of specialists counting digital forms of money for threats identified with intruding on frameworks and divulging private data.

6. **Institution's Loss of Income or Extra Expenses Due to:**
   i. **Security Breach:** Business interference coverage covers the loss of salary and additional costs coming about because of business interference as a result of a security incident or an unexpected or spontaneous outage. Regularly, the protected element must fulfill a holding up period or meeting a damage threshold before the coverage will apply.
   ii. **Security Breaches of Contingent Third Parties:** Many policies presently perceive the relationship of organizations and contain unexpected business interference arrangements. Unforeseen business interference coverage covers an insured's loss of salary and additional cost because of a security incident that intrudes on the administration of an element not possessed, worked, or constrained by the insured however that is depended upon to lead business. For instance, if a ransomware assault forestalls a finance administration from processing a foundation's finance and the organization causes expenses to physically give the finance, the additional expense of doing so would be secured.
   iii. **Framework Failures:** A framework failure refers to any unexpected and spontaneous outage of a computer system or framework. A framework failure may happen, for instance, after an association actualizes a framework fix that ends up being incongruent with existing capacities bringing about a spontaneous outage.

7. **Data Replacement Costs Due to a Security Breach:** This coverage applies to the expenses caused by the insured to supplant, reestablish, or recollect computerized resources from setting up accounts or from partially or completely coordinating electronic information records as a result of their deception, modification, or pulverization from an organization operations security disruption.
8. **Deceptive Fraud Transfer:** It is a kind of cybercrime that happens when an individual is fooled into moving assets to an unauthorized individual or record. In such conditions, cybercriminals consistently hack into an association's organization to send official-looking association messages to instigate organization representatives to move assets to an indicated genuine record. Deceptive fraud transfer schemes are like social engineering hoaxes (Figure 7.2).

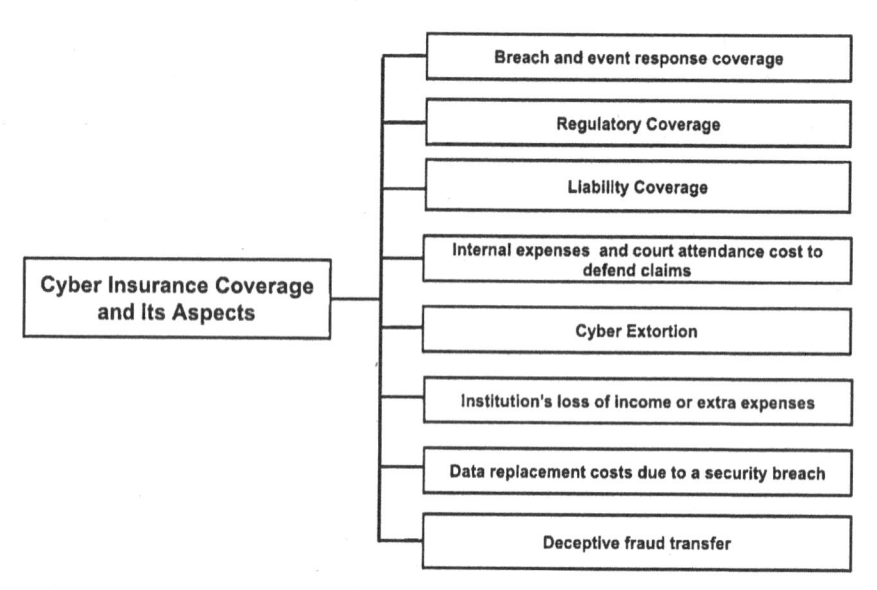

FIGURE 7.2 Cyber insurance coverage and its aspects.

Cyber insurance policies may incorporate a few or the entirety of the accompanying key segments:

1. **Media in the Control of Others:** Cyber insurance may cover decoded media in the consideration or control of third-party processors.

2. **Incidents Happening During the Policy Time Frame However Found Subsequently:** Coverage under a cyber insurance policy is set off, partially, by an insured's report of a claim to the carrier. Contingent upon the phrasing of the policy, it might cover incidents that happened during the policy time frame yet were found after the lapse of the policy time frame. Under a claims-made and announced form,' for instance, a claim must be made and received by the insurer during the strategy time frame or, if applicable, during any extended reporting period that may broaden the disclosure time frame for a while past the termination of the approach.
3. **Coverage for Privacy Breaches Other Than Electronic or Computer Related:** In addition to a breach of a computer system, individual information might be undermined when paper records are lost, taken, or inappropriately dealt with, bringing about an unapproved exposure. For instance, a security breach may happen when individual information paper records are not appropriately discarded.

7.7 DATA PROTECTED BY CYBER INSURANCE

Cyber insurance policies protect a few or the entirety of the accompanying kinds of information:

1. **Personal Health Information (PHI):** It is any data that contains exclusively recognizable health data and incorporates any aspect of a patient's clinical record. The information may be the health status of the patient, game plan of medical services, or payment and installment for medical care. Government law, similar to health insurance portability and accountability act of 1996 (HIPAA), presents different necessities with respect to the treatment of PHI.
2. **Personally Identifiable Information (PII):** It is the information that permits the identity of an individual to whom the information applies to be reasonably derived by either quick or atypical strategies. PII for the most part incorporates a person's name, address, phone number, Social Security number, account numbers, balances, histories, and passwords. It incorporates data that is dependent upon the family educational rights and privacy act (FERPA).

3. **Payment Card Information (PCI):** This incorporates the individual data held by a payment card brand to handle a payment card transaction. PCI likewise alludes all the more extensively to the PCI. PCI-DSS (or the payment card industry data security standards) alludes to the principles, rules, rules, and guidelines required by the payment card brand or the PCI-DSSs Council as for information security, exposure, and treatment of protected data.
4. **Confidential Third-Party/Research Information:** Sensitive third-party information, for example, proprietary innovations, plans, estimates, strategies, formulas, and records that are in the consideration, guardianship, or control of an insured might be viewed as private or protected data. A security incident may happen under a cyber insurance policy when there is an unapproved exposure of secret or confidential third-party data. For instance, if a professor's research database containing data on the social practices of people who have AIDS is breached and the information is taken, insurance may take care of the expense of cases made against the organization and the professor by those whose data is abused.
5. **Data Hosting Outsourced Electronic Processing or Data Storage:** Organizations frequently depend on the third-parties, for example, cloud suppliers and data centers, to perform basic business operations. Coverage under a cyber insurance policy may reach out to a computer system worked by a third party to help the insured. Cyber insurance policies may likewise stretch out inclusion to a common computer system which is a computer system, other than an insured's system, worked to serve an insured by a third-party under created concurrence with an insurance and may consolidate such systems as data facilitating, cloud administrations or registering, co-location, data back-up, data storage, data handling, platforms, programming, and infrastructure-as-a-service (Figure 7.3).

7.8 LOSSES THAT ARE NOT COVERED UNDER CYBER INSURANCE

Like other protection strategies, cyber insurance policies bar inclusion for specific losses. Exclusions are normally gathered in a part of the policy, however, inclusion may likewise be barred or restricted by the definitions or policy language. Normal prohibitions incorporate cases owing to or emerging from war, activity of a nuclear facility, intentionally beguiling or criminal acts, breach of agreement, taking restrictive developments, biased exchange practices, and strategic policies. Insurance carriers exclude these risks since they are reluctant to safeguard these risks, since inclusion may exist in another policy, or on the grounds that inclusion might be against the public policy. Cyber insurance policies likewise normally avoid inclusion for any episode or claim that emerges from or depends on a determined, purposeful, intentional, pernicious, deceitful, deceptive, or criminal act or omission committed by the insured. The overall purpose of this rejection is to keep the insured from accepting a monetary advantage for submitting an unlawful or exploitative act.

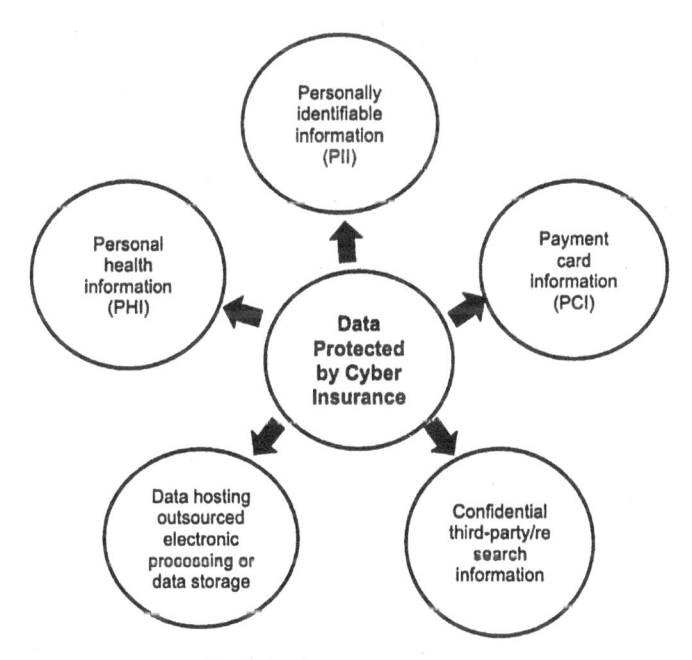

FIGURE 7.3 Data protected by cyber insurance.

In the quick repercussion of a cyber incident, an association with cyber insurance ought to promptly advise its merchant and insurer [98]. Brief warning is liked however regardless, it ought not happen past the prerequisites indicated in the insurance policy. Brief reporting is enthusiastically suggested regardless of whether the association chooses not to document a case. Numerous insurers have a guidance board and other pre-endorsed sellers that the association can use in the aftermath of a breach for remedial, legitimate, and different expenses. To evade delays and augment hierarchical operability, it is suggested that the association have a previous relationship with every merchant or if nothing else have some proportion of knowledge of the preceding a breach. If an association needs to utilize a merchant that is not on an insurer's panel or pre endorsed seller list, this normally is not an issue as long as the insurer is made aware preceding an incident.

7.9 PROCESS OF BUYING CYBER INSURANCE

Numerous associations in industries just as foundations of advanced education face cyber risk exposure. All things considered, higher education foundations warehouse volumes of sensitive data and are also defenseless to losses identified with business interferences. Since not all cyber risks can be mitigated through powerful methods or specialized controls, cyber risk exposure is a practical method for restricting the extent of an educational establishment's exposure. While there is no by and large government or state law ordering that advanced education establishments convey cyber insurance, state or university authorities may, in any case, confirm that conveying such inclusion is to the greatest advantage of the university. Benchmarking the acquisition of cyber insurance normally depends on correlations with associations of comparable industry and income size. Insurance brokers may have benchmarking data dependent on customer information. Different sources are through reviews or companion establishments. Best practices for acquiring cyber liability insurance quote incorporate the following:

1. **Working with an Accomplished Broker:** A compelling merchant ought to have a solid, far-reaching handle of the extent of an association's cyber risk. The person must comprehend and clarify how this risk might be evaluated and ought to be fit for giving proposals on insurance carriers or policies that may be a solid match for the association. It is critical to get fitting inclusion and favorable pricing. Choosing an intermediary with cyber insurance aptitude might be fundamental for securing cyber insurance

that tends to the foundation's needs. Since insurance agencies use policy forms for a wide range of associations, schools and universities must review policy language to search for terms and arrangements that are indistinct in the advanced education setting and address these preceding buying the policy.
2. **Lead a Security Risk Evaluation to Decrease Premium:** A risk assessment gives more prominent straightforwardness into the association's network protection controls and enables the association to distinguish weaknesses and possibly make changes to territories needing improvement, which, if appropriately actualized, could bring about an excellent decrease.
3. **Execute Security Controls That Reduce Premium:** Cohesive and interconnected corporate practices outfitted around individuals, procedures, and innovation-related cybersecurity enhancements diminish risk and can prompt lower charges. Limiting organization admittance to certain clients is a case of an answer that impacts every one of the three categories:
    i. A corporate policy commands the network access limitation;
    ii. The restriction is executed by information security or information technology team; and
    iii. One of the policy's core objectives is to limit human error.

7.10 TRADEOFFS OF SELF-INSURING WITH RESPECT TO BUYING CYBER LIABILITY INSURANCE

Organizations, working with an accomplished insurance intermediary and risk management experts, must assess their own cyber risk appetite and what steps they mean to take to alleviate those risks. Cyber insurance is not a solution, just for all risk alleviation policies and techniques, and it has impediments. For instance, it is not expected to cover the robbery of cash/reserves. Cyber insurance likewise by and large does not cover the effect on institutional reputation, the depreciation of a trading name, or the loss of IP. The managerial weight to manage the occurrence or prosecute an insurance claim is likewise not secured. The cyber insurance market can be unstable. As a rule, buying cyber insurance requires more exertion in reacting to the insurer's requirement for data than purchasing different sorts of insurance, and there is as yet a shortage of information other than that identified with PII-related breaches.

7.11 IMPLICATIONS OF CYBER INSURANCE ON COMPUTING SYSTEMS AND PROCESSES

Cyber insurance may be thought of as a safety net that sits alongside and complements organizational cybersecurity controls. It is not planned to fill in for technical, process, or human-related cybersecurity like ongoing training and awareness of end-users of cybersecurity protocols. Therefore, if organizational frameworks or procedures are exploited or in any case neglect to forestall a cyber-attack, cyber insurance limits the subsequent monetary effect. Key organizational cybersecurity stakeholders (the chief information security officers (CISO), chief information officers (CIOs), chief procurement officers, chief financial officers, general counsel, chief revenue officer) should be fully engaged in all cyber insurance-related purchase discussions, along with discussions related to prescient investigation, predictive analysis and other measurement or similar devices given by insurance brokers, transporters, or outside consultants. These partners are best situated to comprehend the extent of their affiliation's cyber risk and give recommendations on how much risk transfer may be vital. Different ramifications of buying a cyber liability insurance policy incorporate the following [98]:

1. **Improved Online Protection Through Pre-Break Administrations:** Many cyber insurance policies incorporate free or limited administrations that can be utilized by policyholders before a breach happens so as to diminish the probability or seriousness of a future cyber incident. These administrations can incorporate online security appraisals, admittance to cybersecurity expertise through consulting services or white papers, and cybersecurity awareness training programs for users. Numerous insurers additionally keep up lists of pre-affirmed merchants or outsider accomplices who give discounted services, for example, tabletop activities, response planning, and consistent evaluations. Utilizing pre-breach benefits that are incorporated as a major aspect of a cyber insurance policy can be an economic route for an association to improve.
2. **Help to Establishments in Settling on Security Choices:** Some insurance brokers and carriers have in-house staff who give security upgrade suggestions on issues, for example, third-party vendor exposure, estimating authoritative consistency with voluntary

frameworks, for example, NIST, and ISO, and measuring the possible budgetary effect of business interference losses.
3. **Adding Cyber Insurance to Insurance Prerequisites Forced on Contract Accomplices:** For associations with a broad supply chain or dependence on third-party merchants, requiring cyber insurance might be exceptionally fitting. Advanced education establishments may add cyber insurance to contracts with accomplices or might find that the agreement accomplice progressively necessitates that the organization keep up cyber insurance.
4. **Insurer Requirements to Encrypt Portable Media/Computing Devices:** Insurers do not really need encryption of information at rest or in process, yet associations that can show this ability to insurers either on an insurance application or during the underwriting process may lead the insurers to consider the organization safer, which may bring about premium discount.

7.12 INSTITUTIONS' KNOWLEDGE REGARDING THE INSURANCE CLAIM PROCESS

Usually, three parties are involved in the claims handling process, i.e., the insured association's defense counsel, broker claims help, and the insurer's claims team. During the Insurance claims procedure, the association's insurance broker gives related help and legal assistance. The intermediary may also assist with settlement techniques and Cyber Insurance reacting to coverage questions. A critical factor for associations when settling on insurers is claims-paying history. An advantage of cyber insurance is admittance to individuals who are proficient in the different laws identified with cyber incidents. Accidental mismanaging of an event by the association may not renounce the insurance out and out, yet best practices direct captivating the insurance agency during the time spent tending to the occasion that prompted the potential case [98]. Following an incident prompting a case, the organization and insurance agency will assess what turned out badly and think about choices to relieve the risks later on. The insurance agency may ask for certain procedures, yet its capacity to command is a component of its eagerness to proceed as the insurer and whether the market gives the insurer leverage. In numerous occasions, breaches happen in a nation or purview without legitimate prerequisites to notify. In such cases, the organization may do what is fitting or lessen expected commitment irrespective of the fact that the foundation has no legal need to act. The expense of these activities might be secured by insurance despite the fact that not lawfully needed.

7.13 CYBER LIABILITY INSURANCE

According to the International Risk Management Institute (IRMI), cyber liability may be defined as an insurance policy design that is responsible for providing coverage to consumers of technology services or products [99]. The main objective of Cyber Liability Insurance is to cover liability and property losses that may occur when a business engages in various electronic activities. These activities may range from selling on the internet to collecting data within its internal electronic network. Businesses and associations associated with the web are susceptible to cyberattacks. The consequences may be in the form of data breaches. Some bad actors that are responsible for cyber-attacks are as follows:

1. **Disgruntled Former Employees:** Many workers hack into a former employer's system as an act of revenge for their wrongful termination (considered). If the employer is negligent towards changing logins or following good security practices, these attacks can happen very easily.
2. **Human Error:** Negligent employees may leave their cell phones unlocked and unprotected taxi or click on links associated with emails from a party they are not familiar with. There may be lazy employees or managers that use the same password for many workplace programs. Such situations encourage data breaches and ransomware attacks, as it is very easy to vandalize the computer network.
3. **Mobile Devices:** Access systems using smart devices puts data in the systems at risk, hence proper safeguards must be ensured. Employees may also access systems in a shared Wi-Fi hotspot. Script kiddies can seize credentials and easily access the system's network.
4. **Ransomware:** A popular method of generating income for script kiddies and low-level hackers is Ransomware. Once this malware is introduced into a system, it locks (encrypts) the data within. The hacker may then demand a hefty amount for ransom to provide the key that can decrypt the system's data.

5. **Coordinated Attacks:** These attacks are usually launched by a group of hackers in a foreign country with the primary goal of stealing sensitive information. They target organizations that store personal data, related to healthcare or financial organizations. The stolen data is then sold to others for using it for various purposes like opening credit accounts in the name of the individual whose information was taken. This is ordinarily known as identity theft.

7.14 CYBER LIABILITY COVERAGE

Cyber liability is responsible for providing financial coverage for expenses associated with a data breach. These costs can amass quickly once an information breach is found and announced:

1. **Client Loss:** Customers that are influenced by an information breach are probably going to quit working with you since the association neglected to protect their delicate information.
2. **Business Disruption:** According to organizations working on information assurance, the disruption of doing business commonly represents around 40% of the total expenses of an information breach.
3. **Regulatory Fines:** Not only will the customers punish an organization after a data breach, various government agencies may also join the party. The Federal Communications Commission (FCC), FTC, Health, and Human Services (HHS), and state authorities, may impose fines and expensive tasks that the corresponding organization must perform once an information breach has been accounted for.
4. **Legal Costs:** Entrepreneurs must comprehend that benevolent clients can immediately turn out to be disagreeable when their bank account and credit cards are enduring an attack. When large health insurers are attacked, substantial lawsuits may be filed within hours of the data-breach disclosure.
5. **Public Relations (PR):** Once the business's reputation has been corroded, it is costly to recruit a PR firm to rehabilitate the company's image.
6. **Direct Financial Loss:** Following a data breach, attackers can access an organization's financial accounts and may straightaway

transfer accessible assets to accounts they control. A complete Cyber Liability strategy will take care of first-party expenses and third-party expenses in case of an information breach. First-party costs are the consequence of damage to sensitive data and any affected systems. Third-party expenses result from an organization's risk for clients who might be monetarily influenced by an information breach (Figure 7.4).

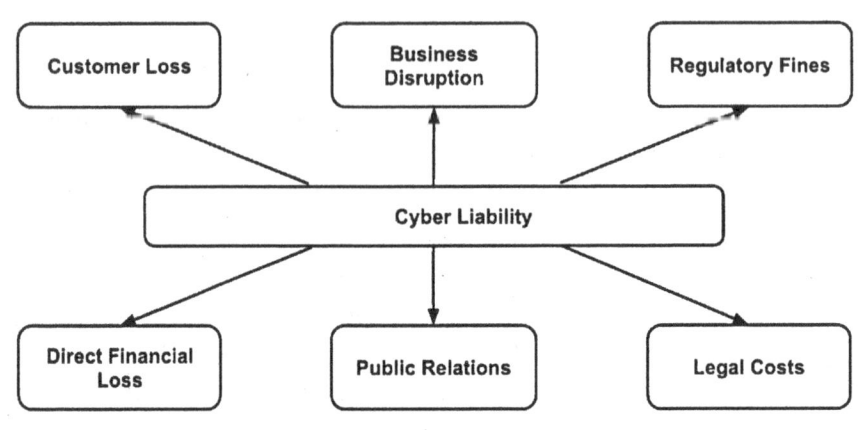

FIGURE 7.4 Cyber liability coverage.

In this chapter, we identified yet another important domain pertaining to cyber risks, i.e., cyber insurance. We highlighted the need for Cyber Insurance and its relationship with Cyber Risks. Apart from that, we also discussed its types, advantages, disadvantages, its coverage, and policies. The types of data that are protected, the kind of information that is not protected by cyber insurance have also been mentioned. Finally, Cyber Liability Insurance has also been discussed briefly.

7.15 SUMMARY AND REVIEW

- Cyber risk refers to the risk of financial harm to an organization which may be attributed to the failure or disruption of its computer systems. Alternatively, it may be defined as the risk that companies face from their handling data and depending on technology in their everyday tasks.

- The duty of care risk analysis standard (DoCRA) gives practices and standards to help balance consistency, security, and business goals when creating security controls.
- The different types of cyber insurance are network security, data loss and restoration, theft, and fraud, business interruption, forensic investigation, extortion reputation insurance, and information privacy.
- Regulatory coverage is responsible for reimbursing the costs of defending an action by regulators due to a privacy breach, but there is no limitation as to what caused the privacy breach.
- Liability insurance is responsible for protecting the policyholder and insured individuals from the risks of liabilities forced by claims and comparable cases.
- Common types of cyber insurance liability coverages include privacy liability, security liability, and multimedia liability.
- Foundation's loss of income happens mainly in light of security breach, security breaches of contingent third gatherings, and system malfunctions and failures.
- Data protected by cyber insurance may be in the form of personal health information (PHI), personally identifiable information (PII), payment card information (PCI), confidential third-party/research information, and information facilitating outsourced electronic processing or information storage.
- The International Risk Management Institute (IRMI), characterizes cyber liability as an insurance policy intended to give coverage to consumers of products or technological services. The coverage is proposed to cover obligation and property losses that occur when a business takes part in different electronic exercises.
- Cyber liability gives budgetary coverage to costs identified with an information breach. These costs can collect quickly once an information breach is found and announced.

QUESTIONS TO PONDER

1. How does cyber-insurance have the potential to be a market solution?
2. What is the use of duty of care risk analysis standard (DoCRA) with respect to Cyber Insurance?

3. How is cyber insurance related to cyber risks?
4. Are there any sorts of events that are explicitly rejected from coverage? In what situations does cyber insurance not cover you?
5. How long after a breach happens do you need to announce it without losing coverage?
6. Would I be able to protect my business without buying Cyber Liability Insurance?
7. How good would be my cyber insurance if a worker caused the attack?
8. By what means can a cyber insurance vendor help lessen an organization's network security risk?

KEYWORDS

- **computer networks**
- **cyber liability insurance**
- **cybersecurity**
- **duty of care risk analysis standard**
- **intrusion-detection systems**
- **network security risk**

CHAPTER 8

Introduction to Cybersecurity Policies

In the previous sections, we discussed ethics, laws, and risks in the domain of cybersecurity. Another significant addition is the topic of policies. While ethics is deemed as the code of responsible behavior, the law is usually seen as an outcome of ethics. Risks related to cybersecurity require a significant level of management, policies could be one way to do that. Hence, policies are made to achieve some goals. In this chapter, we will get acquainted with the overall idea of policies in cyberspace, people who are associated with cybersecurity policies like policymakers and policy audiences, different types of policies, and some very famous cybersecurity policy frameworks.

8.1 CYBERSECURITY POLICIES

Cybersecurity is a global issue. Risk management at the enterprise level is a big concern, hence every employee in an organization including IT professionals and managers must be concerned about security. One of the effective ways of educating employees on the significance of security is a cybersecurity strategy that clarifies an individual's responsibilities for protecting IT systems and data. Cybersecurity policies are responsible for setting the standards of behavior for activities like encrypting email attachments and restricting the use of social media. Cybersecurity policies are significant as cyberattacks and data breaches are potentially costly [100]. Many times, employees are the weak links in an organization's security. This is because employees may share passwords, click on malignant URLs and links, use unapproved cloud applications, and disregard to encode confidential records. All this may lead to catastrophic effects in cyberspace. However, improved cybersecurity policies may be useful for a better understanding of how to maintain the security of data and applications by the employees and consultants. Cybersecurity policies are a must in organizations and public companies that are associated with industries like healthcare, finance, or insurance. In these organizations, cost as well as reputation are at stake, and there are heavy penalties paid if their security infrastructure and procedures are found to be inadequate. Cybersecurity policies promote the public image and credibility of an organization. Often customers, partners, shareholders, and prospective employees demand evidence that the organization is capable of protecting its sensitive data. It might be hard for an association to demonstrate its capacity if it does not have a cybersecurity policy.

> **Definition 8.1: Cybersecurity Policy**

> A cybersecurity policy is the statement of responsible decision-makers about the protection methodology of an organization's vital physical and data resources. It may be thought of as a document that describes a company's security controls and activities. Although security policies do not determine technological solutions, they characterize sets of expectations and conditions that may assist in protecting assets along with its proficiency to organize business. Ideally, a security policy may be defined as the essential manner by which organization prospects for security are converted into explicit, quantifiable objectives. These policies are used to assist users in building, installing, and maintaining systems, so that they do not have to make those decisions by themselves. Policies are written to offer guidance about the kind of behavior and resource uses may be required in acceptable from the employees of an organization. It also states what is forbidden to wield.

Cybersecurity procedures define the rules for the number of employees, consultants, partners, board members, and other end-users who access online applications and web assets, send information over communication networks, and in any case practice dependable security. Usually, the first part of a cybersecurity policy defines the overall security expectations, jobs, and duties in the association. Stakeholders incorporate outside consultants, IT staff, financial staff, etc., [101]. This section characterizes the jobs and obligations or information responsibility along with accountability of the policy. The policy may also include sections with respect to other areas of cybersecurity. These sections may highlight requirements for antivirus software or the use of cloud applications. In order to incorporate guidelines and procedures for large organizations or those in regulated industries, a cybersecurity policy may be several pages long. In the case of small organizations, it may be only a few pages long and must cover fundamental security best practices. Such practices may incorporate:

- Rules for utilizing email encryption;
- Steps for accessing work applications remotely;
- Rules for making and securing passwords;
- Rules on the use of social media.

Cybersecurity policies must be irrespective of their length and must highlight the areas of primary importance in the organization. It might incorporate security for the most delicate or managed information, or security to address the reasons for earlier information breaches. Risk analysis may be used to highlight areas to prioritize in the policy. The policy must be simple and easy to read. It must also encompass technical information for referred to reports, particularly if that information requires periodic updating. In a situation where policy must specify employees encrypting all personally identifiable information (PII), it does not have to highlight the particular encryption programming to utilize or the means for encoding the information.

8.2 CYBERSECURITY POLICY MAKERS

Security policy development is undertaken by different entities in an organization who collectively provide inputs for the policymaking process. Anyone who has a stake in the security policy must be involved in the process of policymaking. This is because policies must be developed according to the requirements of various people in an association. Creation of policy is witnessed by the following entities [102]:

- **Board:** Company board members need to present their expertise for reviewing existing policies and suggesting new ones. Keeping in mind the running conditions of business of the organization, policies can be drafted.
- **IT Team:** Since policy involves developing standards around the usage of computer systems, especially security controls, IT team members happen to be the biggest consumers of the policy information in an organization.

- **Legal Team:** This team ensures the legal points in the document and also guides a specific purpose of appropriateness in the organization.
- **Human Resources (HR) Team:** This team is responsible for obtaining a certified Terms and Conditions certificate from every representative that they have perused and perceived the specified policy. This is on the grounds that the HR group manages prize and discipline-related issues of workers to execute discipline.

Security policymakers must know that picking a policy and validating it are two different things. Choosing a policy is a simple task with respect to analyzing the effectiveness of it. Security policy development approach in an organization may be explained as in Figure 8.1.

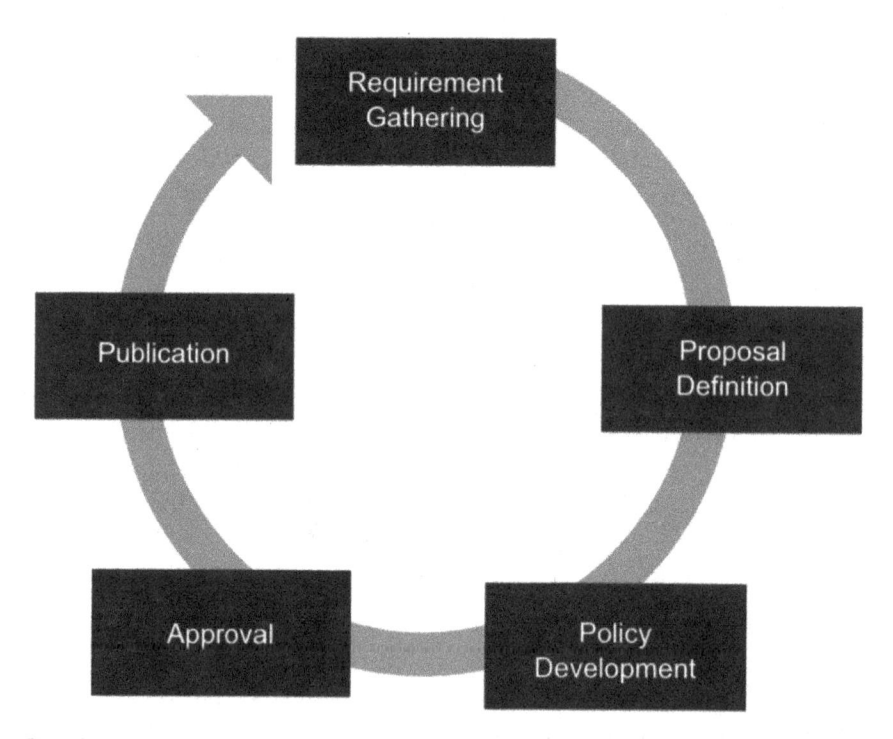

FIGURE 8.1 Security policy development stages.

The IT department, often the Chief Information Officer (CIO) or Chief Information Security Office, is fundamentally liable for all information security policies. On the other hand, stakeholders add to the policy, contingent upon their aptitude and jobs inside the association. Some key partners who are probably going to take an interest in policy creation depending on their roles are as follows [102]:

- C-level business executives are responsible for defining specific requirements for security, followed by the resources present to develop cybersecurity policy. Inadequacy of resources is an obstacle in writing a policy since it cannot be implemented. It is also perceived as a waste of personnel time.
- The legal department is responsible for ensuring that the policy meets legitimate necessities and conforms to government guidelines.
- The HR department explains and enforces employee policies. HR personnel are responsible for confirming that employees have read the policy and disciplining those who are found to violate it.
- Procurement departments appraise cloud services vendors and other relevant service providers. Additionally, they manage cloud services contracts. Procurement personnel is also responsible for verifying whether a cloud provider's security is in par with the organization's cybersecurity policies. They are also responsible for verifying the adequacy of other outsourced important services.
- Board members from public organizations and affiliations are responsible for reviewing and approving policies. Their main responsibilities are policy creation depending on the needs of the organization.

In situations where personnel are invited to participate in policy development, it is important to consider who is the most critical to the success of the policy. A department manager or business executive who enforces policy or provides resources for implementing it may be an ideal participant.

8.3 UPDATING AND AUDITING CYBERSECURITY PROCEDURES

A policy may be defined as a guiding principle that may be used to set direction in an organization. A policy is a course of action for guiding and influencing decisions. Policies are used for making decisions with respect to specific circumstances. Policymaking must take into account the overall framework of objectives, goals, and management philosophies as stated by the senior management. A procedure is a specific method of achieving something. It ought to be planned as a progression of steps to be followed as a steady and monotonous methodology or cycle to achieve a final product.

Since technology is consistently changing, there is a need to update cybersecurity procedures regularly, ideally once a year. Also, establishing an annual review and update process and involving key stakeholders is a must. When reviewing an information security policy, an organization must contrast the policy's rules and the real acts of the association. A policy audit or survey can be helpful in pinpointing rules that no longer address current work processes. An audit may also assist in identifying where better enforcement of the cybersecurity policy is needed. Three important policy audit goals are as follows:

- Comparing the organization's cybersecurity policy to actual practices;
- Determining the organization's exposure to internal threats; and
- Evaluating the risk of external security threats.

An updated cybersecurity policy may be an essential security resource for all organizations without which it is possible for end-users to make mistakes that may result in data breaches. An irresponsible attitude may cost an organization significantly in terms of fees, fines, loss of public trust, settlements, and brand degradation. Hence it is mandatory to create and maintain policies that can help prevent such adverse outcomes.

8.4 CYBERSECURITY POLICY AUDIENCE

Cybersecurity policy holds true for all senior administration, workers, investors, experts, and service providers who use organization resources. The security policy must be comprehensible to its audience. This may be achieved by ensuring that the policies are concise and readable. This will make it easier for employees to adhere to the policies and fulfill their defined roles. Figure 8.2 gives an idea about the cybersecurity policy audience.

In other words, the security policy incorporates all employees, hardware, software, consultants, service providers who use company assets. These assets may be in the form of computer networks data information or any information that may prove to be valuable to the business.

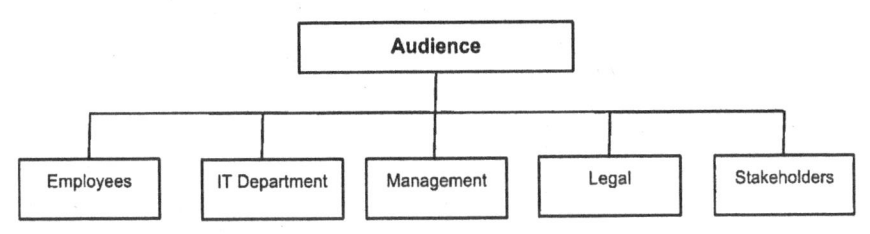

FIGURE 8.2 Cybersecurity policy audience.

8.5 CYBERSECURITY POLICY CLASSIFICATION

Each association commonly has three policies. The first one is drafted on paper, second, one is in employees' minds, and the last one is executed. The security policies are essential to the hierarchy of management control that addresses its audience regarding what must be conducted by the specified terms and requirements of an organization. The policy generally requires what must be done. However, it does not address how it should be done. Security policies may be informative, regulative, and advisory. They may be subdivided into the following categories as:

1. **Physical Security:** It mandates what protection mechanisms must be enforced to safeguard the physical asset of both employees and management. It applies to the prevailing facilities that include doors, entry points, surveillance, alarm, etc.
2. **Personnel Management:** These are responsible for telling their workers how to lead or work every day and conduct business exercises in a safe way. It may take into account password management, confidential information security, etc., and it may apply to individual employees.
3. **Hardware and Software:** It specifies the administrator about the type of technology that may be used. It also defines how network control ought to be arranged and applies to framework and system managers (Figure 8.3).

8.6 CYBERSECURITY POLICY AUDIT

Security documents may be thought of as living documents. Hence, it is mandatory to update them at specific intervals with respect to altering business and customer specifications. After the establishment of policies, they become operational, following which an audit may be performed. Audits are usually performed by outsiders or insider agencies for comparing existing practices to the intentions of the policy. Security policy audits to help the organization to see better the threat the association is prone to and the adequacy of the current protection. The primary goal of the security audit is to bring all the security policies as closely as possible. Security auditing may be thought of as a high-level investigation of the current project progress. It also analyzes the company posture by testing information security related to existing policy compliance. Auditing is performed from time to time because it is necessary to contrast existing practices against a security strategy. This may be useful for substantiating and verifying the effectiveness of security measures. Security reviews are helpful for giving the organization confirmation of best practices that already exist. A successful security audit focuses on:

- Comparing security policy with the actual practice in place;
- Determining an organization's exposure to threats internally;
- Determining the exposure of an organization externally.

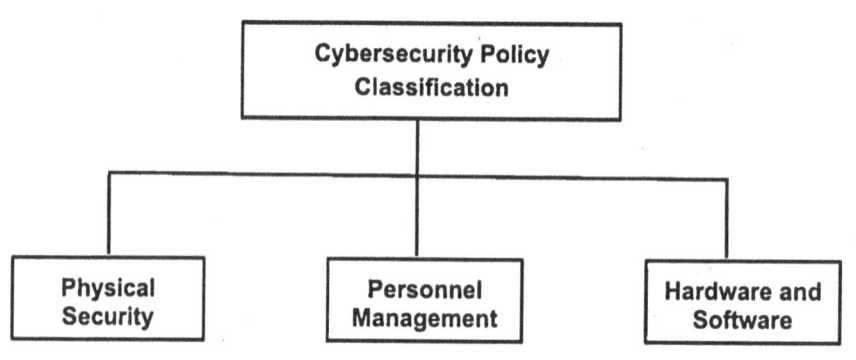

FIGURE 8.3 Cybersecurity policy classification.

8.7 CYBERSECURITY POLICY ENFORCEMENT

Upholding security policies may ensure consistency with the standards and practices directed by the organization [102]. This is because policy, procedures do not work if they are violated. Enforcement may be arguably the most significant aspect of a company since it is responsible for dissuading anyone from deliberately, accidentally violating policy rules. System administrator-level enforcement guarantees appropriate support and forestalls privilege escalation, and employees level enforcement monitors whether assurance of daily working activities actually comply with the policies. Nonetheless, there is a need to keep up an appropriate balance between positive and negative reinforcement. The best employees that abide by rules are rewarded from time to time for increasing their motivation and boosting up their morale high in positive enforcement. On the other hand, in negative enforcement, strict compliance of policies takes the form of threat to the employees.

8.8 CYBERSECURITY POLICY AWARENESS

Company employees are often thought of as soft targets to be compromised [102]. This is because the human elements are the least predictable and easiest to exploit. Trusted employees may be framed to provide valuable information about a company. Therefore, one of the best methods to combat this exposure of information by employees is cybersecurity training and education. It is necessary to make sure that the employees can understand what kind of information they should share and what they must not. This will make the company less vulnerable to risks and cyber-attacks. A good security awareness program must be conducted periodically. This program must highlight all the current security policies that are commanded to be compliant from employees' end. Moreover, awareness programs must coordinate correspondence and suggestions to workers about what they ought to and not ought to share with outsiders. Hence security policy awareness training and education is necessary for mitigating the threat of information leakage.

8.9 WRITING AND DEVELOPING EFFECTIVE CYBERSECURITY POLICIES

Businesses these days run their primary operational processes using cyberinfrastructure. As we know, cyber threats and attacks are rising, therefore companies are also investing in better security systems. No matter how strong the current cybersecurity plan seems, the reality is that all organizations can possibly be attacked. Nowadays, breaches and cyber-attacks are very common [14–16, 23]. Therefore, there is a need to prepare cybersecurity policies. One approach to ensure cybersecurity is to implement the most effective technologies. However, technologies are only effective to a particular extent. Technologies are limited to companies and people who are associated with them. Therefore, policy setting and enforcement is a paramount objective for CIOs and Chief Security Officers. To write and develop effective cybersecurity policies, the following must be considered [103].

1. **Understanding Own Security:** Organizations utilize diverse third-party items in various parts of their operations. Using an off-the-shelf policy for such products is very common, in spite of the fact that it is not the ideal route for the board to comprehend network security. Instead, organizations must find out what their internal team thinks about their security. Essentially, the policy usually comprises mandates made by IT professionals and management. These two parties must analyze the details and need to arrive at a typical decision about the content of the policy. Making time as a group to examine policies may assist organizations in understanding the types of information to work with. Information can be collected and stored. Moreover, organizations may segregate private information from non-private information. Normally, businesses rely on a security industry guidelines report as the benchmark for making their policies. This permits associations to compose a security policy that will be acknowledged not only by their internal teams but also by external auditors and others.

2. **Checking the Compliance:** Using a security industry standards document is extremely effective in aligning an organization's policy to the recognized standards. Additionally, it helps understand all the security compliance requirements in the industry. Cybersecurity regulations are also suggested by the federal government for organizations to follow. In a situation where business deals with health information, the organizations' policy must take into account measures pertaining to technical, physical, and administrative domains for securing it. The organization needs to stay HIPAA compliant. If credit card information is requested from customers, understanding the payment card industry (PCI) Security Standards will help ensure that the organization is compliant. Knowing these standards will help develop, structure, and implement policies

in the best way possible. If someone is involved in government contracts, it assists with understanding the International traffic in arms regulations (ITAR) and export administration regulations (EAR). These guidelines are responsible for providing guidance on securing defense, civilian, and military information. Specific companies associated with industries like finance, insurance healthcare, must adhere to regulatory compliance concerning IT security. Companies in these industries must annually review security compliance requirements and regularly update their security policies and practices.

3. **Considering the Infrastructure:** A well-planned cybersecurity policy must highlight the systems a business uses to protect its basic and client information. An organization needs to work with its IT team to understand the company's capability. This will help evade potential cyber-attacks. The policy may explain which programs will be used for security and identify how updates will be made to seal all potential vulnerabilities. It will also help users understand how data may be backed up. If possible, the policy must also state clearly the cloud servers that are being used for storage. Having this information in policy is critical, as it shows how everything has been taken into consideration. Further, it assists customers, partners, and clients understand the measures an organization has in place to deal with data loss and to mitigate an attack. The cybersecurity policy should include information on controls such as:

   i. Which security programs will be implemented (e.g., protection of endpoints in a layered security environment. This may be achieved by anti-malware, antivirus, anti-exploit software).

   ii. Updates and patches that may be applied for limiting the attack surface and plugging up application vulnerabilities (e.g., setting the frequency for operating system or browser for application updates).

   iii. Backing up of data (e.g., enabling multi-factor authentication along with automated backup to an encrypted cloud server with).

   Also, the policy ought to plainly recognize jobs and obligations. This may include who:
   a. will be issuing the policy?
   b. will be responsible for its maintenance?

    c. will be responsible for enforcing the policy?
    d. will be training users on security awareness?
    e. will respond to and resolve security incidents?
    f. all(users) will have admin rights and controls?

4. **Accountability and its Significance:** Accountability is one of the most important aspects of ensuring policy [13]. An attack is demanding as it takes time and team effort to manage it. Accountability ensures that people become responsible for contacting customers and fixing the problem. Accountability measures of an association may likewise incorporate an alternate course of action for cyber-attacks. For example, in a situation where an attack happens and the chief security technician is away, there must be another person to handle the attack. Alternatively, there must be some contact or backup to manage the attack. It is also recommended to incorporate contact information for clients and customers. This information may be useful in the aftermath of an attack. Clients and customers can be reached out to for questions or assistance. The management must create and follow a schedule for reviewing the company's cyber risk. This would ensure accountability in all many vulnerable areas. In the long run, it may also assist in managing the reputation of the organization. It may also keep the business running when the organization gets attacked.

5. **Considering the Employees:** It is necessary to consider the acceptable use conditions for employees while drafting cybersecurity policies. The smallest of cyber-attacks may be the result of a simple mistake or error made by an employee knowingly or unknowingly. Therefore, an organization must clearly state the best practices for using the company's resources and tools. The employees need to understand the best password management practices. The association ought to likewise have a protocol that workers can use to describe security incidents. The use of social media can also be regulated since it is one of the common sources of phishing scams. For remote workers, organizations must ensure that they understand how to use networks. It is necessary to comply with the stated guidelines. The employees may be asked to not share their credentials and avoid using public networks. The employees must also be aware of the retributive activity for any individual that neglects to adhere to the security rules. Employees

must have sufficient knowledge on how to use work equipment like computers and portable storage devices [11]. They can be taught how to identify scams and spams that they might encounter online. The cybersecurity policy must clearly communicate best practices for users so as to restrict the potential for attacks and enhance the harm. The policies ought to likewise permit representatives the appropriate level of freedom they require to be productive. While banning all Internet and social media usage would certainly help keep the organization safe from online attacks, it is probable that it might lead to counter productiveness. Acceptable use guidelines might include:

i. Knowledge regarding detection of social engineering tactics and other scams;
ii. Knowledge regarding acceptable Internet usage;
iii. Information on how remote workers should access the network;
iv. Use of social media;
v. Knowledge on password management systems;
vi. Knowledge on reporting security incidents.

Additionally, the policy must also state what happens in a situation where users fail to comply with the guidelines. For example, if an employee is responsible for a breach might, he might have to undergo training once again owing to the negligence. The employee may be terminated if the breach is proved to be an internal attack.

6. **Updating Software and Systems:** Adopting a Centralized policy-making in IT is similar to 'push' methodology as it forces new security updates onto a user's device as the device connects to the network. On the other hand, a 'pull' methodology, notifies the user when a new security patch is available and also gives them the option to load this new software onto their device. The rising volume and velocity of cyberattacks demand tougher guidelines. This is also necessary because many users never bother updating their devices.

7. **Conducting Top-to-Bottom Security Audits:** It is necessary for organizations to direct a careful security review of its IT resources and practices. The audit will be responsible for reviewing the security practices and policies of the central IT systems, as well as the

end-user departments and at the edges of your enterprise, which incorporate automated machines and the internet of things (IoT) that an organization could be employing at remote manufacturing plants. The audit must take into account the software and hardware techniques the organization needs to ensure better security. It should also consider remote site personnel habits and compliance with security policies.

8. **Social Engineering:** As part of the end-to-end IT audit, the organization must include social engineering, which reviews whether the employees are exposing loopholes in an organization by exhibiting vulnerability when requested classified data. Social engineering techniques range from adversaries impersonating as important personnel asking for employee credentials to being a user who surrenders passwords and vital information by clicking on attachments that are a part of phishing emails.

9. **Demanding Audits from Vendors and Business Partners:** Companies must design policies in a way that demand customary security review reports from vendors they are thinking about before contracts are agreed upon. Following this, vendors may become a part of their service level agreements (SLA) [11], and must be expected to convey security review document details regarding a yearly premise.

10. **Providing New and Continuing Security Education:** Cybersecurity education must be made mandatory for every new employee orientation. The new employees must sign off that they have read and understood the training. It is advised to introduce a refresher course on practical cybersecurity which can be taught annually. Along these lines, representatives can recall security practices and policies and will likewise know about any policy augmentations or changes in the future.

11. **Watching the Edge:** Manufacturing 4.0 and other remote computing methodologies are moving to compute away from data centers. Thus, manufacturers with remote plants at a specific location may have manufacturing personnel operating automated robots and production analytics with local servers in the plant. It is necessary to maintain hardware and software security in these devices. The devices in this case must be privately controlled under-acknowledged cybersecurity procedures and policies by

workforce who are responsible for carrying out such jobs without an IT background. Since it requires training non-IT personnel in IT security policies and practices, it may be deemed as a security exposure.

12. **Performing Regular Data Backups:** In a situation where an organization's data is held hostage or is somehow compromised due to a ransomware attack, a data backup may assist the organization in rolling back to the previous day's data with little loss. Although the policy is simple, for companies that do not perform data backups, it might be very useful. Corporate ITs must mandate the requirement for data backups and disaster recovery. These must be tested on an annual basis, so as to ensure better security for an organization.

13. **Physically Securing the Information Assets:** Regardless of whether programming, equipment, and system security are set up, it may not support a lot of servers are left unsecured on manufacturing floors and in business units. Physical security, which is accessible just to the workforce with trusted status or security clearance, is crucial [78]. Security policies and practices must address the physical just as the visual aspects of data.

14. **Informing the Board and Chief Executive Officer (CEO):** A successful cybersecurity strategy is one where an organization does not have to explain to the CEO or the board clarifying how a cyber breach occurred and what may be done to mitigate it. Unfortunately, great security systems are invisible, because they never give any problems. Subsequently, CSO's CIOs, managers, and others with security obligations need to plainly clarify practice, approaches, policies, cybersecurity technologies, in plain language to everyone in the organization, including the technical and non-technical staff. If the non-technical people in an organization cannot understand why a certain policy is being enacted or ask for a sizeable investment for a cybersecurity technology, the organization may have trouble making case, unless the association is enduring a humiliating security breach that could end professions and put the whole organization's endurance at risk.

Figure 8.4 summarizes how to effectively develop a cybersecurity policy.

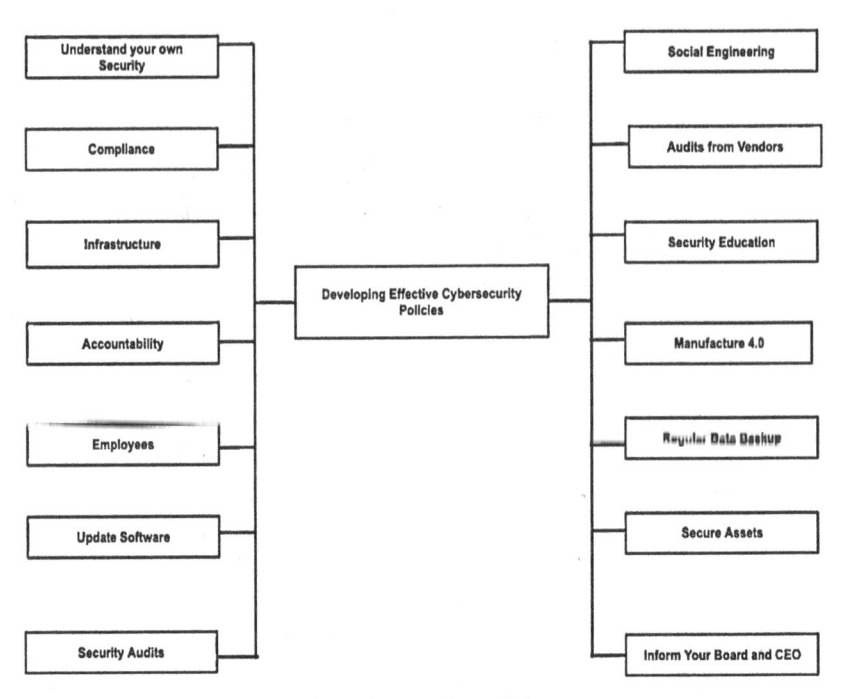

FIGURE 8.4 Developing effective cybersecurity policies.

8.10 TYPES OF CYBERSECURITY POLICIES

Over time most organizations build and manage security programs. To establish the foundation for a security program, companies may first designate an employee to be responsible for cybersecurity. This employee will be responsible for drafting a security program which starts with the process of creating a plan. The objective of this plan is to manage the company's risk using security technologies and auditable work processes. The plan also incorporates documented policies and procedures. Security programs are incomplete without policies and procedures [104]. Cybersecurity policies can be of many types, some of which are discussed in subsections.

8.10.1 ACCEPTABLE USE POLICY (AUP)

An acceptable use policy (AUP), acceptable usage policy or fair use policy, might be characterized as a collection of rules applied by a designer, creator, proprietor, or administrator of a network or website. The AUP restricts the manners by which the system, site or framework might be utilized and may have guidelines on how to use these. AUP records are composed for website proprietors, internet service providers (ISPs), schools, enterprises, universities, and businesses. AUP intends to diminish the potential for lawful activity that a client may take, and frequently with little possibility of implementation. Acceptable use policies are important to the foundation of information security policies. New members of an organization are usually asked to sign an AUP before they are given access to its information systems. Therefore, an AUP must be concise and clear, yet at the same time, it should cover important questions on users, what they are allowed to do with the IT systems and what they are not allowed to do with IT systems. The users must be referred to the comprehensive version of the security policy. The AUP must clearly define the kind of sanctions that will be applied in case a user breaks the AUP. The best way to measure and evaluate this policy is by conducting regular audits. AUPs must specify imperatives and practices that a worker utilizing organizational IT resources must consent to for accessing the corporate network. AUP is basically a standard onboarding policy for new employees who must read and sign the agreement before being provided a network identification. It is recommended that organizations IT, security, legal, and HR departments examine what is encompassed in this policy.

8.10.2 ACCESS CONTROL POLICY (ACP)

Access control pertains to deducing the activities permitted to legitimate users, considering each endeavor by a client to get to an asset in the framework. Usually, complete access is granted only after successful authentication of the user. Nonetheless, numerous frameworks require more modern and complex control. Along with the authentication mechanism, access control deals with how authorizations are organized. In some cases, an authorization may mimic the structure of the organization. In other cases, it may be reflective of the sensitivity level of various documents along with the clearance level of the user acquiring those documents. In order to implement an access control system, organizations need to consider three abstractions: access control approaches, models, and methodologies. Access control policies (ACP) might be characterized as high-level necessities that determine how access is overseen and who may access data under what conditions. The ACP features the access available to workers regarding an organization's data and. ACP incorporates access control standards such as the National Institute of Standards and Technology (NIST's) Access Control and Implementation Guides. The policy also mentions network access controls, complexity of corporate passwords, standards for user access, operating system software controls, etc. Supplementary items discuss various monitoring techniques deployed in corporate systems, information on securing unattended workstations, and information on how to remove access when employees leave the organization.

8.10.3 CHANGE MANAGEMENT POLICY

Change Management may be defined as a formal process used for making changes to IT systems. Change management is responsible for increasing awareness and understanding of suggested changes across an organization. It also ensures that all the modifications made have been thought through and that the change minimizes negative impact on customers and services. Change management takes into account the following steps:

1. **Planning:** Planning the change, which typically includes the execution design, agenda, correspondence plan, test plan, and roll back arrangement.
2. **Evaluating:** Evaluate the change, which is concerned with determining the risk based on the priority level of service and the nature of the proposed change. It is likewise used to decide the change type, and the change process to utilize.
3. **Reviewing:** Survey change plan with peers and additionally Change Advisory Board as suitable to the change type.
4. **Approving:** Obtain approval of the change by the executives or other appropriate change authority as dictated by change type.
5. **Communicating:** Convey about changes with the suitable gatherings (targeted or campus-wide).
6. **Implementing:** Implement the change.
7. **Documenting:** Document the change and any audit, endorsement data and information related to approval.
8. **Post-Change Reviewing:** Survey the change with an eye to future enhancements.

8.10.4 INFORMATION SECURITY POLICY

The primary objective of Information security policies is to ensure that all information technology (IT) users that are part of the organization comply with the rules and guidelines with respect to security of information. This information is stored digitally in the network. Information Security policies are high-level policies that take into account multiple security controls. It is issued by the company for ensuring that all employees who rely on IT assets within the organization, or its networks, observe the stated rules and guidelines. Often employees are asked by the organization to sign a document acknowledging that they will be reading it. This policy is intended for workers to understand that they might be considered responsible with respect to sensitivity of the corporate information and IT assets, considering some specific rules. Every organization is responsible for protecting its data and controlling its distribution both within and outside the organization. For data protection, several techniques may be adopted ranging from encryption of information to authorization by a third party or institution. Placing restrictions on the distribution of data with reference to a classification system laid out in the information security policy may also be useful.

8.10.5 INCIDENT RESPONSE (IR) POLICY

Incident response (IR) policies are responsible for providing an understanding about dealing with certain incidents. It is required that the policy identifies an IR team, who must be notified of issues. The IR team must also have the knowledge and skills to deal with the issues effectively. Some of these issues are viruses, critical system failures, issues relating to unauthorized access, unauthorized changes to systems or data, denial or disruptions of service, or attempts to breach the policies, etc. Incidents that are of criminal nature should likewise be indicated in the approach alongside the data on law enforcement with respect to investigation [105]. An IR policy must also outline who is responsible for specific tasks in situations of a crisis. It may also include related information like:

- Who will be responsible for the investigation or analysis of incidents to determine the occurrence of the incident as well as the resulting consequences?

- Which individuals or departments will be responsible for fixing specific issues and restoring the system?
- Handling of certain incidents;
- References to other documentation.

IR policies are likewise liable for referencing the consequence on what clients are supposed to do once they identified a potential threat. These steps must be clear and concise to evade any confusion. Once an individual gets information about an issue, his/her supervisor or department must be informed of the same. The IR team is then approached. While awaiting the team's arrival, the scene of the incident should be vacated, and those on the scene should not touch anything as this may tamper with the evidence. The users must document who was present in the area during the incident and what all was observed. The IR team usually includes a contingency plan in order to address how a company must deal with intrusions and other such incidents. The contingency plan is responsible for addressing how the company will continue to function during the investigation, like a situation where due to forensic examination, the critical servers are taken offline. Sometimes, such situations are supported by backup equipment that can conveniently replace the servers or other devices. This is done so that employees can still perform their jobs and customers can still make purchases. By having such practices in place, any investigation can avoid negatively impacting normal business practices.

8.10.6 REMOTE ACCESS POLICY

Remote access policy is a document that highlights and mentions acceptable methods of remotely associating with the internal network of an organization. It is useful in large organizations since the networks are dispersed geographically. Sometimes these networks extend into insecure network locations like public networks and unmanaged home networks. The policy considers every single accessible strategy for remotely accessing internal assets:

- Dial-in (serial line internet protocol (IP), point-to-point protocol);
- Integrated service digital network/frame relay;
- Telnet access from the internet;
- Cable modem.

The remote access policy is responsible for defining standards for connecting to the organizational network. It also defines security principles for computers that are permitted to associate with the organizational network. This policy is necessary for organizations that have scattered networks that are capable of extending into insecure network locations.

8.10.7 EMAIL/COMMUNICATION POLICY

A company's email policy may be defined as a document that specifies how representatives can utilize the business' selected electronic correspondence medium. This policy incorporates emails, blogs, social media, and chat technologies. This policy is responsible for providing rules to representatives on what is viewed as the acceptable and unacceptable utilization of any corporate correspondence technology. A viable email strategy will support positive, gainful correspondences while shielding an organization from lawful obligation, reputation damage, and security breaches. The following are some guidelines for email policy.

- Emails are for business use;
- Emails are company property;
- Email policy should address company network and security.

8.10.8 DISASTER RECOVERY POLICY

An association's disaster recovery plan fuses both cybersecurity and IT groups' ideas. This approach is created as a feature of the bigger business coherence plan. The CISO and teams manage an incident through the IR policy. The business continuity plan (BCP) gets activated in a situation where an event has a significant business impact. The essential goal of this policy is to decide the key methods and resources basic for maintaining minimum business capacities after a catastrophe. Some of the best practices to consider for this policy are:

1. **Preparing for Physical Disasters:** Many companies offer alternative workspaces like desks, computer hardware, and phone/Internet access so that business can continue running.
2. **Protecting Against Cyber Disasters:** Cyber disasters typically include major incidents like ransomware attacks and data loss

from breaches. For protection against these, the layers of security must include a firewall with Intrusion Detection System/Intrusion Prevention System and solid host-based endpoint protection.
3. Ensuring redundancy to eliminate single point of failure.
4. **Assigning Responsibilities:** Assigning responsibilities to people who will lead initiatives like purchasing new equipment, coordinating alternative office space, and communicating with senior management, clients, and the press is a must.
5. Testing and reviewing the disaster recovery plan periodically.

8.10.9 BUSINESS CONTINUITY PLAN (BCP)

Business continuity planning (BCP) may be defined as the process involved in the creation of a system that focuses on preventing and recovering from potential threats to an organization. This policy is responsible for the protection of personnel and assets so that they can function efficiently in the event of a disaster. BCP is usually drafted in advance and is based on the input of key stakeholders and personnel. It defines all the risks that can influence the organization's tasks, making it a significant piece of the association's risk management strategy. Risk may incorporate catastrophic events or natural disasters like fire, flood, or weather-related events, and cyber-attacks. After identification of the risks, the plan must:

- Determine how the risks can affect operations;
- Implement safeguards and procedures for mitigate the risks;
- Test procedures to ensure they function;
- Review the process to ensure that it is up to date.

BCPs are an indispensable part of any business. Threats and disruptions may lead to loss of revenue and higher costs, the consequence of which is a drop in profitability. Businesses cannot rely on insurance alone since it cannot cover all the costs and the customers who move to the competition. The BCP is responsible for coordinating efforts across the organization. It uses the disaster recovery plan for restoring equipment, applications, and information regarded necessary for business coherence. BCP's are exceptional to every business. This is because they describe how the organization will operate in the situation of an emergency.

Figure 8.5 depicts the types of cybersecurity policies.

Introduction to Cybersecurity Policies 351

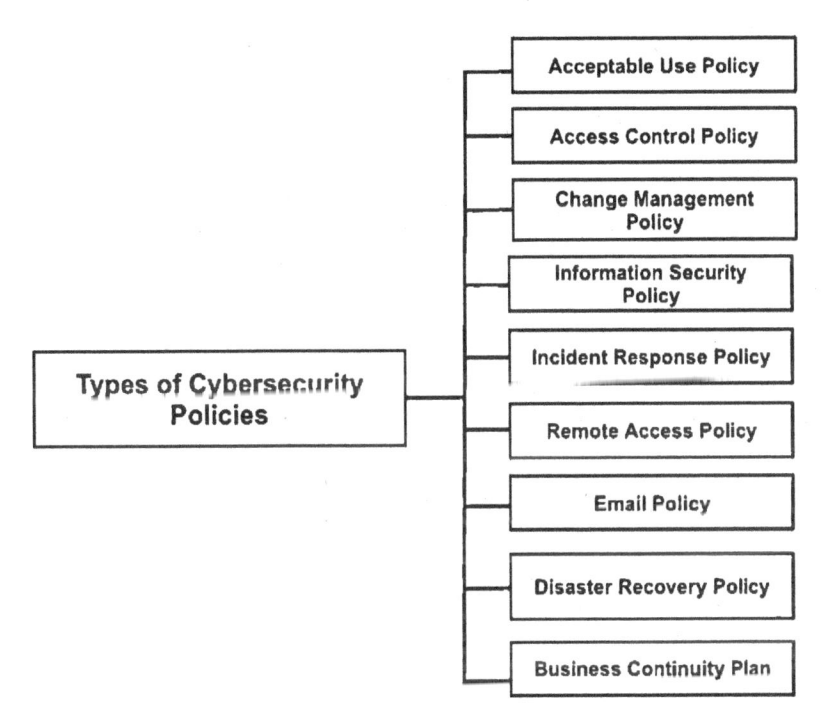

FIGURE 8.5 Types of cybersecurity policies.

8.11 CYBERSECURITY FRAMEWORKS (CSFS)

A cybersecurity framework (CSF) may be defined as a collection of reported methodologies used to characterize policies and procedures with respect to the usage and continuous management of data security controls in an enterprise environment. Since CSFs are prioritized, flexible, and cost-effective, they help in promoting the security and flexibility of critical infrastructure (CI) and different sectors critical to the economy and national security. Frameworks have been around for quite a while. In financial accounting, frameworks have been used for assisting accountants to monitor monetary exchanges. An accounting framework is worked around ideas like resources, liabilities, expenses, and controls. Cybersecurity structures adopt the framework strategy to the work of securing resources. The system is created for giving security managers a reliable and systematic way to mitigate cyber risk despite the complex environment. CSFs are usually compulsory, or if nothing else firmly encouraged, for organizations that need to conform to state, industry, and worldwide cybersecurity guidelines. Frameworks can be of a few sorts like:

1. **Control Frameworks:** These are responsible for developing a basic strategy for the security team. They provide a baseline set of controls and assess the current technical state. These frameworks prioritize control implementation.
2. **Program Frameworks:** These assess the state of the security program. They are responsible for building a comprehensive security program. They also measure program security/competitive analysis. These frameworks also assure simplified communication between the security team and business leaders.
3. **Risk Frameworks:** These characterize key process steps to survey/oversee risk. They structure programs for risk management. Additionally, they can identify, measure, and quantify risk. They prioritize security activities.

While there are several CSFs that are specific to organizations, an important question raised is related to the viability of these frameworks. Have these frameworks considered all security issues of the organization? Are these frameworks feasible? Can they be updated and upgraded in the future? Thus there is a need to measure the success of these frameworks.

During implementation and the steady-state, an organization must measure how well the security systems perform with respect to the chosen framework. An organization's cybersecurity mission, goals, and objectives need to align with its larger mission, goals, and objectives. Therefore, any framework must be capable of translating the security necessities from the most noteworthy strategic degree of administration to the basic level of technical execution. This translation can be performed by specific security programs based on security processes and controls. So as to quantify the advancement and nature of a security program, estimating it on all levels and assuring that programs, controls, and governance do what they are expected to do is necessary. This concept is illustrated in Figure 8.6. Although the purpose is to achieve 100% compliance of security controls, but wrong controls to meet program or governance objectives are picked, it leads to a failing security program.

An organization must be capable of differentiating between the implementation phase of its framework and the final steady-state. During implementation, it may not be expected to meet all desired requirements, therefore, an implementation plan might be needed that offers a moving target to quantify the advancement of your framework implementation after some time. Such estimation may assist in detecting deviations from the intended objective and may also be capable of detecting defects in the process. This is because many requirements could have been overlooked. Practically, with respect to security programs, steady-state is a relative term. Security is never static. There may be changes in the threat landscape, IT infrastructure, or business objectives. All these changes lead to the need to continually assess and adapt to the security program. It is equally essential to comprehend that any framework chosen ought to be applied judicially. Few organizations require the same level of security, hence not every organization has to follow the same steps. It is perfectly valid to comply with all aspects of a framework in an organization. However, the decision must be based on a complete understanding of the prevalent risks. Based on the risks, there may be a need to append requirements beyond what is offered by the selected framework.

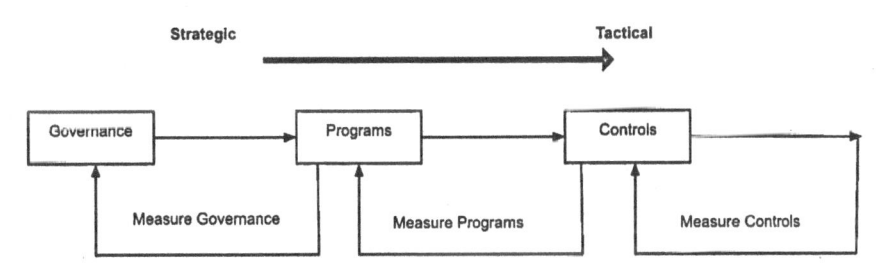

FIGURE 8.6 Conceptual illustration of security governance, programs, and controls.

Since CSFs vary and are not one size fits all, there may be several layers to a mature cybersecurity model. Consequently, an organization may have to consider several methods and measures based on the levels of requirement. The main concern in a business level is about the monetary risk to the business and cost of security, while on a program level, compliance, and processes are the concerns. Technical (security controls) level, deals with reliability and performance.

1. **Measuring Success of Governance:** Reviewing programs to guarantee they meet administration goals:
   i. Auditing;
   ii. Comparing against industry performance;
   iii. Analyzing Security cost and impact.

2. **Measuring Success of Programs:** Reviewing controls against program requirements:
   i. Auditing;
   ii. Comparing against industry benchmarks;
   iii. Comparing against framework.
3. **Measuring Success of Controls:** Reviewing the success of controls is different since controls are supposed to be measured against themselves, i.e., performance indicators. Often, measuring controls is undertaken by security metrics, tools, and benchmarks:
   i. Assessing compliance of security architecture. This can be achieved by performing regular comparison against predefined best practices;
   ii. Estimating consistency of IT resources utilizing tools like vulnerability scanners or pen testing;
   iii. Assessing compliance of information assets using tools for data loss prevention;
   iv. Assessing consistency of workforce through surveys, activities, and security measurements. This can be achieved by measuring click rates with respect to email phishing exercises;
   v. Establishing systematic reporting of security metrics. This can be achieved by basic measures such as number of incidents or the meantime to identify, contain, and respond;
   vi. Performing an Incident analysis followed by post mortem.

Regularly assessing and quantifying security metrics provides an organization with valuable and adequate information to assess and validate security maturity. It may also be used for detecting and correcting deviations. 100% security is never guaranteed, and organizations are not prone to cyberattacks. Any organization could be a possible target for cyber-attacks irrespective of its size or location. Healthcare organizations are at a greater risk than any other industry. It is necessary to introduce frameworks specific to the organization. This is because different organizations have different needs, baselines, and objectives.

8.12 SOME CYBERSECURITY FRAMEWORKS (CSFS)

We have familiarized ourselves with the concept of CSFs and their requirements in the Cyberspace. While these frameworks vary according to the organizations they are deployed in, there are some frameworks that have been considered very effective over the last few decades. In this section, we will be discussing the frameworks in detail.

8.12.1 NATIONAL INSTITUTE OF STANDARDS AND TECHNOLOGY (NIST) CYBERSECURITY FRAMEWORK (CSF)

The NIST CSF presented a policy framework for computer security orchestration which depicts how private sector associations in the United States survey and improve their capacity for preventing, detecting, and responding to cyberattacks [107]. The framework is available in many different languages and provides a high-level classification of cybersecurity outcomes and methodologies to assess and manage those outcomes [106]. The NIST CSF is intended for individual organizations and different associations for assessing the risks they face. The NIST CSF highlights three main components:

1. **The Framework Core:** It consists of a collection of desired cybersecurity exercises and results utilizing basic language that is straightforward and easy to comprehend. This core is liable for directing associations in overseeing and decreasing cybersecurity risk while complimenting their current cybersecurity and risk management approaches.
2. **The Framework Profile:** This defines an organization's unique alignment of its organizational objectives, necessities, risk appetite, and assets concerning the ideal results of the Framework Core. Framework Profiles are necessary for identifying and prioritizing opportunities that may assist in improving security standards and mitigating risk at an organization.
3. **The Framework Implementation Tiers:** The implementation tiers are responsible for providing context on how an organization views cybersecurity risk management. It also assists them in determining appropriate levels of precision with respect to risk analysis and management. It is also an effective communication tool for discussing budget, risk appetite and mission priority.

The NIST CSF underpins five functions. These are further subdivided into a total of 23 categories. Every category is supported by various subcategories of cybersecurity results and security controls. There are a total of 108 subcategories in all. They are as follows:

1. **Identify:** This category is responsible for developing the organizational understanding for managing cybersecurity risk to systems, assets, data, and capabilities.
    i. **Asset Management (ID.AM):** Organizations driven by IT rely on devices, systems, data, and facilities for achieving business purposes. These devices are identified and managed consistently because of their contribution in achieving business objectives and planning organization's risk strategy.
    ii. **Business Environment (ID.BE):** Understanding and prioritizing information pertaining to the organization's objectives, mission, stakeholders, and activities is necessary for managing risks. This information may be useful for cybersecurity roles, responsibilities, and important decisions.
    iii. **Governance (ID.GV):** Managing and monitoring an organization's legal, risk, operational, and environmental requirements can be performed by processes, policies, and procedures. The management must be informed of these and the prevailing cybersecurity risks.
    iv. **Risk Assessment (ID.RA):** The association comprehends the cybersecurity risks to organizational activities. The operations may include organizational assets, functions, mission, reputation, and individuals.
    v. **Risk Management Strategy (ID.RM):** The priorities, constraints, assumptions, and risk tolerances of the organization are defined and are necessary for operational risk decisions.
    vi. **Supply Chain Risk Management (ID.SC):** It is associated with Risk Management Strategy. The priorities, constraints, assumptions, and risk tolerances of the organization are defined for operational risk decisions related with overseeing supply chain risk. The association has well-defined processes for identifying, assessing, and managing supply chain risks.

    In summary, identify corresponds to making a list of all equipment, software, and data that may be used, including laptops, smartphones, tablets, and point-of-sale devices. Creating and sharing an organization cybersecurity strategy that covers:
    a. Roles and responsibilities for anyone working for the organization who accesses sensitive data;

b. Necessary steps that may assist in protecting against an attack and limiting the damage if one occurs.
2. **Protect:** This category is responsible for developing and implementing the suitable shields to guarantee the conveyance of basic CI services.
    i. **Access Control (PR.AC):** Only authorized users, processes, and devices may access assets and associated facilities along with authorized activities and transactions.
    ii. **Awareness and Training (PR.AT):** Cybersecurity awareness training is given to the association's staff and accomplices, with the goal that they may be trained adequately. The training ensures that they can perform their information security-related duties and responsibilities efficiently with respect to related agreements, procedures, and policies.
    iii. **Data Security (PR.DS):** Information and records must be managed consistently complying with the organization's risk strategy. The goal is to ensure the confidentiality, integrity, and availability of data.
    iv. **Information Protection Processes and Procedures (PR.IP):** Security policies are developed taking into account the objectives, roles, scope, purpose, coordination, responsibilities, management commitment, and coordination of the organization's personnel. It also focuses on processes and procedures that must be maintained and used to manage the protection of information, systems, and assets.
    v. **Maintenance (PR.MA):** Industrial control and information system components must be maintained and repaired consistently. This must likewise consent to the individual procedures and policies.
    vi. **Protective Technology (PR.PT):** Consistently managing technical security solutions for ensuring the security and resilience of systems and assets, must also comply with policies, procedures, and agreements.

    Some of the policies that define this aspect of the NIST CSF are as follows:
    a. Controlling who logs on to the network and uses computers, networks, and other devices;
    b. Using security software for data protection;

      c. Encrypting sensitive data, at rest and in transit;
      d. Conducting regular backups of data;
      e. Updating security software routinely and furthermore automating those updates;
      f. Having formal policies for securely discarding electronic records and old tools and devices;
      g. Preparing every individual who utilizes your computers, gadgets, and network about cybersecurity;
      h. Helping employees understand their personal risk and their crucial role in the workplace.

3. **Detect:** This category is responsible for developing and implementing the appropriate activities for identifying the occurrence of a cybersecurity event.
   i. **Anomalies and Events (DE.AE):** Anomalous activity must be detected timely followed by analyzing the potential impact of the events.
   ii. **Security Continuous Monitoring (DE.CM):** Monitoring the information system and assets at discrete intervals for identifying cybersecurity events and verifying the effectiveness of protective measures is necessary.
   iii. **Detection Processes (DE.DP):** Testing and maintaining detection processes and procedures may ensure timely and adequate awareness of anomalous events.

   Some approaches that compare to this part of the NIST cybersecurity structure are:
         a. Monitoring information systems with respect to software, devices (external) and unauthorized personnel access;
         b. Investigating any abnormal activities on the network or by staff;
         c. Checking network for unauthorized users or connections.

4. **Respond:** This category is responsible for developing and implementing the proper exercises to take action with respect to a distinguished cybersecurity incident.
   i. **Response Planning (RS.RP):** Execution and maintenance of response processes and procedures must be conducted for timely response to detected cybersecurity events.
   ii. **Communications (RS.CO):** Coordinating response activity with internal and external stakeholders may ensure external assistance from law enforcement offices.

iii. **Analysis (RS.AN):** Conducting analysis for efficient response and recovery activities is necessary.
iv. **Mitigation (RS.MI):** Specific Activities must be performed to target prevention of expansion of an event, mitigating its effects, and eradicating the incident.
v. **Improvements (RS.IM):** Lessons learned from current and previous activities can be used to improvise organizational response activities.

   Accordingly, it is important to have a response plan that can:
   a. Notify workers, clients, and others whose information might be at risk.
   b. Keep business operations up and running.
   c. report the incidents to law enforcement and other authorities.
   d. Investigate and contain an attack.
   e. Update cybersecurity policy and plan based on the lessons learned.
   f. Prepare for inadvertent events that may put data at risk.

5. **Recover:** This category is responsible for developing and implementing the appropriate activities for maintaining plans that can ensure resilience and restore any capabilities or services that might have been impaired following a cybersecurity event.
   i. **Recovery Planning (RC.RP):** Executing and Maintaining Recovery processes and procedures can lead to timely restoration of assets and systems vandalized due to cybersecurity events.
   ii. **Improvements (RC.IM):** Incorporating lessons learned into future activities is beneficial for improving recovery planning and processes.
   iii. **Communications (RC.CO):** Coordinating centers like sellers, proprietors of attacking frameworks, casualties, and ISPs and other external and internal parties may coordinate restoration activities.

      After attack objectives must include:
      a. Repairing and restoring the equipment and parts of the network that were damaged due to the attack;
      b. Keeping employees and customers informed of the response and recovery activities.

Profiles may be regarded as an association's novel arrangement to their business prerequisites and targets, risk appetite, and assets against the ideal results in the Framework Core. Profiles are used to optimize the CSF to best serve an organization. Since profiles are voluntary frameworks, there is no correct or incorrect method to utilize it. Moreover, organizations have varying organizational understanding, security requirements, cybersecurity risks management techniques, risk tolerance, appropriate safeguards, etc. In this way, cybersecurity prerequisites, practices, targets, and working can be planned to subcategories in the Framework Core for creating the current profile. These requirements and objectives need to be weighed against the current state for gaining an understanding of the security gaps. Following the completion of the cybersecurity risk assessment process, organizations need to create a prioritized implementation plan. This plan must consider the size of the gap, cost of protective technologies and size of the gap.

The NIST CSF, depicts four implementation tiers. Higher tiers indicate that the organization's risk management framework (RMF) is closer to the characteristics mentioned in the framework. The four tiers are:

- **Tier 1:** Partial;
- **Tier 2:** Risk informed;
- **Tier 3:** Repeatable; and
- **Tier 4:** Adaptable.

Tiers do not necessarily represent maturity levels. This is because they are based on the organization's requirements and need to identify their preferred tier. Hence tiers are selected for meeting organizational goals, reducing cybersecurity risk to a tolerable level, and feasible implementation at a financial and operational level.

The following are some benefits of the NIST CSF:

1. **Superior and Unbiased Cybersecurity:** The NIST CSF highlights the combined reviews and experience of information security professionals and subject matter experts. It is widely accepted in the industry and is considered a best practice. It is comprehensive and includes an in-depth set of controls. Cybersecurity leaders and practitioners are concerned with holding up an organization against cyber-attacks, which is their top priority. NIST CSF plays a major role in that mission.
2. **Enabling Long-Term Cybersecurity and Risk Management:** The CSF takes an organization out of the one-off review consistency attitude, and into a more versatile and responsive stance. A

more grounded procedure over the long haul is continuous consistency. Although risk management using policies is a challenging task, using the right tools can ensure an efficient and continuous compliance approach and the framework can be used with ease.
3. **Ripple Effects Across Supply Chains and Vendor Lists:** Cybersecurity practices and posture is turning into a considerable selling point. It is also capable of enabling and enables faster and secure business growth.
4. **Overcome Any Issues Among Technical and Business Side Partners:** The CSF adopts a risk-based approach, which is well understood by executives. The risk management approach promotes cybersecurity management integrated with business goals which results in effective communication and better decision-making throughout the organization. This will also ensure that security budgets are distributed efficiently. A shared security vocabulary for both business and technical stakeholders, makes the risk management process simple due to improved communication throughout the organization.
5. **Flexibility and Adaptability of the Framework:** The CSF is one of the most flexible frameworks given its risk-based, outcomes-driven approach. Numerous businesses and framework firms in the finance, energy, transportation, and other small and medium-sized enterprises have adopted the CSF. As it is a voluntary framework, it is extremely customizable. The Core Functions are intuitive and collective with the Implementation Tiers. The Profiles act as blueprints for speeding adoption and provide ongoing guidance.
6. **Built for Future Guideline and Consistency Necessities:** With changes in laws and regulations being prevalent, organizations must cope up with changing policies and procedures. Organizations that execute policy Frameworks are in a better position when new laws and regulations emerge. CSF can be used as a foundation for compliance standard guidelines when new regulations are introduced. This has tremendous impact in private industry and CI, and is believed to impact other industries in near future as well. Compliance requirements are rising across industries and geographies, hence there is a need for a reliable foundation on which a cybersecurity program can be built. The framework will not only allow new guidelines to be introduced, but would also encourage updating the existing standards.

The NIST CSF may be depicted as in Figure 8.7.

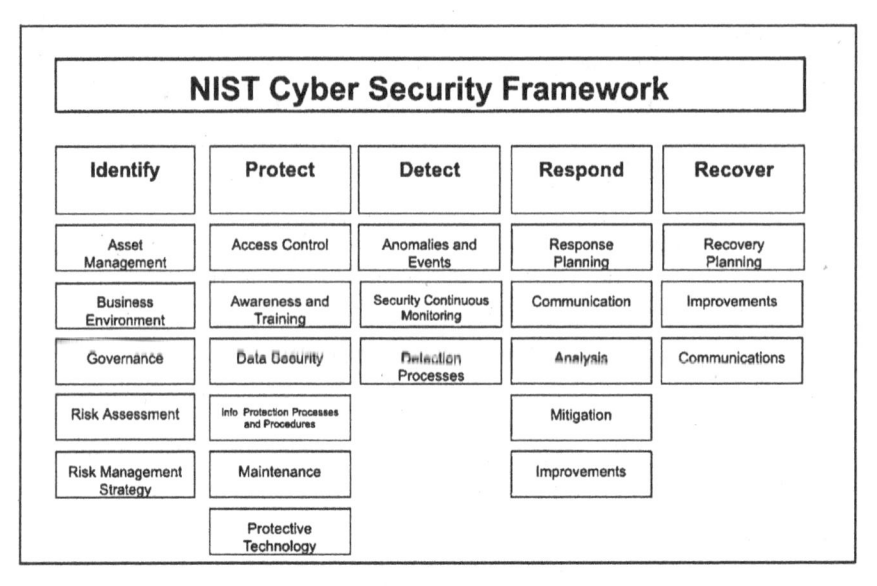

FIGURE 8.7 NIST cybersecurity framework.

> **Key changes made to the NIST cybersecurity framework:**
> - The framework now incorporates authentication and supply chain risk management;
> - Federal agencies have adopted and leveraged the CSF;
> - The CSF is presently embraced internationally;
> - Translation has made the framework more accessible T;
> - The framework can assist small businesses.

8.12.2 CONTROL OBJECTIVES FOR INFORMATION AND RELATED TECHNOLOGY (COBIT) FRAMEWORK

Control objectives for information and related technology (COBIT) framework was introduced by the ISACA (Information Systems Audit and Control Association). It is responsible for IT governance and management [108]. COBIT also serves as a supportive tool for managers. COBIT permits connecting the significant gap between technical issues, business risks, and control requirements and incorporates thoroughly recognized guidelines that may be applied to any organization in any industry. Additionally, COBIT offers quality, control, and dependability of data frameworks in an association. This is a significant aspect of many modern businesses. The various components of COBIT include:

1. **Framework:** It is responsible for assembling the objectives of IT governance in order to bring in the best practices in IT processes and sectors by connecting business necessities.
2. **Process Descriptions:** It is a reference model which may also be thought of as a common language for each person in the association. The process description integrates planning, building, operating, and monitoring the IT processes.
3. **Control Objectives:** These include a total list of necessities that have been considered by the administration for successful IT business control.
4. **Maturity Models:** These are responsible for accessing the development and the capacity of each process while tending to the weaknesses.
5. **Management Guidelines:** These assist in better-assigning duties, estimating exhibitions, concurring on basic targets, and outlining better interrelationships with each different process.

ISACA's globally accepted framework COBIT 5 is responsible for providing an end-to-end business view of the governance of the enterprise IT that reflects the central role of IT in creating value for enterprises. COBIT5 may be thought of as a comprehensive framework that helps enterprises in making ideal incentives from IT by keeping up a balance between acknowledging benefits and improving risk levels and asset use. COBIT 5v empowers data and related technology to be represented and overseen in a complete manner for the entire enterprise by taking end-to-end businesses and functional areas of responsibility, considering IT-related interests of internal and external stakeholders. COBIT 5 methodologies are generic and may be applied to various enterprises irrespective of their sizes. The COBIT 5 principles are discussed in subsections (Figure 8.8) [108, 109, 128].

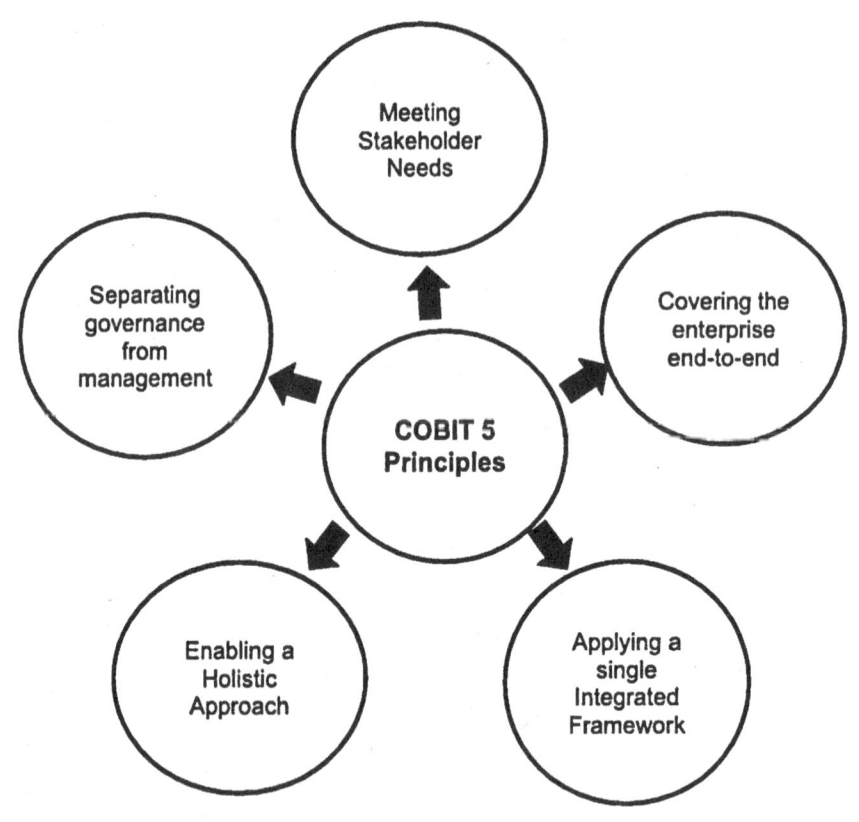

FIGURE 8.8 COBIT 5 principles.

8.12.2.1 MEETING STAKEHOLDERS NEEDS

The need for enterprises is to create value for stakeholders (Figure 8.9) [128].

Conveying enterprise stakeholder value makes appreciable governance and management of IT assets mandatory [109]. COBIT 5 is significantly responsible for it since it provides an extensive framework to assist enterprises in achieving their goals and delivering value. COBIT 5 vouches for effective governance and management of enterprise IT. Stakeholder needs must be transformed into enterprises' actionable strategy. COBIT 5 is responsible for translating stakeholders' needs to specific, actionable, and customized goals with respect to the enterprise, IT-related goals, and enabler goals (Figure 8.10) [128].

Introduction to Cybersecurity Policies

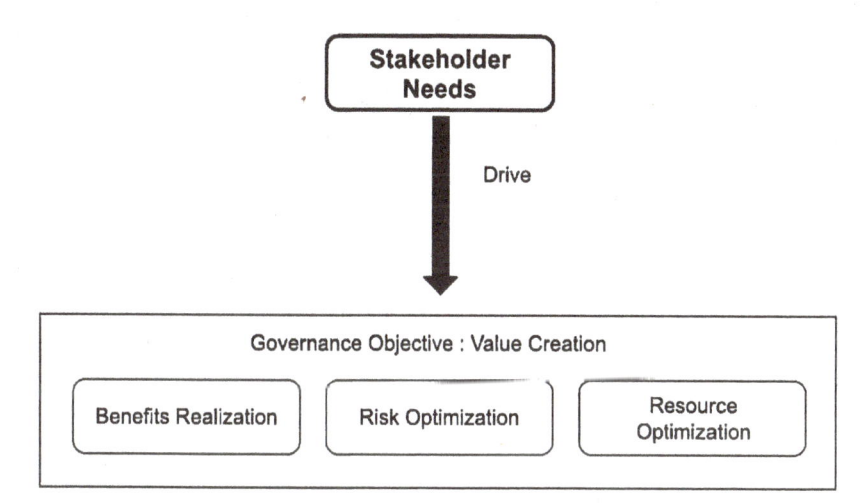

FIGURE 8.9 Meeting stakeholder needs.

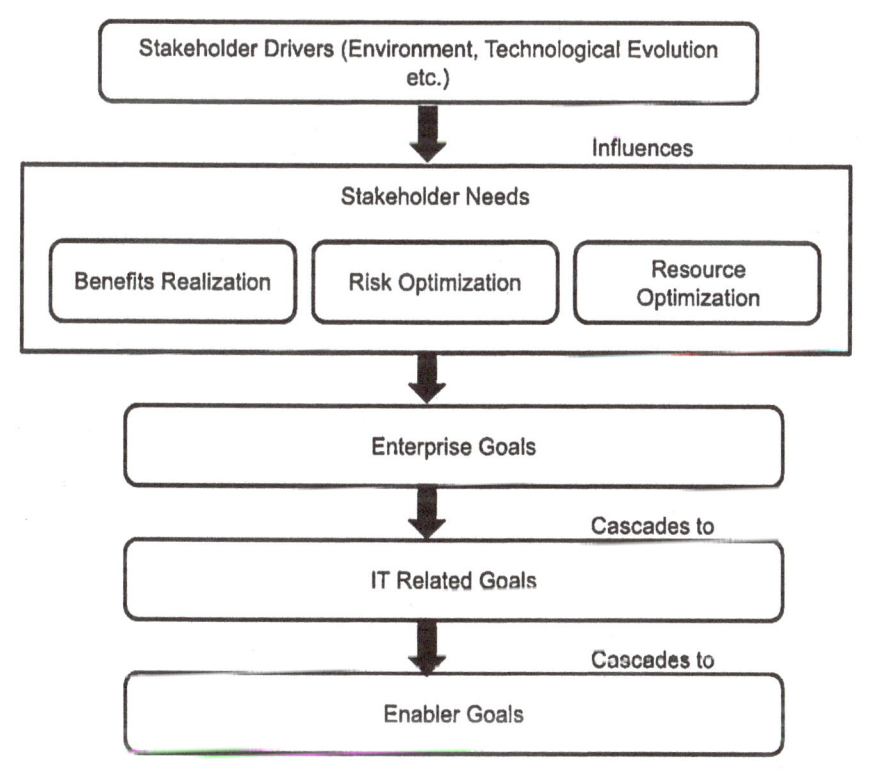

FIGURE 8.10 Meeting stakeholder needs and goals.

8.12.2.2 COVERING THE ENTERPRISE END-TO-END

The governance is responsible for valuing creation, but the key components of a governance system include benefits realization, risk optimization, and resource optimization, which in turn influence the roles, activities, and relationships (Figure 8.11) [128].

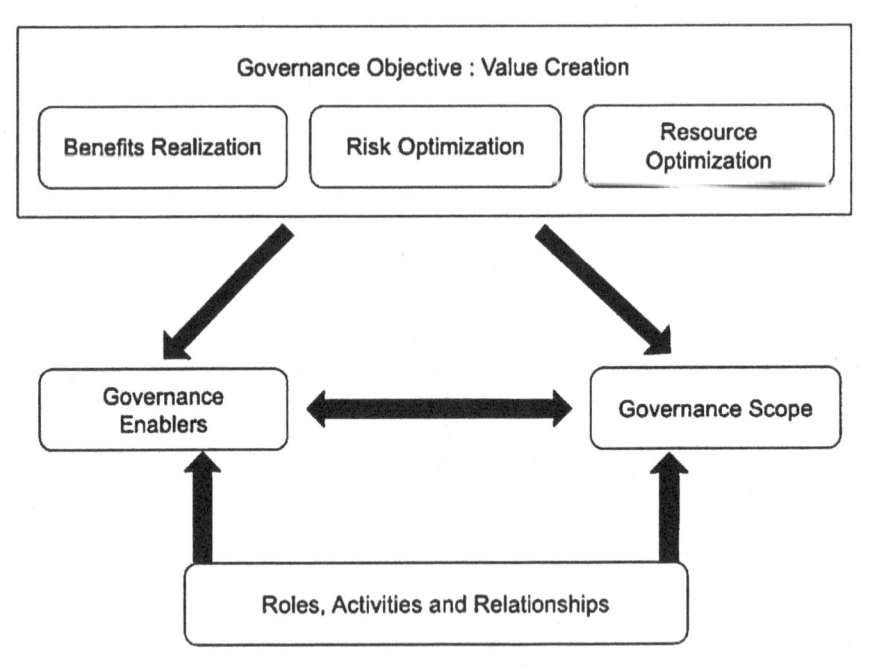

FIGURE 8.11 Covering the enterprise end-to-end.

The roles, activities, and relationships follow a loop trend wherein stakeholders, governing bodies, management, and operations are all dependent on each other. The same has been illustrated in Figure 8.12 [128].

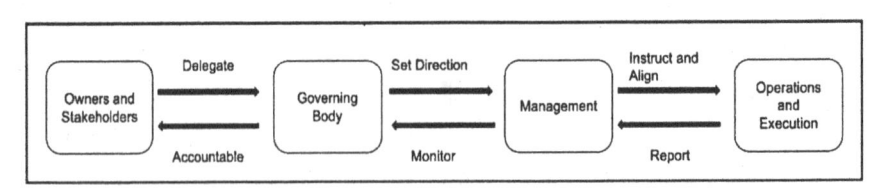

FIGURE 8.12 Roles, activities, and relationships.

8.12.2.3 APPLYING A SINGLE INTEGRATED FRAMEWORK

COBIT 5 lines up with the most recent applicable different standards and ventures utilized by enterprises. This permits the enterprise to utilize COBIT 5 as the larger administration and management framework integrator ISACA plans an ability to encourage COBIT client planning of practices and exercises to third-party references.

8.12.2.4 ENABLING A HOLISTIC APPROACH

COBIT 5 characterizes a lot of empowering influences to help the usage of extensive administration and management of enterprise IT. COBIT 5 empowering agents are factors that exclusively and collectively impact if something will work, driven by goals cascade. They are described by the COBIT 5 framework in seven categories (Figure 8.13) [128].

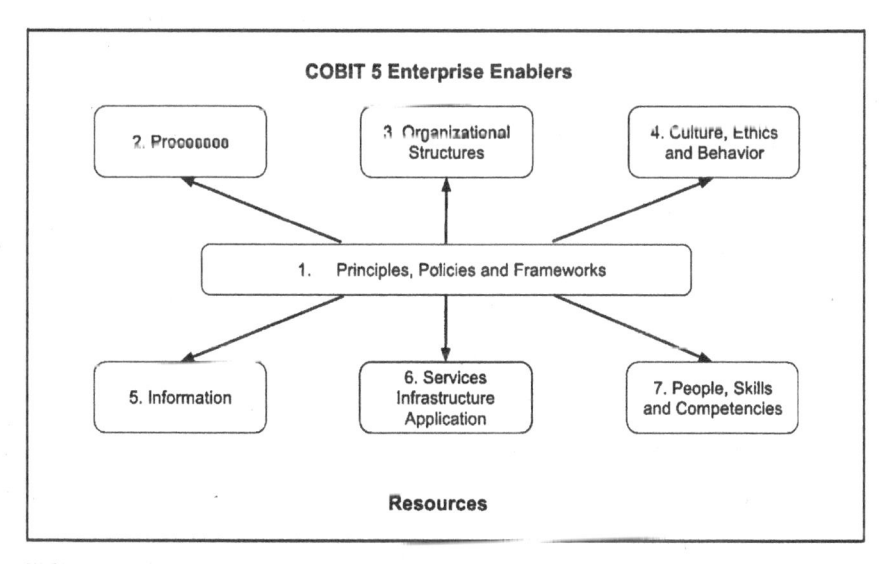

FIGURE 8.13 COBIT 5 enterprise enablers.

- Principles, policies, and frameworks necessary to translate the desired behavior into practical guidance for day-to-day management.
- For achieving certain objectives, and producing a set of outputs to achieve the overall goals, an organized set of practices and activities are defined by the processes.

- Organizational structures are the key decision-making entities in an organization.
- Culture, ethics, behavior of individuals, and organizations.
- Information is pervasive throughout the environment. A well-governed and operational organization rely on information. At the operational level, information is very critical.
- Services infrastructure applications include infrastructure, applications, and technology for providing IT processing and services to the enterprise people.
- Skills and competencies are connected and necessary for the completion of all activities and making correct decisions.

8.12.2.5 SEPARATING GOVERNANCE FROM MANAGEMENT

Administration guarantees that the venture targets are accomplished by assessing partner needs, conditions, and alternatives, and setting direction through prioritization and decision making. It also assists in monitoring performance and progress against agreed objectives. COBIT is not prescriptive but advocates that organizations implement governance and management processes such that key areas are covered (Figure 8.14) [128].

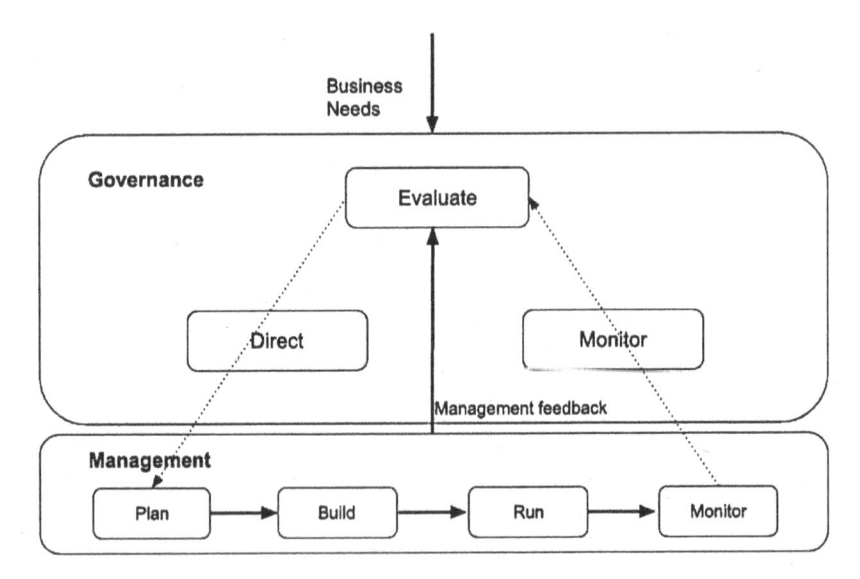

FIGURE 8.14 Separating governance from management.

> **Key changes made to COBIT:**
> - New processes for data, projects, and compliance;
> - Updates to cybersecurity and privacy;
> - Updated linkages to all relevant standards, guidelines, regulations, and best practices.

8.12.3 PAYMENT CARD INDUSTRY DATA SECURITY STANDARD (PCI DSS) FRAMEWORK

PCI DSS or the Payment Card Industry Data Security Standard is a compliance framework that relies on an industry-mandated set of standards. The primary objective of the PCI DSS framework is to keep consumers' card data safe while it is used by merchants and service providers [110]. The framework was founded by credit card companies American Express, Visa, Discover, and Mastercard, and is managed by the PCI Security Standards Council (PCI SSC). If organizations do not comply with the standards, the card companies that control these standards may impose fines against these organizations. PCI DSS is applicable to all organizations, along with stores, banks, merchants, processors, developers, etc., who are responsible for processing, or transmitting cardholder data. Approval of DSS consistency may not be compulsory under a characterized limit for annual transactions. It may also depend on which payment cards one intends to accept at the place of business. PCI SSC founding members have their own compliance program for protecting their cardholders' data. These compliance programs must be contacted directly for specific requirements.

The PCI Data Security Standard defines 12 requirements with respect to compliance. These prerequisites are composed of six legitimately related groups called control objectives. The six groups are:

- Establishing and supporting secure network and systems;
- Protecting cardholder data;
- Maintaining vulnerability management programs;
- Implementing strong access control measures;
- Regularly monitoring and testing networks; and
- Maintaining an information security policy.

PCI DSS splits these six requirements into many sub-requirements differently, however, the 12 high-level requirements have not been altered ever since the standard was established. Each requirement/sub-requirement may be detailed into three sections:

1. **Requirement Declaration:** It provides a detailed description of the requirement. PCI DSS is supported utilizing legitimate usage of the necessities.
2. **Testing Processes:** It is used for obtaining confirmation about the proper implementation of the processes and methodologies conducted by the assessor.
3. **Guidance:** It details the core objectives of the requirement and respective content that may aid the correct definition of the prerequisite.

The 12 requirements essential to build and maintain a protected network and frameworks can be summed up as follows:

1. **Introduce and Support Firewall Arrangement to Secure Cardholder Information:** The firewall is responsible for scanning all network traffic and blocking untrusted networks from accessing the system.
2. **Change Seller Provided Defaults for System Passwords and Other Security Boundaries:** Passwords may be easily retrieved by adversaries using public information. This can allow malicious individuals to gain unauthorized access to systems.
3. **Protect Stored Cardholder Data:** Cardholder data must be protected using a combination of masking, hashing, encryption, and truncation techniques.
4. **Encrypt Transmission of Cardholder Data Over Open, Public Networks:** Using trusted keys and certificates for strong encryption, can significantly reduce the risk of being targeted by adversaries through hacking.
5. **Protect All Systems Against Malware and Perform Regular Updates of Antivirus Software:** There are several ways for malware to enter a network. Some of these include Web use, worker email, cell phones, or storage devices. Up-to-date anti-virus software or supplemental anti-malware software may be capable of reducing the risk of exploitation via malware.
6. **Develop and Maintain Secure Systems and Applications:** Weaknesses and loopholes in systems and applications allow malicious

individuals to gain unauthorized access. Hence, security patches must be immediately installed to fix the vulnerability. This may also be instrumental in preventing exploitation and compromise of sensitive data.

7. **Limit Access to Cardholder Information to Just Authorized Staff:** Restrict access to cardholder data must be implemented. Data must be accessed only on a need-to-know basis.
8. **Recognize and Verify Admittance to Framework Components:** A unique identifier must be associated with each person who accesses the system components. This is essential for guaranteeing accountability.
9. **Restrict Physical Access to Cardholder Information:** Physical access to cardholder information or frameworks that hold this information must be secure to forestall unauthorized access or expulsion of information.
10. **Track and Observe All Access to Cardholder Information and System Assets:** Logging methodologies ought to be set up to track client activities that are critical to prevent, identify or limit the effect of information compromises.
11. **Test Security Frameworks and Methods Consistently:** Every day new vulnerabilities are continuously being discovered. Therefore, it is necessary to test software, systems, and processes, and software for identifying vulnerabilities, which if not timely addressed, may lead to undesirable cybersecurity events.
12. **Keeping up a Data Security Strategy for All Work Forces:** A solid security strategy incorporates making personnel realize the sensitivity of information and their obligation to ensure it.

Figure 8.15 is an illustration of the PCI DSS CSF [111].

Key changes made to PCI DSS:
- Change management execution and documentation; all applicable PCI DSS prerequisites must be actualized on all new or changed frameworks and networks.
- Execute multi-factor authentication for any administrator access to the cardholder data environment (CDE).
- Quarterly administration survey of strategy and process consistency with staff.
- Respond and report inadequacy of any basic security controls in an ideal manner.

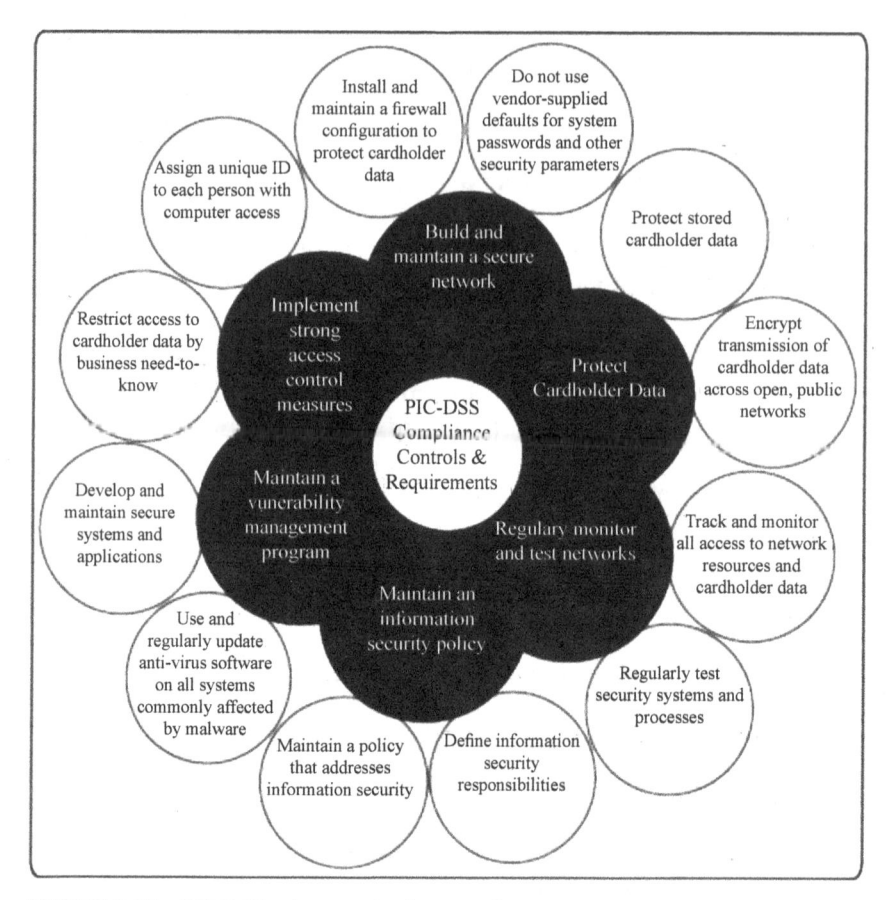

FIGURE 8.15 PCI DSS cybersecurity framework.

8.12.4 ISO/IEC 27000 SERIES

The ISO/IEC 27000-series defines information security standards distributed together by the International Organization for Standardization (ISO) and the International Electrotechnical Commission (IEC) [112]. The series aims to provide best practice recommendations on information security management, which is basically the management of information risks through information security controls with respect to the overall information security management system (ISMS). Data security breaches are normal and are perhaps the greatest risks that associations face.

Confidential information finds use over all sectors of businesses nowadays. This increases the value of data legitimately as well as illegitimately. Countless incidents occur every month, ranging from cybercriminals accessing databases with sensitive information employees losing personal data. Following the information breach, associations experience the ill effects of money-related and reputational harm, which can be devastating. Therefore, organizations invest heavily in securing their systems and networks. ISO 27001 guidelines are adopted by organizations for effective security. These guidelines are applicable to associations of every kind imaginable. Since the framework is broad and scalable, it may be implemented in business of any size. The series is deliberately broad in scope, and covers more than just privacy, confidentiality, and IT/technical/ cybersecurity issues. Thus, it is applicable to organizations of all shapes and sizes and encourages scalability. All organizations are encouraged to assess their information risks, before treating them according to their needs. This may be performed using the guidance and suggestions if necessary. Because of the dynamic nature of information risk and security, and since security is not 100% achievable, the ISMS system additionally fuses constant feedback and improvement exercises to react to changes in the threats, weaknesses, or effects of incidents.

The ISO 27000 series describes how to implement an ISMS. An ISMS may be defined as a systematic approach to risk management, which supports measures for tending to the three pillars of data security, i.e., people, processes, and technology. The series incorporates 46 individual standards, which also includes ISO 27000. Some of the standards in the series are as follows:

1. **ISO 27001:** This is the predominant specification in the ISO 27000 series. It defines the implementation requirements for an ISMS. It is the only standard in the series against which organizations can be audited and certified. This is because it contains an overview of everything that must be done for achieving compliance. These are detailed in each of the following standards.
2. **ISO 27002:** This is an additional standard that provides details about the information security controls that have been chosen by the organizations for the purpose of implementation. Organizations are required to adopt controls based on their requirements and relevance. Usually, organizations adopt controls that may be

useful during risk assessments. Annex A of ISO 27001 lists these controls. ISO 27002 provides a comprehensive overview of these controls describing the objectives, working, and implementation of these controls.

3. **ISO 27017 and ISO 27018:** Introduced in 2015, these standards describe how organizations should ensure the protection of sensitive information in the Cloud. With organizations migrating much of their sensitive data to the Cloud, these guidelines are much in demand. ISO 27017 is a code of practice. It provides detailed information on how to apply Annex A controls to the information stored in the Cloud. Under ISO 27001, organizations have the alternative to treat these as a different set of controls. So, a set of controls can be chosen from Annex A for 'normal' data. Similarly, a set of controls from ISO 27017 can be chosen for data in the Cloud. ISO 27018 works in a similar fashion, however, it takes into account personal data that is stored online.

4. **ISO 27701:** This is the most recent standard in the ISO 27000 series. It underpins what organizations must do for implementing a PIMS (privacy information management system). GDPR trains associations to grasp suitable organizational and technical measures for securing individual data. However, GDPR does not mention how to do it. Hence, this standard was introduced in response to that. ISO 27701 fills that gap, essentially bolting privacy processing controls onto ISO 27001.

8.12.5   CIS CRITICAL SECURITY CONTROLS

The Centre for Internet Security (CIS) Critical Security Controls are a set of laid-out actions for enhancing security in cyberspace that are based on a defense-in-depth set of specific and actionable best practices. The idea is to eradicate the most widely recognized cyber-attacks [113]. CIS Controls organize and focus on few activities that incredibly lessen cybersecurity risk. The CIS Controls are capable of minimizing identity theft, denial of service, privacy loss, cyber threats, the risk of data breaches, corporate espionage, data leaks, and theft of intellectual property (IP). The CIS Controls may assist in answering questions related to:

- The most critical domains for setting up a risk management program;

- Different defensive advances that may give the best value;
- Tracking of risk management program maturity;
- Identifying root causes of attacks;
- Tools that can solve problems;
- CIS controls planning to an association's administrative and compliance systems.

The critical controls are intended to assist associations with shielding their frameworks and information from known attack vectors. It can likewise be a powerful guide for organizations that do not yet have a cognizant security program.

The overview of each critical control is as follows:

1. **Inventory and Control of Hardware Assets:** It is a known fact that attackers are ceaselessly monitoring for new and potentially weak frameworks so as to find a way into a target's network. Attackers are especially intrigued by devices that go back and forth from the enterprise network, like laptops or Bring-Your-Own-Devices (BYOD). This is because these devices usually do not install security updates or may have already been compromised. Following the detection, attackers can exploit this equipment and access an association. Additionally, it may also be used to launch cyber-attacks. This control mandates associations to oversee hardware devices on their network to guarantee just approved devices approach sensitive areas. Managed control of all devices is critical in arranging and executing framework reinforcement, IR, and recuperation [114].

2. **Inventory and Control of Software Assets:** Attackers are also on the lookout for vulnerabilities that can be remotely exploited in software. EternalBlue, is a popular vulnerability that is known to the old versions of Windows operating systems. This vulnerability was capable of launching the WannaCry ransomware which is a computer worm. Attackers can conveniently distribute malicious software through emails, phishing scams, websites, and third-party sites that may seem trustworthy. When the content on an exploitable or compromised machine is accessed by a victim user, attackers can easily gain unauthorized access and install malicious software. If the defenders do not have sufficient knowledge or control on this malicious software, they cannot secure their assets. This results in exposure to sensitive data and data breaches. Once a

machine is compromised within a network, it can be used to initiate further attacks. The Inventory and Control of software assets can mitigate this risk by ensuring that associations effectively deal with all software on the network. Consequently, just authorized software can be introduced and executed.

3. **Continuous Vulnerability Management:** For vulnerability management and vulnerability assessment, cyber defenders must take in a steady stream of new data, software updates, patches, security warnings, and so forth. Attackers may utilize this data as it is to exploit holes between the presence of new weaknesses and their remediation to attack targets. So as to limit this risk, this control expects associations to consistently secure, evaluate, and take action on new data so as to determine weaknesses, remediate them and limit the window of opportunity. Without monitoring for weaknesses and proactively tending to issues, associations face a significant risk of compromise.

4. **Controlled Use of Administrative Privileges:** This control is based on the principle of least privileges and other access control methods. It is liable for making processes and tools for controlling, preventing, and tracking risks. It is also responsible for assigning and configuring administrative privileges. This may be significantly useful in reducing the misuse of administrative privileges, which is a commonly used attack method of attack for breaking into an organization. Attackers may also use social engineering methods to trick victims into downloading and opening malicious files that automatically run. If the administrative privileges are assigned to the victim, the adversary can take over the victim's machine, install malware and steal sensitive data.

5. **Secure Configuration for Hardware and Software on Workstations, Laptops, Servers, and Mobile Devices:** In order to promote deployment and usage of operating systems and applications, default configurations focus mainly on ease rather than security. Default passwords, basic controls, outdated protocols and default passwords may be easily exploited when left in the default state. Therefore, there is a need to establish, implement, and actively manage the security configuration of devices like laptops, servers, and mobiles. It is equally important to manage the security configurations of workstations using configuration management

to ensure that attackers cannot exploit vulnerable services and settings. This will mitigate a significant amount of risk.
6. **Maintenance, Monitoring, and Analysis of Audit Logs:** Insufficient analysis of security logging has been used by attackers as a means to hide their location as well as for installing malicious software on a victim's machine. Therefore, it is necessary for organizations to gather, oversee, and investigate review logs of incidents to support the identification, detection, and recovery from attacks.
7. **Email and Web Browser Protections:** Due to their technical complexity, web browsers and email clients are common points of attack. They are flexible and are used abundantly. Using social engineering, it is possible to craft content to lure users into taking required actions. This can also lead to loss of valuable data and installation of malicious code. This control is responsible for minimizing the attack surface and opportunities for attackers that are used for manipulating human behavior. Also known as social engineering, these techniques can also witness user interactions with web browsers and email systems.
8. **Malware Defenses:** Malicious software is commonly used for vandalizing systems, devices, and data. Such software can conveniently enter through end-user devices, removable media, cloud services, web pages, user actions and email attachments. Besides, advanced threats are intended to go bypass, evade, avoid, and incapacitate defenses. Optimizing the use of automation may assist organizations in controlling the installation, spreading, and execution of malicious code. This in turn will also encourage rapid updation defenses, data gathering, and appropriate actions.
9. **Limitation and Control of Network Ports, Protocols, and Services:** It is a known fact that attackers continuously search for devices and accessible network services that are vulnerable to exploitation. Poorly configured web servers, DNS (domain name system) servers installed by default, file, and print services mail servers, are some examples of such services. This control is responsible for managing the ongoing operational use of ports, networked devices, services, and protocols, with the objective of minimizing the windows of vulnerability available to attackers.

10. **Data Recovery Capabilities:** If attackers gain access to a machine, they may make significant changes to configuration and software. Subtle modifications to data stored potentially may have a remarkable effect on the organization's ability to operate. In a circumstance where an attacker is found, associations must eliminate all aspects of the attacker from the system. In this manner, associations must utilize processes and tools to appropriately back up valuable data with a validated methodology for ideal recuperation of it.
11. **Secure Configuration for Network Devices, such as Firewalls, Routers, and Switches:** To ensure ease of deployment, default configurations of network infrastructure are encouraged and designed. This may interfere with the security requirements. In default states, pre-installed software, Open services and ports, default passwords, and support for older protocols may be vulnerable. Hence it is necessary for organizations to establish, execute, and manage the security configuration of network infrastructure devices. This may be performed by using configuration management and changing control processes for preventing attackers from exploiting vulnerable services. Securing configuration is a continuous process. This is because hardware and software configuration may have to be performed multiple times. This may prove to be useful for attackers as they can exploit slipping infrastructure after some time as clients request special cases for genuine business needs. When these exceptions are left open after the business is no longer needed, it can open potential attack vectors.
12. **Boundary Defense:** Attackers target systems that can be accessed on the Internet. Nation-state actors and organized crime groups are known to abuse configuration and structural loopholes in the systems' periphery network devices and client machines for gaining unauthorized access into the organizations. Boundary defense controls may be useful in detecting, preventing, and correcting the flow of information being transferred across systems of various trust levels with an attention on security-damaging information.
13. **Data Protection:** Sensitive data may reside in many places. Therefore, protection of that data is of utmost importance. This may be accomplished using a combination of techniques using integrity protection, data loss prevention and encryption. Data protection

is achieved using tools, controls, and processes. The primary aim is to forestall information exfiltration, alleviate the impacts of exfiltrated information and guarantee the protection and integrity of confidential data.

14. **Controlled Access Based on the Need to Know:** There is a degree of assurance related with encoding information. This is because, even if a data breach occurs, plaintext cannot be accessed without adequate resources. Further, it is necessary to put controls in place in order to mitigate the threat of data breaches. Organizations rely on processes and tools for tracking, controlling, preventing, and correcting secure access to basic resources as per access control privileges of individuals, systems, and applications dependent on a need or right previously classified.

15. **Wireless Access Control:** Many data breaches are a result of attackers gaining wireless access to organizations from outside the physical building. This may be done by connecting wirelessly to access points. Public Wi-Fi networks may be useful in conducting man-in-the-middle attacks and may also successfully install backdoors which can possibly reconnect to the network of a target organization. Wireless access controls are processes and tools for tracking, controlling, preventing, and correcting the secure use of wireless local area networks (WLANs), wireless client systems and access points.

16. **Account Monitoring and Control:** Attackers are continuously on the lookout for legitimate but inactive user accounts. These accounts may be used by an attacker for impersonating legitimate users. Hence, detecting such user accounts is a challenging task for the security personnel. This control necessitates active management of a user account throughout its life. It takes into account their creation, use, dormancy, and deletion to minimize opportunities for attackers and attack surface.

17. **Implement a Security Awareness and Training Program:** Cybersecurity is a technical challenge, however the role of employees in an organization plays a significant role in the failure or success of even the best cybersecurity programs. Cybersecurity programs vary across designs, implementations, operations, uses or oversight. Therefore, for every functional job, associations need to recognize security aptitudes, specific knowledge, and abilities required for

supporting the defense of the organization. The organizations are also responsible for developing and executing plans and strategies to assess, identify gaps and remediate through training, planning, policy, and security awareness programs.

18. **Application Software Security:** Attackers may also exploit weaknesses found in web-based and other application programming. These loopholes may be the result of coding errors, logical mistakes, fragmented necessities and inability to test for unusual or unanticipated conditions. So as to relieve this attack vector, associations must deal with the security of all in-house and obtained programming over its life cycle.

19. **Incident Response (IR) and Management:** Developing and implementing an IR infrastructure is useful for organizations to protect their information and reputation. This infrastructure may include plans, defined roles, training, communications, management oversight to quickly discover attacks and then contain the damage, or annihilate the adversary's access and reestablish the integrity of the system, network, and frameworks. Security events are an integral part of every organization. When an incident occurs, it might be too late to design the appropriate procedures, report data collection, manage, address legal procedures and communicate strategy. Therefore, it is important to develop an IR planning prior to a successful attack.

20. **Penetration Tests and Red Team Exercises:** Testing the overall defense of an organization is a good approach for evaluating its security. This is often done by simulating the actions and objectives of an attacker. Adversaries frequently misuse the gap between good defense structure and actual execution, which is usually the window of time between a vulnerability being discovered and its remediation. A successful defensive strategy demands a comprehensive program with appropriate information security policies. It must also incorporate strong technical defenses and proper action by people. Red Team exercises can explore an organization defenses, policies, and processes for improving organizational readiness. It can also improve defenses and inspect current levels of performance (Figure 8.16).

Introduction to Cybersecurity Policies   381

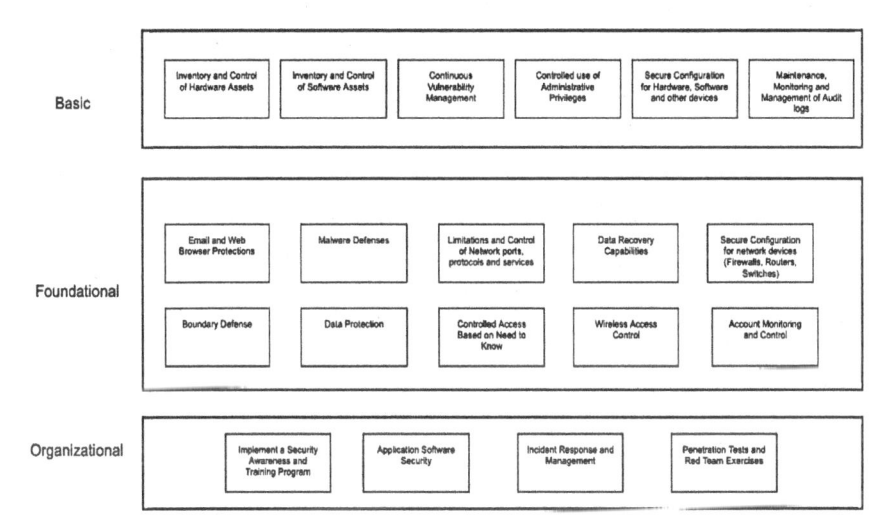

FIGURE 8.16 CIS critical security controls.

Key changes made to CIS critical security controls:
- Improving the consistency and simplifying the wording of each sub-control;
- Implementing one ask per sub-control;
- Bringing more focus on authentication, encryption, and application whitelisting;
- Accounting for enhancements in security innovation and rising security issues;
- Better alignment with other frameworks (such as the NIST CSF);
- Supporting the advancement of related items (e.g., estimations/measurements, execution guides);
- Identifying types of CIS controls (basic, foundational, and organizational).

8.12.6 IASME

IASME Governance is a basic and reasonable information assurance standard. It was designed to improve the cybersecurity of small and medium-sized enterprises (SMEs) [115]. The standard incorporates five Cyber Essentials technical topics and appends additional topics that mostly relate to people and processes, such as:

- Risk evaluation and the supervision;
- Preparing and managing individuals;
- Change management;
- Keeping a track of (monitoring) backup;
- IR and business continuity.

An IASME Governance certificate ensures that an organization is capable of providing assurance to customers, suppliers, and partners that the organization's security has been reviewed by an autonomous third-party and meets current best practices. The IASME Governance Standard is an official data and cybersecurity procedure that is suitable for any organization in any segment and SMEs specifically. It gives a working structure to guarantee data protection from the ever-changing threat landscape. The IASME Governance Standard incorporates clear guidance on good information security practices, so the business realizes where to begin taking safety efforts. The standard is accessible at two levels of affirmation:

1. **IASME Governance Self-Assessment:**
    i. Candidates are required to complete an online questionnaire which includes 160 simple questions about their organization. A Certification Body is responsible for evaluating the same, and also awards the certification if the candidate's answers comply with the standard.
    ii. The evaluation incorporates confirmation to the Cyber Essentials standard.
2. **IASME Governance Audited (or IASME Gold):**
    i. The competitor association is visited by an IASME Certification Body who confirms consistency with the norm and, if fitting, issues accreditation.

The IASME governance standard enables businesses in:

- Identifying risks to their information.
- Applying satisfactory obstructions or controls to decrease the probability or effect of undesirable security breaches.
- Keeping information risk at an acceptable level.
- Using an organized self-evaluation of what they are doing to secure data.
- Proactively verifying that the security controls that are implemented provide the expected degree of data and cybersecurity.

- Being autonomously surveyed by an assessor who will comprehend their business risk and confirm their viability.
- Raising the concerns of data risks in organizations their supply chain.
- Giving themselves, clients, and their supply chain, a degree of confirmation comparable to ISO/IEC 27001 and comparable standards.

The IASME Governance Standard uses a framework to establish the risk profile which considers:

- How information systems are used;
- How redistributed (counting cloud) services are utilized;
- Whether people in an organization use their own equipment for business (BYOD);
- How remote and portable frameworks are utilized;
- Awareness and attitude to the threat environment;
- Assessed estimation of the business' data resources;
- Assessed estimation of the business' IT.

The IASME Governance Standard is tied in with surveying risk to the business data and keeping that risk at an allowable level to an association, clients, and supply chain. IASME Governance Standard is intended to show an equalization of proactive measures and the capacity to be tough even with inadvertent or deliberate data and cybersecurity events.

8.12.7 SERVICE ORGANIZATION AND CONTROL (SOC 2)

SOC 2 or service organization and control is an evaluating system that guarantees service providers safely oversee information to secure the interests of the association and the protection of its customers. It was created by the American Institute of Certified Public Accountants (AICPA). It characterizes models for overseeing client information dependent on five trust administration standards to be specific security, accessibility, handling integrity, confidentiality, and privacy. Unlike the PCI DSS, which has exceptionally rigid prerequisites, SOC 2 reports are one of a kind to every association. In accordance with explicit strategic approaches, each plans its own controls to conform to at least one of the trust standards. These inside reports give significant information about how a service provider oversees data. There are two sorts of SOC reports:

1. **Type I:** This type details a vendor's systems and determines if their design is appropriate with respect to relevant trust principles. A type I report is based on trust services criteria (TSC) chosen by a company followed by some other controls. These controls are placed to mitigate security risks. A type 1 report is also responsible for reviewing security controls based on a specified point in time.
2. **Type II:** This type discusses the operational viability of those frameworks. A sort II report highlights how those controls are organized and actualized over some undefined time frame.

SOC 2 certification is issued by outside auditors. These auditors are responsible for assessing the degree to which a vendor follows at least one of the five trust standards. These standards depend on the frameworks and methodologies set up. Trust standards are analyzed as follows:

1. **Security:** The security principal is responsible for inspecting the protection of framework assets against unapproved access. Access controls assist in preventing misuse of software, potential system abuse, unauthorized removal of data, disclosure of information, and improper alteration of information. IT security tools such as network and web application firewalls (WAFs), multi-factor authentication, and intrusion detection and prevention assist in preventing security breaches which may otherwise result in unauthorized access of systems and data.
2. **Availability:** The availability principle pays attention to the accessibility of the system, services, and products, as detailed in an SLA. Although the principle does not address system functionality and usability, it takes into account security-related standards that may have affected availability. Monitoring network availability, performance, site failover, incident handling, and security are critical in this context.
3. **Processing Integrity:** The processing integrity principle examines whether or not a system achieves its purpose by analyzing whether it conveys the perfect information at the perfect cost at the correct time. It is necessary to ensure that data processing is complete, authorized, accurate, valid, and timely. However, processing integrity is not the same as data integrity. If data is erroneous before being input into a system, the processing entity is not responsible for detecting it. A combination of tracking and quality affirmation strategies, can help guarantee processing integrity.

4. **Confidentiality:** Data might be viewed as confidential if its access and revelation is limited to a predefined set of people or associations. Encrypting data can protect its confidentiality during transmission. Network and application firewalls, along with meticulous access controls, can be useful for safeguarding information being processed or stored on computer systems.
5. **Privacy:** The privacy principle highlights system's disclosure, collection, retention, disposal, and use of personal information with respect to an organization's privacy notice, as well as with criteria set forth in the AICPA's generally accepted privacy principles (GAPP). PII details information that can distinguish an individual. It may incorporate information like name, address, Social Security number, etc. Other personal data related to sexuality, religion, health, and race is also considered sensitive. Hence, extra levels of protection may be demanded. Controls might be utilized to shield PII from unapproved access (Figure 8.17).

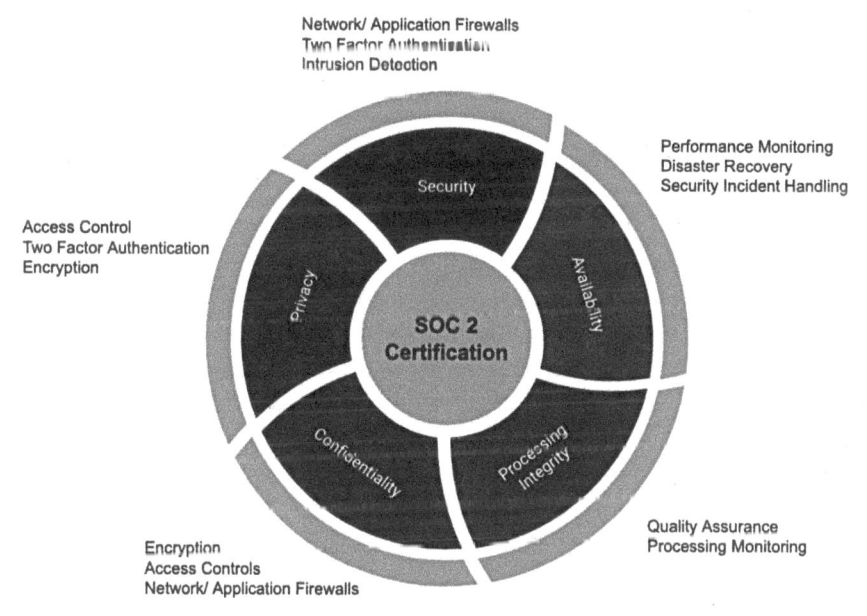

FIGURE 8.17 SOC principles.

Four regions of security practices that are basic to meeting SOC 2 consistency are as follows:

1. **Monitoring the Known and Unknown:** Achieving SOC 2 compliance makes it necessary to establish methodologies and practices with required degrees of oversight over the association. Frameworks might be valuable in observing irregular framework action, authorized and unauthorized system configuration changes, and client access levels. One of the ways to achieve it is by baselining it with normal activities in the cloud environment. This can provide significant information about the abnormal activities in the cloud environment. Continuously monitoring the activities can assist in detecting potential threats that may have raised from internal and external sources.
2. **Anomaly Alerts:** When a security incident happens, it is necessary to manifest that adequate alerting procedures are in place. This is done to show that in an occasion including unapproved access to client information, The organization is capable of responding to the attack and taking corrective timely measures. Many times, the problem with alerting is that a lot of false positives may occur. Therefore, in order to combat this, there is a need to process the alarms only when activity deviates from the normal that has been defined for the unique environment. SOC 2 mandates organizations to set up alarms for any exercises that bring about unauthorized:
   i.   Modification or exposure of configurations, data, and controls;
   ii. Activities involving File transfers;
   iii. Privileged filesystem, account, or login access.
3. **Detailed Audit Trails:** Discovering the main driver of an attack with regards to response is yet another necessary task. Audit trails are a common way of attaining insight that one needs to carry out security operations since they are capable of providing the necessary cloud context, providing the what, who, where, when, and how of a security incident so that quick and informed decisions can be made. Review trails give profound experiences into:
   i.   Expansion, change, or evacuation of key framework components;
   ii. Unauthorized changes of information and system designs;
   iii. The breadth of attack impact and the point of source.
4. **Actionable Forensics:** Customers need assurance that the organization has the ability to take corrective action on alerts before customer data is exposed worldwide. Systems must analyze the MTTD (mean time to detect), security associations ought to be similarly

compulsive about cutting MTTR (mean time to remediate). Since decisions can only be as good as intelligence, actionable data is used to make informed decisions. This comes as host-based checking, where there is greater perceivability:
i. Where an attack began?
ii. Where it is headed out to?
iii. What parts of the framework it affected?
iv. The nature of the impact.
v. What its best course of action maybe?

The Canadian Institute of Chartered Accountants (CICA) and AICPA designed the GAPP, which is a global privacy framework. The primary objective of GAPP is to assist organizations in creating an effective privacy program. The privacy program must be capable of addressing privacy risks, business opportunities and obligations. The following principles are considered globally recognized good practices:

1. **Management:** The entity must be capable of defining, documenting, communicating, and assigning accountability for its privacy policies and procedures. It may incorporate questions like:
   i. Collection of PII and the reason behind it;
   ii. The source of the PII;
   iii. The departments responsible for collecting the information;
   iv. The departments that use the information;
   v. Accessing;
   vi. PII being provided to external businesses.

2. **Notice:** An entity must be capable of providing notice about its privacy policies and procedures and identifying the motive behind a collection of personal information, its use, if it can be retained or disclosed. Customers/service organizations may want information on why their information is needed, how it may be used, and how long the information will be retained. If it is not required by the company to store data for over 7 days, then strategies must guarantee that the data is appropriately taken out from the framework after that assigned timeframe.

3. **Choice and Consent:** The entity must be capable of describing the choices available to the individual and must obtain implicit or explicit consent with respect to the assortment, use, and exposure of individual data.

4. **Collection:** The entity must be capable of gathering individual data just for the reasons distinguished in the notification.
5. **Use, Retention, and Disposal:** The entity should limit the use of personal information and comply with the purposes mentioned in the notice and for which the individual has provided implicit or explicit consent. Similarly, if the information is no longer needed, it may be disposed of. Factors for retention periods may include, but are not limited to current activities, access requests, legal/regulatory requirements, the advantage of storing data, assortment purpose, client verification necessities, likely legitimate risks, industry guidelines, and contractual obligations. Other considerations may also incorporate whether the removal process will happen on-site or off-site. The removal process ought to be irreversible. Removal Options could be physically delivering data on electronic storage devices unreadable by means of squashing, destroying, and so forth. For non-disposable devices like systems and tablets, overwriting, cleaning, or deleting, burning or shredding paper records may be considered.
6. **Access:** The entity must be equipped for giving people admittance to their own data for audit and update.
7. **Divulgence to Third Parties:** The entity must be capable of revealing individual data to outsiders just for the reasons distinguished in the notification and with the implicit or explicit assent of the person.
8. **Security for Privacy:** The entity must be capable of protecting personal information from unauthorized access. Data breaches may be due to various reasons ranging from social engineering to lost devices. Conducting a PII storage inventory may help identify the weakest link in the storage practices. This must be followed by reviewing electronic and physical means of storage. For example, making sure that physical files are kept in a secured location as papers lying on desks can be easily acquired. A layered approach must be used for the electronic storage of information. Different devices have different security measures like wireless monitoring, firewalls, encryption, etc. A combination of such measures can increase privacy. USBs, hard drives, and CDs are some other storage devices.
9. **Quality:** The entity must be capable of maintaining complete, correct, and important individual data for the reasons recognized in the notification.

10. Monitoring and Enforcement: The organization must be capable of monitoring compliance with its privacy policies and procedures. The procedures must be capable of addressing privacy-related disputes and complaints.

8.12.8 HITRUST CSF

The health information trust alliance (HITRUST) set up a common security framework (CSF) that is robust and scalable. The HITRUST CSF structure can be utilized by all associations that are creating, accessing, storing, or exchanging sensitive and/or regulated data [116]. The CSF details a prescriptive set of controls that are responsible for coordinating the necessities of various guidelines and standards. Organizations may procure the CSF framework from HITRUST or buy access to the myCSF tool to decide the particular CSF prerequisites that apply to the association. The HITRUST CSF is responsible for assisting organizations in addressing challenges through a framework that is flexible, comprehensive, and supports scalable security and privacy controls. The structure brags of a risk-based methodology and records government and state guidelines, norms, and frameworks. The HITRUST CSF is responsible for:

- Including, harmonizing, and cross-referencing existing, globally recognized standards, regulations, and business requirements. These incorporate ISO, EU GDPR, NIST, and PCI.
- Scaling controls based on type, size, and complexity of an organization.
- Providing prescriptive requirements to ensure clarity.
- Following a risk-based methodology offering different degrees of execution prerequisites dictated by explicit risk limits.
- Allowing for the adoption of alternate controls, when necessary.
- Evolving as per user input and changing conditions in the norms and administrative condition on an annual basis.
- Providing a unified approach for managing data protection compliance.

The HITRUST myCSF is an administration, risk, compliance, and consistency tool worked by HITRUST that is utilized by associations to evaluate compliance with different guidelines and structures. A HITRUST myCSF appraisal is customized to every association's unique framework and factors so every evaluation is unique to an association. HITRUST necessities depend on ISO 27001 and applied to the healthcare industry. Every one of the controls characterized by HITRUST has three diverse usage levels related to them. The execution levels build off of one another. This implies a level 3 execution incorporates the entirety of the level 1 and level 2 usage too. Execution levels are based upon three special risk factors:

- Association factors, similar to the kind of association or the size of the association;
- Framework factors, similar to internet connections, number of records, or the utilization of cell phones in the association; and
- Regulatory factors, like state or specialized industry requirements.

The HITRUST CSF, is primarily concerned with security and privacy. The framework has been introduced specifically for the healthcare industry. The framework hierarchy is similar to that of ISO 27001/27001. There are 14 control categories that may be further subdivided into 46 control objectives that have 149 controls. For every one of the 149 controls, there are up to three execution levels in which one must meet for each risk factor involved. Risk factors may be organizational, system, and regulatory. There are a total of 845 requirement statements that span over each implementation level. This makes the HITRUST CSF very scalable (Figure 8.18).

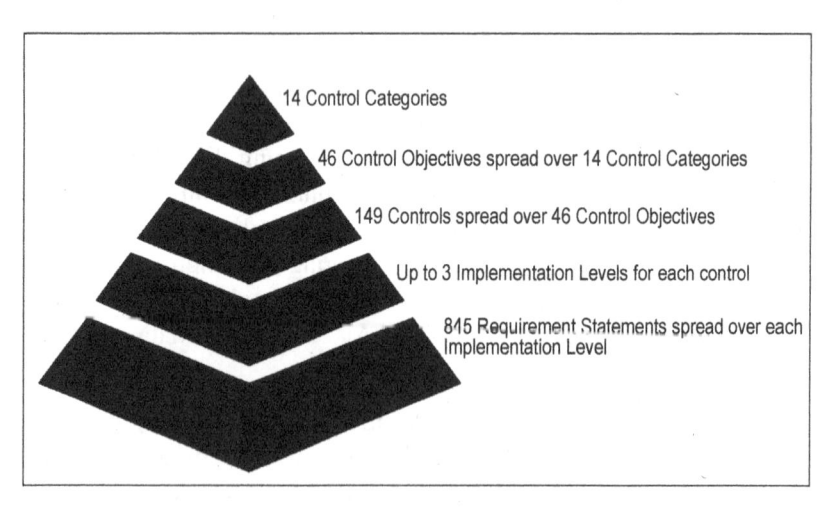

FIGURE 8.18 HITRUST common security framework.

While HIPAA and HITRUST are both confined to the healthcare industry, they are not entirely the same.

HIPAA (For covered entities and business associates)	HITRUST CSF (For IT service providers, doctors' offices, hospitals, health insurance plans, and other medical facilities)
Includes:	Includes:
• Administrative, physical, and technical safeguards. • Organizational requirements and policies and procedures and documentation requirements.	• HIPAA security rule. • PCI/DSS. • Control objectives for information and related technology (COBIT). • National Institute of Standards and Technology (NIST) risk management framework. • International Organization for Standardization (ISO). • Federal Trade Commissions (FTC) red flag rule. • Center for Medicare and Medicaid services addressable risk safeguards (CMS ARS) Federal and state regulations

In this chapter, we familiarized ourselves with the concept of Cybersecurity policies, how they are written, by whom they are written, and how they can be updated. Moreover, relevant terminologies topics like cyber policy audience, cyber policy classification, cyber policy audit, cyber policy enforcement, cyber policy awareness have also been discussed in detail. Developing and creating policies could be challenging, and this chapter also discusses how to effectively write and develop policies. Apart from that, the chapter discusses and reviews several cybersecurity policy frameworks like NIST CSF, ISO Standard series, PCI/DSS, HITRUST CSF, etc.

8.13 SUMMARY AND REVIEW

- A cybersecurity policy is the announcement of dependable decision-makers about the protection component of a company's crucial physical and information assets. It may be thought of as a document that describes a company's security controls and activities.

- Cybersecurity techniques characterize the standards for the quantity of representatives, specialists, accomplices, board individuals, and other end-clients who access online applications and web assets, send information over networks, and in any case practice responsible security.
- An updated cybersecurity policy might be a fundamental security asset for all associations. Without an updated cybersecurity strategy, end clients may commit errors and cause information breaches. An imprudent methodology can cost an association generously in fines, legitimate charges, settlements, loss of public trust, and reputation degradation. Establishing and supporting policy can help prevent these adverse outcomes.
- The primary goal of the security audit is to bring all the security policies as closely as could reasonably be expected. Security reviewing is an elevated level examination of the current venture progress and company posture on test data security identified with existing strategy consistency.
- A cybersecurity framework (CSF) may be defined as a series of documented processes used to define policies and procedures regarding the implementation and ongoing management of information security controls in an enterprise environment.
- The NIST CSF gives a policy framework of computer assistance and direction for how private sector associations in the United States can evaluate and improve their capacity to forestall, distinguish, and react to cyberattacks.
- Control objectives for information and related technology (COBIT) structure was made by the ISACA (Information Systems Audit and Control Association). It is liable for IT administration and support. It was intended to be a steady tool for administrators.
- PCI DSS or the Payment Card Industry Data Security Standard is a compliance system dependent on an industry-mandated collection of principles proposed to guard customers' card information when it is utilized with vendors and service providers.
- The ISO/IEC 27000 series aims to provide best practice recommendations on information security management, which is basically the management of information risks through information security

controls with respect to the overall information security management system (ISMS).
- The Centre for Internet Security (CIS) Critical Security Controls are a set of laid out activities for cybersecurity that structure a defense-in-depth collection of explicit and actionable best practices. The aim is to mitigate the most common cyber-attacks.
- IASME Governance is a basic and reasonable Information Assurance standard. It was designed to improve the cybersecurity of small and medium-sized enterprises (SMEs).
- SOC 2 or Service Organization and Control is an inspecting method that guarantees service providers safely oversee information to secure the interests of the association and the privacy of its customers. It was created by the American Institute of Certified Public Accountants (AICPA).
- The health information trust Alliance (HITRUST) has set up a common security framework (CSF) that can be utilized by all associations that make, access, store, or trade confidential and/or regulated information.

QUESTIONS TO PONDER

1. How is 'utilizing,' 'adopting,' and 'executing' a Framework different?
2. For what reason should an association utilize the Framework?
3. Can you deduce any relationship between risk management and policy?
4. Does the NIST Framework apply just to critical infrastructure organizations?
5. Is the NIST Framework preventive against cyber-attacks?
6. For what reason do medicinal and healthcare services need a security framework?
7. If an organization utilizes an outsider to make credit card exchanges, would they say they are as yet answerable for client information under PCI compliance?
8. Who uses CIS Controls? How are they updated?
9. How has the ISO 27000 grown over the last few years?

KEYWORDS

- **cybersecurity**
- **human resources**
- **information technology**
- **organization's security**
- **personally identifiable information**
- **risk management**

Bibliography

1. Paulsen, C., & Byers, R., (2019). *Glossary of Key Information Security Terms*. Retrieved from: https://csrc.nist.gov/publications/detail/nistir/7298/rev-3/final (accessed on 25 February 2021).
2. Gibson, W., (1996). Civilization and the edge of popular culture, an interview to bob Catterall. *CITY, 5, 6,* 174–177.
3. Department of Homeland Security Authorization Act for Fiscal Year-2006, Part 3, 109th Congress First Session, Report, May 2005.
4. Cyberethics (2019). Retrieved from: https://en.wikipedia.org/wiki/Cyberethics (accessed on 25 February 2021).
5. Pusey, P., & Sadera, W. A., (2011). Cyberethics, cyber safety, and cybersecurity: Preservice teacher knowledge, preparedness, and the need for teacher education to make a difference. *Journal of Digital Learning in Teacher Education, 28*(2), 82–88. http://etec.ctlt.ubc.ca/510wiki/Cyberethics#cite_note-1 (accessed on 25 February 2021).
6. *Know the Rules of Cyberethics,* (n.d.). Retrieved from: https://www.cisecurity.org/daily-tip/know-the-rules-of-cyber-ethics/ (accessed on 25 February 2021).
7. Christoph, S., & Pavan, D., (2018). *Cyberethics: Serving Humanity with Values*. Globethics.net Global 17.
8. Morgan, C., & Vezza, A., (2016). Retrieved from: https://www.iab.org/ (accessed on 25 February 2021).
9. The Computer Ethics Institute, (n.d.). Retrieved from: Computerethicsinstitute.org. (n.d.). Retrieved March 12, 2021, from https://computerethicsinstitute.org/publications/tencommandments.html (accessed on 25 February 2021).
10. Vallor, S., & Rewak, W., (2018). *An Introduction to Cybersecurity Ethics*. Santa Clara University.
11. Priyadarshini, I., (2018). *Features and Architecture of the Modern Cyber Range: A Qualitative Analysis and Survey*. The University of Delaware.
12. *Grey Hat,* (2019). Retrieved from: https://en.wikipedia.org/wiki/Grey_hat (accessed on 25 February 2021).
13. Priyadarshini, I., & Pattnaik, P. K., (2015). Some issues of accountability framework in data-intensive cloud computing environment. *African Journal of Computing and ICT, 8*(3), 2.
14. Priyadarshini, I., (2019). Introduction on cybersecurity. *Cyber Security in Parallel and Distributed Computing: Concepts, Techniques, Applications and Case Studies,* 1–37.
15. Priyadarshini, I., & Cotton, C., (2019). Internet memes: A novel approach to distinguish humans and bots for authentication. In: *Proceedings of the Future Technologies Conference* (pp. 204–222). Springer, Cham.

16. Priyadarshini, I., Wang, H., & Cotton, C., (2019). Some cyberpsychology techniques to distinguish humans and bots for authentication. In: *Proceedings of the Future Technologies Conference* (pp. 306–323). Springer, Cham.
17. Rogers, C., (2020). *Privacy vs. Security*. Retrieved from: https://lossofprivacy.com/2010/12/privacy-vs-security/ (accessed on 25 February 2021).
18. Hary, G., (2018). *Ethical Issues in Cyberspace and IT Society*. Ritsumeikan Asia Pacific University.
19. Ahmed, T., (2009). *Copyright Infringement in Cyberspace and Network Security: A Threat to E-Commerce*. KIIT Law School, KIIT University.
20. Gerhardt, D., (2006). Plagiarism in cyberspace: Learning the rules of recycling content with a view towards nurturing academic trust in an electronic world. *Richmond Journal of Law and Technology*.
21. Stückelberger, (2013). *Ethics in the Information Society: The Nine 'P's A Discussion Paper for the WSIS+10 Process 2013–2015*. Globalethics.net.
22. Loren, L. P., (2009). The evolving role of for-profit use in copyright law. Lessons from the 1909 act. *Santa Clara Computer and High Tech. L. J., 26*, 255.
23. Priyadarshini, I., (2018). Cybersecurity risks in robotics. In: *Cyber Security and Threats: Concepts, Methodologies, Tools, and Applications* (pp. 1235–1250). IGI Global.
24. Vo, T., Sharma, R., Kumar, R., Son, L. H., Pham, B. T., Tien, B. D., & Le, T., (2020). Crime rate detection using social media of different crime locations and Twitter part-of-speech tagger with Brown clustering. *Journal of Intelligent and Fuzzy Systems*, 1–13. (Preprint).
25. Negin, B., (2004). *Ethics and Information and Communication Technology* (p. 99). Public administration and public policy.
26. Taylor, M. J., & Moynihan, E., (2002). Analyzing IT ethics. *Systems Research and Behavioral Science: The Official Journal of the International Federation for Systems Research, 19*(1), 49–60.
27. Thekkilakattil, A., & Dodig-Crnkovic, G., (2015). Ethics aspects of embedded and cyber-physical systems. In: *2015 IEEE 39th Annual Computer Software and Applications Conference* (Vol. 2, pp. 39–44). IEEE.
28. U.S. Department of Defense (2017). *Department of Defense Directive 3000.09 (Autonomy in Weapon Systems)*. Accessed at: http://www.esd.whs.mil/Portals/54/Documents/DD/issuances/dodd/300009p.pdf (accessed on 25 February 2021).
29. Kelsey, A., (2018). *Russia Prepares for a Future of Making Autonomous Weapons*. C4ISR. Accessed at: https://www.c4isrnet.com/electronic-warfare/2018/06/11/russia-prepares-for-a-future-of-making-autonomous-weapons/ (accessed on 25 February 2021).
30. Rowe, N. C., (2010). The ethics of cyberweapons in warfare. *International Journal of Technoethics (IJT), 1*(1), 20–31.
31. Wolter, P., & Van, C. A., (2009). The precautionary principle in a world of digital dependencies. In: *Computer, 42*(6), 50–56. doi:10.1109/MC.2009.203.
32. Timmermans, J., Stahl, B. C., Ikonen, V., & Bozdag, E., (2010). The ethics of cloud computing: A conceptual review. In: *2010 IEEE Second International Conference on Cloud Computing Technology and Science* (pp. 614–620). IEEE.
33. Popescul, D., & Georgescu, M., (2014). Internet of things-some ethical issues. *The USV Annals of Economics and Public Administration, 13, 2*(18), 208–214.

34. Puri, V., Jha, S., Kumar, R., Priyadarshini, I., Abdel-Basset, M., Elhoseny, M., & Long, H. V., (2019). A hybrid artificial intelligence and internet of things model for generation of renewable resource of energy. *IEEE Access, 7*, 111181–111191.
35. Puri, V., Kumar, R., Van, L. C., Sharma, R., & Priyadarshini, I., (2020). A vital role of blockchain technology toward the internet of vehicles. In: *Handbook of Research on Blockchain Technology* (pp. 407–416). Academic Press.
36. Quek, S. G., Selvachandran, G., Munir, M., Mahmood, T., Ullah, K., Son, L. H., & Priyadarshini, I., (2019). Multi-attribute multi-perception decision-making based on generalized T-spherical fuzzy weighted aggregation operators on neutrosophic sets. *Mathematics, 7*(9), 780.
37. Jha, S., Kumar, R., Abdel-Basset, M., Priyadarshini, I., Sharma, R., & Long, H. V., (2019). Deep learning approach for software maintainability metrics prediction. *IEEE Access, 7*, 61840–61855.
38. Tuan, T. A., Long, H. V., Kumar, R., Priyadarshini, I., & Son, N. T. K., (2019). Performance evaluation of botnet DDoS attack detection using machine learning. *Evolutionary Intelligence*, 1–12.
39. Priyadarshini, I., & Cotton, C., (2020). Intelligence in cyberspace: The road to cyber singularity. *Journal of Theoretical and Artificial Intelligence*. Taylor and Francis.
40. Cooper, I., & Yon, J., (2019). Ethical issues in biometrics. *Sci. Insight, 30*(2), 63–69.
41. Alterman, A., (2003). A piece of yourself: Ethical issues in biometric identification. *Ethics and Information Technology, 5*(3), 139–150.
42. Nuseir, M. T., & Ghandour, A., (2019). Ethical issues in modern business management. *International Journal of Procurement Management, 12*(5), 592–605.
43. Trademark, I., (2015). *Cyber law-Standard, and Legal Issues.*
44. Berner, S., (2003). Cyber-terrorism: Reality or paranoia? *SA Journal of Information Management, 5*(1).
45. Saini, H., Rao, Y. S., & Panda, T. C., (2012). Cyber-crimes and their impacts: A review. *International Journal of Engineering Research and Applications, 2*(2), 202–209.
46. Stückelberger, & Duggal, (2018). *Cyberethics 4.0: Serving Humanity with Values.* Globethics.
47. Fischer, E. A. (2014). Federal laws relating to cybersecurity: Overview of major issues, current laws, and proposed legislation. Retrieved from: https://fas.org/sgp/crs/natsec/R42114.pdf (accessed on 25 February 2021).
48. Cybercrime Laws of The United States-OAS, (n.d.). Retrieved from: https://www.oas.org/juridico/spanish/us_cyb_laws.pdf (accessed on 25 February 2021).
49. Fischer, E. A., (2014). *Federal Laws Relating to Cybersecurity: Overview of Major Issues, Current Laws, and Proposed Legislation.*
50. Dulcinea. *On This Day: Robert Tappan Morris Becomes First Hacker Prosecuted for Spreading Virus.* Retrieved from: http://www.findingdulcinea.com/news/on-this-day/July-August-08/On-this-Day--Robert-Morris-Becomes-First-Hacker-Prosecuted-For-Spreading-Virus.html (accessed on 25 February 2021).
51. Zetter, K., (2017). *Judge Acquits Lori Drew in Cyberbullying Case, Overrules Jury.* Retrieved from: https://www.wired.com/2009/07/drew-court/ (accessed on 25 February 2021).

52. Zetter, K., (2018). *The Most Controversial Hacking Cases of the Past Decade*. Retrieved from: https://www.wired.com/2015/10/cfaa-computer-fraud-abuse-act-most-controversial-computer-hacking-cases/ (accessed on 25 February 2021).
53. Edward Holman & Kelly Singleton (2018). *FTC Announces Settlement with PayPal for Alleged FTC Act and GLBA Violations by Venmo*. Retrieved from: https://www.wsgrdataadvisor.com/2018/03/ftc-venmo-paypal-settlement/ (accessed on 25 February 2021).
54. FairAug, L., (2017). *4 Gramm-Leach-Bliley Tips to Take from FTC's TaxSlayer Case*. Retrieved from: https://www.ftc.gov/news-events/blogs/business-blog/2017/08/4-gramm-leach-bliley-tips-take-ftcs-taxslayer-case (accessed on 25 February 2021).
55. Nicandro Iannacci (n.d.). *Katz v. the United States: The Fourth Amendment Adapts to New Technology*. Retrieved from: https://constitutioncenter.org/blog/katz-v-united-states-the-fourth-amendment-adapts-to-new-technology/ (accessed on 25 February 2021).
56. *The United States vs. Warshak-Case Brief*. (2016). Retrieved from: https://www.quimbee.com/cases/united-states-v-warshak (accessed on 25 February 2021).
57. *Yates vs. the United States-2015*. (2019). Retrieved from: https://en.wikipedia.org/wiki/Yates_v._United_States_(2015) (accessed on 25 February 2021).
58. *Wiest vs. Lynch-Case Brief*. (2019). Retrieved from: https://www.quimbee.com/cases/wiest-v-lynch (accessed on 25 February 2021).
59. Shweta Saraswat (2010). *Former UCLA Medical Center Employee Huping Zhou Sentenced Jail Time for Looking at Private Medical Files*. Retrieved from: https://dailybruin.com/2010/05/05/former-ucla-medical-center-employee-huping-zhou-se/ (accessed on 25 February 2021).
60. *Former Hospital Employee Sentenced for HIPAA Violations*. (2015). Retrieved from: https://www.justice.gov/usao-edtx/pr/former-hospital-employee-sentenced-hipaa-violations (accessed on 25 February 2021).
61. *20 Catastrophic HIPAA Violation Cases to Open Your Eyes*. (2020). Retrieved from: https://www.medprodisposal.com/hipaa/20-catastrophic-hipaa-violation-cases-to-open-your-eyes/ (accessed on 25 February 2021).
62. Garrett Hinck (2019). *Wassenaar Export Controls on Surveillance Tools: New Exemptions for Vulnerability Research*. Retrieved from: https://www.lawfareblog.com/wassenaar-export-controls-surveillance-tools-new-exemptions-vulnerability-research (accessed on 25 February 2021).
63. Lindsey, N., (2020). *Concerns Mount as Israel Eases Rules on Cyber Weapons for Cyber Espionage*. Retrieved from: https://www.cpomagazine.com/cyber-security/concerns-mount-as-israel-eases-rules-on-cyber-weapons-for-cyber-espionage/ (accessed on 25 February 2021).
64. Hannah Natanson & Derek G. Xiao (n.d.). *Email Lists Revealing Students' Private Information Remained Public for Years: News: The Harvard Crimson*. Retrieved from: https://www.thecrimson.com/article/2017/2/21/hcs-emails-public/ (accessed on 25 February 2021).
65. Algar, S., (2018). *Controversial Former Brooklyn Principal Rehired by DOE*. Retrieved from: https://nypost.com/2018/09/14/controversial-former-brooklyn-principal-rehired-by-doe/ (accessed on 25 February 2021).

66. Satariano, A., (2019). *Google Is Fined $57 Million Under Europe's Data Privacy Law*. Retrieved from: https://www.nytimes.com/2019/01/21/technology/google-europe-gdpr-fine.html (accessed on 25 February 2021).
67. Simpson, I., (2018). *Twitter in Violation of GDPR?* Retrieved from: https://sionik.com/insights/2018/10/14/twitter-in-violation-of-gdpr (accessed on 25 February 2021).
68. Keane, S., (2019). *British Airways Faces $230M GDPR Fine for 2018 Data Breach*. Retrieved from: https://www.cnet.com/news/british-airways-faces-record-breaking-230m-gdpr-fine-for-2018-data-breach (accessed on 25 February 2021).
69. Porter, J., (2018). *Google Accused of GDPR Privacy Violations by Seven Countries*. Retrieved from: https://www.theverge.com/2018/11/27/18114111/google-location-tracking-gdpr-challenge-european-deceptive (accessed on 25 February 2021).
70. Olson, P., (2019). *Marriott Faces $124 Million Fine Over Starwood Data Breach*. Retrieved from: https://www.wsj.com/articles/marriott-faces-123-million-fine-over-starwood-data-breach-11562682484 (accessed on 25 February 2021).
71. Brasseur, K., (2020). *CCPA cited in Hanna Andersson/Salesforce Breach Lawsuit*. Retrieved from: https://www.complianceweek.com/data-privacy/ccpa-cited-in-hanna-andersson/salesforce-breach-lawsuit/28410.article (accessed on 25 February 2021).
72. *Google and YouTube will Pay Record $170 Million for Alleged Violations of Children's Privacy Law*. (2019). Retrieved from: https://www.ftc.gov/news-events/press-releases/2019/09/google-youtube-will-pay-record-170-million-alleged-violations (accessed on 25 February 2021).
73. O'Donnell, L., & O'Donnell, L., (n.d.). *Lawsuit Claims Google Collects Minors' Locations, Browsing History*. Retrieved from: https://threatpost.com/lawsuit-claims-google-collects-minors-locations-browsing-history/153134/ (accessed on 25 February 2021).
74. Kalia, A., & McSherry, C., (2015). *Viacom vs. YouTube*. Retrieved from: https://www.eff.org/cases/viacom-v-youtube (accessed on 25 February 2021).
75. Brandom, R., (2020). *Apple Sues Security Vendor for DMCA Violations*. Retrieved from: https://www.theverge.com/2020/1/3/21047275/apple-corellium-dmca-lawsuit-copyright-research-jailbreak-security-copyright (accessed on 25 February 2021).
76. Errick, K., (2020). *Photographer Files Copyright Infringement Case Against Facebook-Tech*. Retrieved from: https://lawstreetmedia.com/tech/photographer-files-copyright-infringement-case-against-facebook/ (accessed on 25 February 2021).
77. Darkin, S., (2020). *PokerStars Founder Isai Scheinberg Faces US Jail After Guilty Plea*. Retrieved from: https://www.casinobeats.com/2020/03/26/pokerstars-founder-isai-scheinberg-faces-us-jail-after-guilty-plea/ (accessed on 25 February 2021).
78. Chatterjee, J. M., Privadarshini, I., & Le, D. N., (2019). Fog computing and its security issues. *Security Designs for the Cloud, IoT, and Social Networking, 59*–76.
79. Popescu, C. R. G., & Popescu, G. N., (2018). Risks of cyber-attacks on financial audit activity. *The Audit Financiar Journal, 16*(149), 140.
80. *Delphi Technique*, (2017). Retrieved from: http://acqnotes.com/acqnote/careerfields/delphi-technique (accessed on 25 February 2021).
81. Linstone, H. A., & Turoff, M., (1975). *The Delphi Method* (pp. 3–12). Reading, MA: Addison-Wesley.

82. Jones, J., (2006). An introduction to factor analysis of information risk (fair). *Norwich Journal of Information Assurance, 2*(1), 67.
83. Card, A. J., Ward, J. R., & Clarkson, P. J., (2012). Beyond FMEA: The structured what-if technique (SWIFT). *Journal of Healthcare Risk Management, 31*(4), 23–29.
84. Gantt, H., & Adamiecki, K., (2015). *Gantt Chart.*
85. Usmani, F., (2020). *What is a Monte Carlo Simulation?* Retrieved from: https://pmstudycircle.com/2015/02/monte-carlo-simulation/ (accessed on 25 February 2021).
86. *Sensitivity Analysis.* (n.d.). Retrieved from: https://project-management-knowledge.com/definitions/s/sensitivity-analysis/ (accessed on 25 February 2021).
87. Pecb. (n.d.). *Risk Assessment with OCTAVE.* Retrieved from: https://pecb.com/whitepaper/risk-assessment-with-octave (accessed on 25 February 2021).
88. Schmoeller, D., (2020). *Pros and Cons of the FAIR Framework.* Retrieved from: https://reciprocitylabs.com/pros-and-cons-of-the-fair-framework/ (accessed on 25 February 2021).
89. Pritam, N., Khari, M., Kumar, R., Jha, S., Priyadarshini, I., Abdel-Basset, M., & Long, H. V., (2019). Assessment of code smell for predicting class change proneness using machine learning. *IEEE Access, 7*, 37414–37425.
90. *Cybersecurity Risk Management Framework (RMF).* (n.d.). Retrieved from: https://aida.mitre.org/cyber-rmf/ (accessed on 25 February 2021).
91. Knowledge Leader, P., (n.d.). *Five Components of the COSO Framework You Need to Know.* Retrieved from: https://info.knowledgeleader.com/bid/161685/what-are-the-five-components-of-the-coso-framework (accessed on 25 February 2021).
92. Marsden, E., (2017). *The ISO 31000 Standard Risk Management: Principles and Guidelines.* Retrieved from: https://risk-engineering.org/ISO-31000-risk-management/ (accessed on 25 February 2021).
93. Justin Whitaker & Jacob Armijo (2020). *Enterprise Cybersecurity Risk Remediation.* Retrieved from: https://www.dayblink.com/enterprise-cybersecurity-risk-remediation/ (accessed on 25 February 2021).
94. Marotta, A., Martinelli, F., Nanni, S., Orlando, A., & Yautsiukhin, A., (2017). Cyber-insurance survey. *Computer Science Review, 24*, 35–61.
95. Meland, P. H., Tondel, I. A., & Solhaug, B., (2015). Mitigating risk with cyber insurance. *IEEE Security and Privacy, 13*(6), 38–43.
96. *Cyber Insurance.* (2020). Retrieved from: https://en.wikipedia.org/wiki/Cyber_insurance (accessed on 25 February 2021).
97. *Schneier on Security,* (n.d.). Retrieved from: https://www.schneier.com/blog/archives/2018/04/cybersecurity_i_1.html (accessed on 25 February 2021).
98. Frequently Asked Questions about Cyber Insurance A Resource, Report by Urmia and Educause, November 2017.
99. Hibberd, G., & Cook, A., (2014). The rise of cyber liability insurance. In: *Cyber Crime and Cyber Terrorism Investigator's Handbook* (pp. 221–230). Syngress.
100. Li, L., He, W., Xu, L., Ash, I., Anwar, M., & Yuan, X., (2019). Investigating the impact of cybersecurity policy awareness on employees' cybersecurity behavior. *International Journal of Information Management, 45*, 13–24.
101. Cavelty, M. D., & Egloff, F. J., (2019). The politics of cybersecurity: Balancing different roles of the state. *St Antony's International Review, 15*(1), 37–57.

102. Kumar, A. (2017). *An Introduction to Cyber Security Policy.* Retrieved from: https://resources.infosecinstitute.com/cyber-security-policy-part-1/#gref (accessed on 25 February 2021).
103. *How to Write an Effective Cybersecurity Policy.* (2019). Retrieved from: https://www.theamegroup.com/write-effective-cybersecurity-policy/ (accessed on 25 February 2021).
104. Hayslip, G., & Hayslip, G., (2018). *9 Policies and Procedures You Need to Know About if You're Starting a New Security Program.* Retrieved from: https://www.csoonline.com/article/3263738/9-policies-and-procedures-you-need-to-know-about-if-youre-starting-a-new-security-program.html (accessed on 25 February 2021).
105. *Incident Response Policy*, (n.d.). Retrieved from: https://www.sciencedirect.com/topics/computer-science/incident-response-policy (accessed on 25 February 2021).
106. Mesker, K., Engineer, I. C., & Chevron, E. T. C., (2014). Adapting NIST cybersecurity framework for risk assessment. In: *NIST Conference*.
107. Shen, L., (2014). The NIST cybersecurity framework: Overview and potential impacts. *SciTech. Lawyer, 10*(4), 16.
108. Ridley, G., Young, J., & Carroll, P. (2004, January). COBIT and its Utilization: A framework from the literature. In: *37th Annual Hawaii International Conference on System Sciences, 2004,* (8pp). IEEE.
109. Garsoux, M. '*COBIT-5 ISACA's New Framework for IT Governance, Risk, Security and Auditing: An Overview.* COBIT5, ISACA.
110. Morse, E. A., & Raval, V., (2008). PCI DSS: Payment card industry data security standards in context. *Computer Law and Security Review, 24*(6), 540–554.
111. Didier, G., (2013). *Demystifying PCI DSS Expert Tips and Explanations to Help You Gain PCI DSS Compliance.* Risk and Compliance Product Manager at Rapid.
112. Disterer, G., (2013). *ISO/IEC 27000, 27001 and 27002 for Information Security Management.*
113. Gonzalez, C. P., Hunting, B., Burgess, J., & Jensen, L. H. *CIS Critical Security Controls.*
114. UpGuard, (2020). *What are the CIS Controls for Effective Cyber Defense?* Retrieved from: https://www.upguard.com/blog/cis-controls (accessed on 25 February 2021).
115. Henson, R., Dresner, D., & Booth, D., (2011). *IASME: Information Security Management Evolution for SMEs.*
116. Donaldson, S. E., Siegel, S. G., Williams, C. K., & Aslam, A., (2015). Cybersecurity frameworks. In: *Enterprise Cybersecurity* (pp. 297–309). A Press, Berkeley, CA.
117. Priyadarshini, I., (2019). Introduction to blockchain technology. *Cyber Security in Parallel and Distributed Computing: Concepts, Techniques, Applications and Case Studies,* pp. 91–107.
118. Priyadarshini, I., & Sarraf, J., (2017). Smart and accountable water distribution for rural development. In: *Proceedings of the 5th International Conference on Frontiers in Intelligent Computing: Theory and Applications* (pp. 251–258). Springer, Singapore.
119. Patro, S. G. K., Mishra, B. K., Panda, S. K., Kumar, R., Long, H. V., Taniar, D., & Priyadarshini, I., (2020). A hybrid action-related k-nearest neighbor (HAR-KNN) approach for recommendation systems. *IEEE Access, 8,* 90978–90991.
120. Ayyasamy, A., Julie, E. G., Robinson, Y. H., Balaji, S., Kumar, R., Thong, P. H., & Priyadarshini, I., (2020). AVRM: Adaptive void recovery mechanism to reduce void nodes in wireless sensor networks. *Peer-to-Peer Networking and Applications,* pp. 1–15.

121. Sarraf, J., Priyadarshini, I., & Pattnaik, P. K., (2016). Real time bus monitoring system. In: *Information Systems Design and Intelligent Applications* (pp. 551–557). Springer, New Delhi.
122. Jha, S., Kumar, R., Priyadarshini, I., Smarandache, F., & Long, H. V., (2019). Neutrosophic image segmentation with dice coefficients. *Measurement, 134*, 762–772.
123. Jha, S., Kumar, R., Chiclana, F., Puri, V., & Priyadarshini, I., (2019). Neutrosophic approach for enhancing quality of signals. *Multimedia Tools and Applications*, pp. 1–32.
124. Priyadarshini, I., Mohanty, P., Kumar, R., Son, L. H., Chau, H. T. M., Nhu, V. H., & Tien Bui, D., (2020). Analysis of outbreak and global impacts of the COVID-19. In: *Healthcare* (Vol. 8, No. 2, p. 148). Multidisciplinary Digital Publishing Institute.
125. Stephen Brown (n.d.). *UNESCO Comprehensive Study on Internet-Related Issues: United Nations Educational, Scientific and Cultural Organization.* Retrieved from: http://www.unesco.org/new/en/communication-and-information/crosscutting-priorities/unesco-internet-study/unesco-comprehensive-study-on-internet-related-issues/ (accessed on 25 February 2021).
126. *1992 ACM Code*, (2019). Retrieved from: https://ethics.acm.org/code-of-ethics/previous-versions/1992-acm-code/ (accessed on 25 February 2021).
127. *The Ethics of Autonomous Weapons Systems*. (n.d.). Retrieved from: https://www.law.upenn.edu/institutes/cerl/conferences/ethicsofweapons/ (accessed on 25 February 2021).
128. Adapted from COBIT 5 ©2012 ISACA. All rights reserved. Used with permission.

Index

A

Access
　control, 90, 346, 357
　policy, 345, 346
　rights, 42, 43, 65
Accidental plagiarism, 51
Account monitoring and control, 379
Accountability, 34, 108, 114, 220, 292, 340
　for decisions by autonomous systems, 114
　its significance, 340
Actionable forensics, 386
Advancements, 64, 106, 124, 127, 211
Advantages of cyber insurance, 308, 310
Affinity diagram, 251
Aggravated identity theft, 158
Algorithms, 20, 39, 93, 103, 108
Ambiguity, 111
American Institute of Certified Public Accountants, 383, 393
Analysis, 249, 254, 270–273, 279, 293, 303, 305, 359
Anomalies and events, 358
Anomaly alerts, 386
Antitrust
　exemption, 184
　laws, 165
Application
　enforcement of group health plan, 207
　software security, 380
Applying
　risk identification tools and techniques, 254
　single integrated framework, 367
Arming employees with knowledge, 97
Artificial
　general intelligence, 119
　intelligence, 77, 133
　stupidity, 117
Assess, 289
Assessing and managing vulnerabilities, 98

Assets, 130, 279
　management (ID.AM), 356
Assigning responsibilities, 350
Assumption analysis, 251
Attributing responsibility to software, 77
Audit controls, 90
Authentication, 4, 90, 158
Authorize, 289
　authorization to operate, 287
　authorizing officials, 287
Automated
　clearing house, 233
　security tools, 41
Autonomous
　unpredictable behavior, 112
　IoT entities, 116
Autonomous weapons systems, 91
Autonomy, 126, 133
Availability, 4, 10, 16, 59, 177, 247, 288, 357, 384
Avoiding
　harm to others, 74
　risks, 281
Awareness and training, 357

B

Barnes, 225
Bisexual, 218
　bisexual, gay, lesbian, transgender, queer, and questioning, 218
Bitcoin, 53, 99–102, 104, 132
Black hats, 25, 26
Blockchain, 99–105, 132
　community, 103
　ethics in voting, 102
　security issues, 100, 106
　technology, 99–102, 104, 132
　transparent trade tracing, 103
Blockchain transactions, 103

Blue hats, 27
Board, 12, 177, 331, 333, 343, 346
Boundary defense, 378
Bow-tie analysis, 263, 303
Brainstorming, 250
Breach and incident response (IR) coverage, 313
Bring your own device, 242, 305
British Airways, 222
Building company culture, 282
Bullying, 11
Business continuity plan (BCP), 349, 350
  disruption, 325
  environment, 356
  interruption insurance, 312

C

California Consumer Privacy Act, 223, 225
Canadian Institute of Chartered Accountants, 387
Cardholder data environment, 371
Case 1: identifying common healthcare security threats, 83
Case 2: addressing data security issues in healthcare, 85
Case 3: dealing with security breach in a healthcare organization, 86
Case 4: minimizing security threats in a healthcare organization, 87
Case 5: improving data security in healthcare organizations, 88
Case Study 1: The United States vs. Robert Morris, 189
Case Study 10: The United States vs. Huping Zhou, 209
Case Study 11: The United States vs. Joshua Hippler, 209
Case Study 12: Nurse outs STD patient to man's girlfriend, man sues, 210
Case Study 13: UCLA HIPAA violations, 210
Case Study 14: Cloud-based HIPAA trouble, 210
Case Study 15: The United States and cybersecurity controls, 213
Case Study 16: Israel's take on cybersecurity tools and weapons, 214
Case Study 17: The Harvard case, 218
Case Study 18: Principle on naming students who failed in a newsletter, 219
Case Study 19: Google fined $57 million, 221
Case Study 2: The United States vs. Lori Drew, 190
Case Study 20: The Twitter GDPR violation case, 222
Case Study 21: The British Airways data breach, 222
Case Study 22: Google location tracking, 223
Case Study 23: The Starwood Hotel data breach, 223
Case Study 24: Barnes vs. Hanna Anderson LLC, 225
Case Study 25: Google and YouTube COPPA violation, 228
Case Study 26: Google's G-Suite COPPA violation, 229
Case Study 27: Viacom vs. YouTube, 231
Case Study 28: Apple DMCA violations, 232
Case Study 29: Photographer Baffoli vs. Facebook, 232
Case Study 3: The United States vs. Aaron Swartz, 192
Case Study 30: The PokerStars case, 234
Case Study 4: GLBA PayPal-Venmo case, 198
Case Study 5: GLBA TaxSlayer, LLC case, 199
Case Study 6: The United States vs. Katz, 202
Case Study 7: The United States vs. Warshak, 203
Case Study 8: Yates vs. The United States, 205
Case Study 9: Wiest vs. Lynch, 206
Cause and effect diagrams, 251
Central processing unit, 76
Challenges in cyber law, 135, 150
Change management policy, 346
Changing nature of threats, 121
Chatbots, 116
Checklists, 250

Index

Chief information security
  officer, 24, 175, 248, 322
  critical security controls, 283, 374, 381
Child
  labor, 128, 130, 133
  pornography, 135, 158, 161, 235, 237
Choice and consent, 387
Client-server overuse, 46
Clinger-Cohen Act, 163, 235
  Information Technology Management
    Reform Act, 173
COBIT, 362–364, 367–369, 391, 392
Collection, 387, 388
Commission Nationale de l'Informatique
  et des Libertés, 221
Committee of sponsoring organizations,
  290, 304
Common
  security framework, 389, 390, 393
  vulnerability scoring system, 299, 302
Communicating, 34, 77, 346
  consultation, 294
  information responsibly, 77
Communications, 163, 166, 173, 179, 180,
  200, 203, 235, 325, 358, 359
  Act of 1934, 166, 180
  Assistance For Law Enforcement Act of
    1994 (CALEA), 172
  Decency Act of 1996, 173
Communities, 59, 60, 78, 120, 174, 279
Community emergency response team, 277
Computer
  Fraud and Abuse Act, 187, 189, 235,
    236
  computer fraud and misuse, 49
  hardware and software, 76, 136, 137,
    153, 155
  networks, 3, 26, 308, 328, 334
  Security Act of 1987, 163, 235
  systems, 3, 11, 16, 54, 69, 74, 82, 132,
    134, 159, 161, 163, 175, 235, 307,
    326, 331, 385
Conducting
  regular audits, 88
  top-to-bottom security audits, 341
Confidential third-party/research
  information, 318

Confidentiality, 4, 10, 12, 16, 76, 134, 147,
  188, 195, 196, 198, 215, 220, 247, 288,
  357, 373, 383, 385
Confronting power to destroy, 120
Considering
  all phases, 88
  employees, 340
  infrastructure, 339
Continuous vulnerability management, 376
Contracts, 137, 147, 148, 154, 294, 309,
  323, 333, 339, 342
Contribution to society and human well
  being, 74
Control, 78–80, 107, 114, 201, 291, 292,
  352, 362, 363, 376, 390–393
  activities, 291
  environment, 291
  frameworks, 352
  objectives, 363, 390
    information and related technology
      (COBIT) framework, 362
  strength, 280
  systems, 69, 78, 79, 113, 132, 166
Controlled
  access, 379
  use of administrative privileges, 376
Cookies, 148, 228
Cooperation, 58, 122
Coordinated attacks, 325
Copyright infringement, 44–47, 65, 143,
  230, 233, 314
Copyrights, 44, 48, 49, 66, 75, 108, 145
Coso risk management framework (RMF),
  290
Counteracting attack attempts, 82
Counterfeit Access Device and Computer
  Fraud and Abuse Act, 163, 170, 235
Counterfeiting, 46
Coverage for privacy breaches, 317
  covering enterprise end-to-end, 366
Creating strong cybersecurity culture, 98
Credit card, 20, 51, 82, 83, 158, 170, 196,
  225, 233, 236, 338, 369, 393
Criminal offenses related to copyright, 161
Critical infrastructure (CI), 164, 174, 178,
  283, 314, 351, 357, 361, 369, 393
Crosswalk, 89

Cryptocurrency, 53, 99–102, 105, 132
Cryptojacking, 100, 101
Crystallization, 141
Curious case of black hats and white hats, 25
Cyber
   attacks and their business significance, 245
   criminality, 63
   extortion, 315
   insurance
      coverage and its aspects, 312
      need for it, 307
   law, 7, 8, 14, 37, 135–137, 139, 140, 148, 150, 152, 153, 155, 157, 163, 188, 234, 235
      intellectual property (IP), 143
      in United States, 163
   liability
      coverage, 325
      insurance, 320, 322, 324, 328
   physical system, 172
   risk, 96, 241–246, 248, 249, 273, 274, 283, 285, 302, 305, 307, 308, 310, 311, 313, 314, 320–322, 326, 328, 340, 351
      insurance as part of risk mitigation, 310
   security research and development (R&D) act, 164, 176
Cyberbullying, 52, 63, 65, 66, 149
Cybercrimes, 26, 65, 88, 101, 128, 133, 135, 136, 140–143, 152, 153, 155, 157, 234, 237, 241, 305
   against
      companies/organizations, 142
      government, 143
      individuals, 142
      property, 143
      society, 143
   categories from the legal perspective, 141, 152
   intellectual property (IP), 62
Cyberethics, 3, 6, 7, 8, 10, 13, 14, 37–39, 41, 52, 60, 65, 67, 128, 139–141, 152, 155, 157, 237
   cyber law, 138, 157

economic ethics, 15
environmental ethics, 15
laws, 14
philosophy, 14
political ethics, 15
religious ethics, 15
societal norms, 15
Cybersecurity
   computer systems ethics, 74
   control systems ethics, 78
   detailed study, 69
   embedded processors ethics, 76
   enhancement act oF 2014, 178
   ethics, 5, 6, 16, 18, 36, 38, 57, 58, 60, 65, 69, 133
      applications, 81
      artificial general intelligence, 116
      autonomous weapons, 91
      biometrics, 124
      blockchain, 98
      cloud, 106
      finance, 95
      intelligence, spying, and secret services, 120
      internet of things (iot), 111
      medical and healthcare industry, 81
   framework, 96, 249, 283, 351, 354, 362, 372, 392
   information sharing act, 182
   internet ethics, 69
   issues in business, 128
   policies, 163, 242, 329, 333, 338, 340, 344, 351
      audience, 334
      audit, 335
      awareness, 337
      classification, 335
      enforcement, 336
      makers, 331
   resource allocation, 22
   risk, 248
      identification and management, 249
      policies, 239
   roles, duties, and interests, 24, 25
   telecommunication systems ethics, 72
Cyberspace, 3–8, 10–16, 18–20, 23, 26, 37–39, 41–43, 45, 49, 52, 56, 57, 59, 64,

65, 69–74, 132, 134–137, 139–141, 143, 151–153, 157, 165, 167, 187, 241, 243, 302, 309, 329, 354, 374

D

Dark
  electronic processing or data storage, 318
  loss and restoration insurance, 312
  protected by cyber insurance, 317
  protection, 120, 222, 378
    commission, 222
  recovery capabilities, 378
  replacement costs, 316
  retention, 149
  security, 357
    breach notification act, 180
    strategy, 371
  web, 97, 225
Decentralized
  exchange, 105
  majority owned, 104
Deceptive fraud transfer, 316
Decision
  making, 91, 92, 252, 266, 283, 289, 361, 368
  tree analysis, 269, 303
Defamation, 142, 149, 308, 312, 314
Defense federal acquisition regulation (DFAR), 180
Defining
  risk baselines, 296
  set hierarchy of rules, 297
Delphi technique, 259
Demanding audits from vendors/business partners, 342
Department of defense, 91, 165
  appropriations act, 171
Description of an event, 255
Design
  patents, 145
  process, 79
Detailed audit trails, 386
Detect, 358
Detection processes, 358
Determining key metrics, 300

Development of secure systems and applications, 370
Devising comprehensive incident response (IR) plans, 98
Difference between piracy and copyright infringement, 47
Difficult
  control, 112
  cybersecurity collaboration, 213
  identification, 112
  making cybersecurity tools, 212
Diffusion, 42
Digital
  certificates help safeguard healthcare data, 90
  financial loss, 325
  forensics, 137
  millennium copyright act (DMCA), 230
  plagiarism, 51
  versatile disc, 56
Disadvantages of, cyber insurance, 311
Disaster recovery policy, 349
Disgruntled former employees, 324
Distributed
  control systems, 78
  denial of service, 243, 308
Distributing responsibility, 282
Divulgence to third parties, 388
Document signing, 90
Documenting, 255, 346
  risk identification process, 255
Domain disputes, 147
Duty of care risk analysis standard, 310, 327, 328

E

E-commerce, 136, 155, 225
Economic power of technology, 64
E-government act of 2002, 164, 176, 235
Elasticity, 106, 133
Electromagnetic, 72, 167, 201
  radiation, 167
Electronic communications
  privacy act, 163, 170, 200, 235
  services, 4
  systems, 4, 39

Email, 16, 17, 29, 53, 70, 72, 82, 96, 97, 100, 137, 142, 149, 202, 217, 218, 222, 224, 228, 229, 237, 247, 314, 329, 331, 349, 354, 370, 377
   communication policy, 349
   security, 90
   web browser protections, 377
Embedded
   processors, 3, 11, 69, 76, 77, 132
   system, 76–78, 134
Emphasizing speed, 283
Employment, 53, 82, 130, 147, 149, 206, 225
Enabling long-term cybersecurity and risk management, 360
Encouraging diverse views, 283
Encrypt transmission of cardholder data, 370
Encryption issues, 41
End-user piracy, 46
Enforcing
   property rights on information, 112
   right to private life, 114
Ensuring
   access to information, 113
   integrity of the information, 113
Enterprise
   risk management, 290
   technology risk assessment, 181
Environmental impacts, 43, 67
   blockchain technology, 103
Equitability, 93
Establishing
   context, 294
   drivers, 276
   formal security and ethical framework, 96
   procedures, 85
Ethereum, 100
Ethical
   benefits or dangers, 104
   challenges
      accountability for cybersecurity, 34
      balancing security with other values, 30
      competing interests and obligations, 32
      data storage and encryption, 33
      faced by cybersecurity professionals, 30, 41
      incident response (IR) activities, 31
      IoT, smart grid, and product design, 34
      network monitoring and user privacy, 32
      security breaches and vulnerabilities, 31
      security research and testing, 35
   issues affecting
      accounting and finance, 129
      cybersecurity, 20
   in cybersecurity
      privacy, 20
      property, 21
   manner, 35
   way, 69
Ethics
   cybersecurity, 16, 26, 41, 64
      professions, 61
   cyberspace and cybersecurity, 3, 37
   regulation and freedom, 64
   versus legal, 139
European
   Commission, 221
   Economic Area, 219, 236
   Union, 44, 219, 222, 236
Evil geniuses
Protection against unintended consequences, 118
Exchange hacks, 100
Exemption from federal and state FOIA law, 184
Expected monetary value, 268, 303
Export administration regulations, 339
External environment, 279
Extortion insurance, 312

F

Facebook, 29, 30, 39, 130, 228, 232, 233
   White Hat program, 30
Failure mode and effect analysis, 266
Fair, 19, 54–56, 59, 61, 63, 65, 66, 87, 117, 120, 146, 155, 170, 278–280, 303, 344
   use doctrine, 55, 65
Family Educational Rights and Privacy Act of 1974 (FERPA), 215
Fault tree analysis, 272, 303
Regulations for electronic records use, 181

Federal
  Advisory Committee Act (FACA), 169
  Bureau of Investigation, 180
  Communications Commission, 166, 325
  Financial Institutions Examination
    Council, 96
  Information Security Management Act,
    164, 175, 177, 235
  Power Act, 166
  Trade Commission, 87, 174, 313, 391
Financial
  establishments, 159, 237
  sector, 15
  services, 197
Flexibility and adaptability of the
  framework, 361
Focusing on human rights and ethical
  behavior (IoT), 116
Food
  FDA (Food and Drug Administration),
    181
  industry, 147
Forensic Investigation Insurance, 312
Privacy Act of 1974, 182
Framework, 283, 287, 304, 315, 355, 360,
  363, 390, 393
  core, 355, 360
  failures, 315
  implementation tiers, 355
  profile, 355
Fraud, 20, 49–51, 65, 82, 87, 101–103,
  135, 136, 142, 143, 157, 158, 169, 170,
  174, 180, 187, 188, 192, 205, 234, 235,
  290, 313, 316, 327
  access devices, 158
  computers, 159
  electronic mail, 159
  radio, television or wire, 159
Fraudulent financial reporting, 129
Freedom of,
  expression and association, 63
  Information Act (FOIA), 168
Frequency, 113, 266, 279, 280, 339
Federal Trade Commission Act, 179
Function Creep, 108

G

Gathering information to identify risks, 253
Gender
  inequality, 11
  oriented, 60
General data protection regulation, 219, 236
Generally accepted privacy principles, 385
Generation sensitive, 60
Global positioning system, 113
Gold, 104, 382
Google, 38, 221–223, 228, 229, 231, 237
Governability, 93
Governance, 356
Gramm-Leach-Bliley act, 194, 235
Gray hats, 25, 26
  ethical point of view, 29

H

Hacker, 88
Hacking, 26, 27, 137, 140, 143, 157, 187,
  193, 235, 284, 307, 308, 312, 370
Hanna Andersson, 225
Harassment, 11, 142, 173
Harddisk loading, 46
Hardware and software, 335, 376
Harmful actions, 42, 43, 52, 65
Hazard and operability, 254
Health
  human services, 86, 207
  information trust alliance, 389
  insurance portability and accountability
    act of 1996 (HIPAA), 206
    HIPAA (for covered entities and
      business associates), 391
    HIPAA administrative simplification,
      207
    HIPAA health insurance reform, 207
    HIPAA tax-related health provisions,
      207
  plans, 391
High-performance computing act, 172
Hitrust CSF, 389–391
Homeland security act (HSA), 164, 174
Honoring
  confidentiality, 76
  property rights (copyrights and patents),
    75

Hospitals, 113, 137, 391
Human
　centered approach, 79
　civilization, 7
　error, 324
　resources, 74, 291, 332, 394
　rights, 11, 57, 64
Humanity, 117
Humidity, 114

I

Identification, 94, 97, 112, 113, 127, 128, 143, 154, 158, 217, 249–256, 261, 276, 278, 284, 291, 300–303, 307, 345, 350, 377
Identify, 66, 73, 276, 284, 356
　mitigate risks, 276
　oriented, 60
　threats, 276
Identity theft, 26, 50, 52, 53, 83, 158, 174, 177, 180, 235, 307, 308, 325, 374
　Assumption Deterrence Act of, 174
　Penalty Enhancement Act, 177
Implement, 249, 283, 289, 296, 346, 350, 369, 372, 379, 381
　cybersecurity framework (CSF), 283
　security awareness and training program, 379
Implications of cyber insurance, 322
Important ethical issues in cybersecurity, 18, 37
Improved online protection, 322
Inadequate disposal of old hardware, 85
Incident response, 21, 42, 98, 181, 215, 243, 284, 313, 380
　management, 380
　plan, 284
　response (IR), 42
　policy, 347
Incorporated intelligence, 112
Individual risk per annum, 266
Industrial control systems, 78, 166
Inequality, 117
Information and communication, 58, 292
　technology, 3, 37, 39, 58, 83, 106, 139, 173, 241, 302, 307, 321, 347, 394

　collection, 10
　commissioner's office, 222
　privacy insurance, 312
　protection processes and procedures, 357
　security management system, 289, 372, 393
　security policy, 334, 347, 369
　systems, 73, 90, 91, 175, 180, 182, 183, 195, 241, 267, 273, 294, 305, 345, 358, 383
　　audit and control association, 362, 392
Infrastructures, 3, 37, 42, 67, 274
Inspectors general, 175
Institutional review boards, 181
Insurance, 82, 87, 89, 176, 181, 200, 206, 207, 221, 236, 285, 302, 307–312, 314, 316–330, 339, 350, 391
Insurer requirements to encrypt portable media, 323
Integrity, 3, 4, 10, 12, 16, 19, 42, 49, 60, 61, 75, 90, 91, 113, 129, 180, 188, 198, 241, 247, 291, 357, 378–380, 383, 384
　controls, 91
Intellectual property, 21, 47, 50, 136, 144, 154, 155, 157, 235, 308, 374
Intelligence reform and terrorism prevention act (IRTPA), 177
Intelligibility and fairness, 120
Interception of wire, oral, or electronic communication, 162
Internal
　costs and court participation cost, 314
　revenue service, 217
International
　Electrotechnical Commission, 372
　Organization for Standardization (ISO), 372, 391
　Risk Management Institute, 324, 327
　standard, 293, 304
Internet, 5, 7, 12, 44, 46, 49, 62, 63, 65, 69–72, 84, 90, 94, 99, 100, 106–108, 111, 112, 115, 132, 133, 136, 137, 141, 142, 144, 145, 148–154, 161, 167, 168, 170–173, 186, 187, 189, 192, 212, 224, 228, 230, 232, 233, 236, 307, 341, 349, 374, 378, 393

Index

etiquette, 134
internet of things (IoT), 34, 111–116, 133, 134, 186, 246, 342
piracy, 46
protocol, 3, 111, 348
service provider, 45, 162, 235, 243, 305, 345
Interviews, 250
Intrusion-detection systems, 308, 328
Inventory and control of hardware assets, 375
IPhone operating system, 232
ISO, 283, 293, 294, 304, 323, 372–374, 383, 389–392, 393
   27001, 283, 373, 374, 390
   27002, 373, 374
   27017, 374
   27018, 374
   27701, 374
   31000 STANDARDS, 293
   IEC 27000 SERIES, 372
IT
   service providers, 391
   team, 331

J

Jurisdiction, 150, 158, 159, 174, 187

K

Key cyber risks and threats, 247
Khalil Shreateh, 29

L

Lead security risk evaluation to decrease premium, 321
Legal
   costs, 325
   framework, 116
   team, 332
Liability, 79, 230, 308, 309, 314, 324, 326–328
Limitation and control of network ports/protocols/services, 377
Litecoin cash, 104
Lognormal curve, 271
Long-term costs, 274

Losses not under cyber insurance, 318
Lost and stolen mobile devices, 84

M

Machine
   design ethics, 79
   intelligence, 117
   vision, 116, 133
Maintenance, monitoring, and analysis of audit logs, 377
Major vulnerabilities-employees and encryption, 82
Making yourself look good online, 71
Malicious activities, 159
Malware, 16, 35, 43, 46, 51, 53, 54, 66, 83, 84, 97, 100, 101, 106, 137, 143, 189, 225, 232, 246, 247, 249, 282, 295, 303, 304, 308, 324, 339, 370, 372, 376
   phishing attempts, 83
   defenses, 377
Management, 164, 177, 235, 254, 258, 259, 287, 289, 324, 327, 346, 356, 363, 387
   guidelines, 363
   managing third-party risks, 98
Marketing, 196, 221, 228
Massachusetts Institute of Technology, 189
Massive open online courses, 58
Matrix, 257, 264, 265, 303
Maturity models, 363
Mean time to
   detect, 386
   remediate, 387
Measuring success of,
   controls, 354
   governance, 353
   programs, 354
Media in control of others, 316
Meeting stakeholders needs, 364
Miniaturization and invisibility, 111
Misdirecting words/digital pictures, 161
Mitigating risk, 281
Mitigation, 359
Mixing in proactive reporting, 301
Mobile devices, 324, 376
Monitor, 289

Monitoring, 32, 98, 106, 127, 183, 292, 294, 358, 384, 386, 389
  defending information systems, 183
  enforcement, 389
  review, 294
  known and unknown, 386
Monte Carlo analysis, 270, 271
Mosaic plagiarism, 51
Multimedia liability, 314

N

National
  cybersecurity
    communications integration center, 179
    protection advancement act (NCPAA) of 2015, 179
  Institute of Standards and Technology, 3, 39, 69, 163, 166, 249, 304, 346, 391
    cybersecurity framework (CSF), 355
  Public Radio, 232
  Science Foundation, 164
  Security Act of 1947, 167
Nature of copyrighted work, 55
Need for cyber
  cyberethics, 10
  laws, 136
  regulation based on cyberethics, 140
Netiquette, 70–72
Network security
  insurance, 312
  risk, 328
Networking and Information Technology Research and Development, 172
New York Attorney General, 228
NIST
  cybersecurity framework, 96, 362
  risk management framework (RMF), 289
Nominal group technique (NGT), 251
Non-discrimination, 63
Non-public personal information, 195
Non-repudiation, 4
Non-waiver of privilege, 184
Normal or bell curve, 271
Norms and values, 123
Not abusing power, 71
Notice, 387

O

Octave, 275–278, 303
Omnibus Crime Control and Safe Streets Act, 168, 201
Omnipresence, 111
Online
  gambling, 158, 234, 235
  medical devices, 84
  service providers, 230
Operational risk identification, 252
Opportunities, 122, 256, 257, 269, 303, 305
Organization's security, 85, 329, 382, 394
Origin of cybersecurity ethics, 56
Other technical safeguards, 90
Outsourcing, 122
Ownership, 108

P

Packaging, 46
Paperwork reduction act of 1995, 163, 235
Patents, 49, 63, 75, 145, 147, 154
Payment card
  data security standard, 318, 369, 392
  industry, 314, 318, 338, 369, 392
  information (PCI), 318
    DSS, 369–372, 383, 392
    Security Standards Council, 369
Pen registers, trap, and trace devices, 162
Penetration tests and red team exercises, 380
People-centered, 59
Performing
  assessment, 277
  continuous threat monitoring, 97
  regular data backups, 343
Personal
  financial information, 10
  health information, 10, 114, 317, 327
  privacy, 42, 65, 140
  identifiable information, 10, 169, 246, 302, 314, 327, 331, 394
Personally identifiable information (PII), 317
Personnel management, 335

Phishing, 5, 16, 51, 82, 83, 85, 90, 97, 137, 142, 155, 243, 247, 282, 295, 304, 308, 340, 342, 354, 375
Physical security, 335
Physically securing information assets, 343
Piracy, 45–47, 57, 62, 63, 65, 143, 155
Plagiarism, 51, 62, 65, 66, 314
Planning, 195, 346
Plant patents, 145
Policy for IoT safety, security, and privacy, 116
Posse comitatus act of 1879, 165
Post-change reviewing, 346
Potential for loss of life, 266
Preparing for,
    octave, 277
    physical disasters, 349
Preservation, 61, 140
Pressure, 20, 121, 129
Principles of cyberethics, 12, 139
Prioritizing cybersecurity risks, 283
Privacy, 8, 9, 10–12, 15, 16, 19–21, 30, 32, 36, 38, 42, 57, 61–63, 71, 80, 84, 87, 102–104, 107–109, 113, 114, 120, 125, 126, 128, 130, 133, 136, 137, 140, 148, 151, 152, 171, 172, 177, 179, 180, 197–202, 208–210, 218, 219, 221–225, 227, 229, 236, 237, 307, 308, 312, 314, 317, 327, 369, 373, 374, 383, 385, 387–390, 393
    act of 1974, 169, 236
    privacy and security, 8–11, 61, 62, 172
    concerns, 80
    information management system, 374
    liability, 314
Probability
    impact, 263
    consequence matrix, 263, 303
Probable loss magnitude, 280
Procedural cyber laws, 162, 235
Process
    buying cyber insurance, 320
    descriptions, 363
Processing integrity, 384
Professional services, 88
Profile assets, 276
Program frameworks, 352

Project
    management body of knowledge, 251
    risks, 252, 258, 265, 293, 303, 304
        identification, 252
Promoting ethical use of IoT technologies, 115
Proof-of-work, 105
Proprietary information, 184
Protect stored cardholder data, 370
Protecting against cyber disasters, 349
Protections for sharing and receiving information, 183
Protective technology (PR.PT), 357
Prototype, 304
Providing new/continuing security education, 342
Public relations, 308, 325
Purpose and character of the use, 55
Putting the Risk Prioritization Framework to Use, 299

Q

Qualitative risk analysis, 258, 259, 263–265, 273, 303
Quality, 388
    safety, 248
Quantitative risk analysis (QRA), 264, 273, 303
Questionnaire, 260
Questions to ponder, 38, 66, 133, 154, 236, 304, 327, 393

R

Racist robots, 117
Racketeer Influenced and Corrupt Organizations Act (RICO), 169
Radio-frequency identification, 113
Ransomware, 5, 31, 53, 65, 66, 82, 101, 207, 247, 249, 284, 295, 303, 304, 315, 324, 343, 349, 375
    attacks, 53, 65, 101, 324, 349
Re-assess/re-evaluate, 297
Recognize and verify admittance to framework components, 371
Recovery planning, 359
Red hats, 26

Reexamining the network, 87
Regulations for use of electronic records in clinical investigations, 181
Regulatory
  coverage, 313
  fines, 325
Reliability, 3, 22, 30, 33, 93, 172, 181, 195, 291, 353
Reliable and useful metrics, 300
Remediation techniques, 295, 304
Remember the human, 70
Remote access policy, 348, 349
Reporting the breach, 86
Reputation insurance, 312
Requirement declaration, 370
Research and development, 48, 82, 164, 235
Respecting other people's privacy, 71, 75
Response planning, 358
Responsibility, 58, 79, 93, 204
Restrict physical access to cardholder information, 371
Results of questionnaire four, 260
Revenue offsets, 208
Ripple effects across supply chains and vendor lists, 361
Risk
  analysis, 249, 257–259, 271, 273, 274, 279, 293, 302, 310, 327, 328, 355
  assessment, 79, 175, 185, 199, 245, 246, 262, 264, 267, 271, 273–276, 278, 280, 287, 296, 297, 303, 321, 360, 374
    frameworks, 275
    matrix, 264
  evaluation, 250, 293
  frameworks, 352
  identification, 249, 256, 293, 298, 303
    using SWOT analysis, 256
  management, 23, 96, 175, 243, 245, 246, 248, 249, 252, 254, 256, 258–260, 273, 278, 280–282, 284, 285, 287, 290–295, 298, 299, 301–304, 309, 310, 321, 350, 352, 355, 360–362, 373–375, 391, 393, 394
    framework, 280, 286, 287, 290, 292, 304, 360, 391
  strategy, 356
  monitoring, 250
  prioritization, 296, 298, 299
  remediation, 241, 295–302
    process, 296, 298–300
  treatment, 250, 293
Robot rights, 118
Roles and responsibilities, 42

S

Safety, 79, 195, 254
Sale restrictions, 41
Salesforce.com, 225
Sarbanes-Oxley, 204, 235
  Act of 2002, 204
Scrubbing personal information before sharing, 183
Search engine, 29
  optimization, 29
Securing networks, 42, 67
  secure configuration for network devices, 378
Securities and Exchange Commission, 204, 313
Security, 7, 8, 9, 10, 26, 29, 32, 35, 37, 43, 51, 53, 90, 117, 164, 167, 173, 174, 177, 183, 186, 199, 208, 212, 218, 224, 235, 274, 282, 283, 297, 298, 310, 314, 315, 317, 331, 332, 335, 336, 338, 343, 344, 347, 353, 357, 358, 369, 370, 374, 380, 384, 385, 388, 392, 393
  breach, 32, 315
    contingent third parties, 315
  continuous monitoring, 358
  for privacy, 388
  liability, 314
Self-determination, 107
Self-awareness, 274
Self-plagiarism, 51
Semiconductor, 137
Sensitivity analysis, 271, 272, 303
Separating governance from management, 368
Service
  level agreement, 342
  organization and control, 383
Service-oriented architecture, 107
Setting strict access, 88

Sexual
  abuse of minors, 161
  exploitation of children, 160
Sharing
  expert knowledge, 71
  information, 283
  information, 87
  receiving cyber threat indicators, 183
Short message service, 52
Singularity, 118
Smith-Mundt Act, 167
SOC 2, 383–386, 393
Social
  engineering, 100, 342
  exclusion, 127, 133
  media, 11, 15, 52, 61, 63, 70, 100, 102, 104, 112, 128–130, 148, 149, 243, 329, 331, 340, 341, 349
    ethical issues, 130
  networking, 63, 128, 133, 190
Societal and technological ecosystem, 116
Software flaws, 100
Sources of ethics, 13
Specific ethical issues in cybersecurity, 25, 43
Speed, 30, 103, 117, 231, 284
Staff or insider threats, 83
Stakeholders, 17, 30, 32, 33, 35, 42, 62, 98, 123, 129, 178, 251, 253, 256, 274, 286, 294, 322–334, 350, 356, 358, 361, 363, 364, 366
Stalking, 11
Standard deviation, 268
State Department Basic Authorities Act of 1956, 167
Strategic risk identification, 252
Strength weakness opportunity and threat, 254
Strict personal device regulations, 86
Structured
  query language, 244
  what-if technique, 261
Substantive cybercrime laws, 157
Superior and unbiased cybersecurity, 360
Supply chain, 82, 174, 246, 323, 356, 361, 362, 383
  risk management, 356

Sustainable development, 116
Sustainable development, 64
Swift analysis, 261
SWOT, 254, 256, 257, 303, 305
  analysis, 256, 257, 305
  example, 305
  matrix, 256

T

Technological infrastructure, 73
Technology, 3, 39, 49, 59, 69, 73, 111, 121, 130, 163, 172, 175, 178, 189, 249, 304, 346, 391
  privacy, 130
Telecommunications, 3, 11, 39, 72, 73, 158, 170, 172, 173, 237
  systems, 69, 132, 134, 166
  networks, 3, 11, 39
Terminate, 285
Terrorism Risk Insurance Act of 2002, 176
Test security frameworks, 371
Testing processes, 370
Theft and fraud insurance, 312
Third-party requests, 30
Threats, 61, 256, 257, 269, 279, 284, 303, 305, 350
  capability, 280
  event frequency, 280
Three-point estimate, 267, 268
Threshold, 193, 315
Tolerate, 285
Traceability, 93
Trade secrets, 20, 21, 48, 75, 146, 147
Trademark, 48, 66, 137, 145–148, 154
Tradeoffs of self-insuring with respect to buying cyber liability insurance, 321
Training employees, 282
Transfer, 285
Transferring risk, 281
Transmission security, 91
Transparency, 23, 30, 38, 63, 65, 93, 108, 123, 186, 197, 200, 211, 220, 222, 223, 295, 300
  disclosure and accountability, 23
Transportation
  importation of obscene matters, 160
  obscene matters, 160

Triangular curve, 271
Trust services criteria, 384
Types of
  cyber insurances, 312
  cybercrime laws, 157
  cybersecurity policies, 344, 350
  ethical issues in cybersecurity, 42

U

U.S. Information and Educational Exchange Act of 1948, 167
Ubiquity, 111
Ultra-connectivity, 112
Unauthorized publication, 162
Understanding
  broader impacts of cybersecurity practice, 35
  identifying risks, 253
  network map, 87
  own security, 338
Unemployment, 116
Unfair methods of competition, 165
Uniform curve, 271
United Kingdom, 212, 236
United States, 44, 54, 66, 145, 150, 157–160, 163–165, 167–169, 174, 175, 177, 178, 186–190, 192, 194, 200, 202, 203, 205, 208, 209, 212, 213, 215, 216, 223, 224, 230–232, 234–236, 355, 392
  computer emergency readiness team, 175
Unlawful Internet Gambling Enforcement Act of 2006 (UIGEA), 233
Unrestricted access to computers, 85
Unsecured mobile devices, 84
Updating
  auditing cybersecurity procedures, 333
  software and systems, 341
User and device authentication/access control, 90
Utility patents, 145

V

Vacuum, 78, 122, 125
Values in ethics, 57

Vendors, 84
Virtual private network (VPN) encryption, 87
Vulnerability, 17, 23, 24, 26, 28–32, 34, 35, 38, 41, 67, 96, 98, 103, 109, 170, 174, 181, 212, 213, 215, 244, 245, 249, 275, 280, 282, 297–303, 342, 354, 369, 371, 375–377, 380
  disclosure, 41, 67, 215

W

War zone, 91, 93
Wassenaar arrangement, 210, 212, 236, 237
  countries, 212
Watching the edge, 342
Web
  application firewalls, 384
  filter, 97
  server security, 90
White hats, 25–27
WHO, 248
Wire communication, 4
Wireless access control, 379
Wireless local area networks, 379
World Intellectual Property Organization (WIPO), 230, 236
Writing and developing effective cybersecurity policies, 337

Y

Yates, 205, 206
Young
  adults, 63
  people, 63
  users, 227
Youngsters' dress organization, 225
YouTube, 228, 229, 231

Z

Zero day
  exploits, 35
  specific burndown, 300
Zero-footprint, 97